PROGRESS IN CIRCADIAN RHYTHM RESEARCH

Progress in Circadian Rhythm Research

Anne-Laure Léglise
Editor

Nova Biomedical Books
New York

For permission to use material from this book please contact us:
Telephone 631-231-7269; Fax 631-231-8175
Web Site: http://www.novapublishers.com

Library of Congress Cataloging-in-Publication Data

Progress in circadian rhythm research / Anne-Laure Léglise, editor.
p. ; cm.
Includes bibliographical references and index.
ISBN-13: 978-1-60021-876-7 (hardcover)
ISBN-10: 1-60021-876-8 (hardcover)
1. Circadian rhythms. I. Léglise, Anne-Laure.
[DNLM: 1. Circadian Rhythm. 2. Chronobiology Disorders. QT 167 P964 2007]
QP84.6.P79 2007
612'.022--dc22 2007026861

Published by Nova Science Publishers, Inc. ✦ New York

Contents

Preface

A circadian rhythm is a roughly-24-hour cycle in the physiological processes of living beings, including plants, animals, fungi and cyanobacteria. The term "circadian", comes from the Latin circa, "around", and dies, "day", meaning literally "about a day." The formal study of biological temporal rhythms such as daily, weekly, seasonal, and annual rhythms, is called chronobiology. In a strict sense, circadian rhythms are endogenously generated, although they can be modulated by external cues such as sunlight and temperature. Most of a person's body systems demonstrate circadian variations. The body systems with the most prominent circadian variations are the sleep-wake cycle, the temperature regulation system, and the endocrine system. The malfunctioning of a person's circadian system, or biological clock, causes circadian rhythm disorders. Circadian rhythms and their relation to health outcomes in the globalized hyper world of the 21st century are now understood to be extremely important. This new book presents the latest research from around the world in this dynamic field.

Chapter I - Oncology is a domain of particular interest in chronobiology research, focusing the efforts of a wide range of scientists. A great deal of data concerning the association between biological rhythms and cancer is now available, associating genetics, cell biology, animal experimentation, epidemiology, and clinical research. Despite these efforts, little has been done to associate biological rhythmicity and therapeutics, although some teams have shown the advantage of such an approach on survival and toxicity control. There have been two major approaches: chronochemotherapy and melatonin supplementation. Preliminary data suggest that improving circadian rhythms may improve global cancer prognosis.

Several groups have studied psychosocial aspects in relation to cancer, initially intending to improve overall survival, with equivocal results. Presently such surveys focus on the quality of life and nursing management, although improved study design may show an impact on survival after all.

The authors present here the rationale for a new type of investigation, based partially on Ernest Rossi's suggested impact of hypnosis on immunity and gene expression, and then an original protocol being conducted among palliative care cancer patients. This is prospective randomized trial testing the impact of Ericksonian hypnosis on immunity and biological rhythms. Biological parameters and activity/rest rhythms are measured before inclusion in the trial and after two months; hypnosis intervention is planned once a week for 8 weeks. In the second arm, patients receive a training CD for relaxation that they are told to use *ad libitum*. Using suggestions aimed at behavioral changes and personnal insights, the authors hope to obtain an immunitary rebound thanks to the amplification of ultradian and circadian rest/activity rhythms. The principal measurements include rhythms of salivary melatonin, cortisol and leptin, rest/activity cycle measured by actimetry, temperature. Levels of interleukins IL-1, IL-2, IL-6, IL-15, INF-γ and TNF-α and lymphocytes counts are included. Circulating expression of clock genes are measured in parallel with the one of the interleukin receptor genes. Other genes concerning hypnosability and some brain memory managment functions will also be studied. Overall survival and quality of life are secondary goals. A multicenter inclusion is planned; the authors intend to compare the effect of different therapists, and if significant differences are observed, to search for ways to improve hypnosis managment.

Last inclusions are forecast for 2009, for a total of 100 patients per arm, while an interim analysis should give us a good estimate of the final results after 30 patients are accrued in each arm.

Chapter II - The 24h cycle of day and night is the most powerful synchroniser (*Zeitgeber*) of the human circadian system thereby regulating a wide range of behavioural, physiological and metabolic rhythms. A common feature of varied forms of mental health conditions are circadian rhythm disruption of hormonal and sleep/wake cycles. Significantly, various neuropsychiatric disorders ranging from schizophrenia and depression to age-related medical conditions like Alzheimer's and dementia can be improved with specific light treatments that are designed to act as *Zeitgeber* for the circadian clock. In recent years our understanding of the circadian timing system has changed profoundly. Less than a decade ago is was thought that the hypothalamic suprachiasmatic nuclei (SCN) imposed a 24h temporal structure on the entire body, but the discovery of tissue and organ specific circadian oscillators has radically changed this view. It now appears that the SCN does not drive circadian rhythms in the periphery but acts to coordinate and perhaps sustain their activity. This has led to the hypothesis that pathologies may arise from or are exacerbated by sustained internal desynchronisation – the inappropriate alignment of physiological rhythms in different organ systems over the 24h day. If this is indeed the case, then this might explain why bright light given at the appropriate time can alleviate some of the symptoms associated with mental health conditions. Light might be acting as a strong synchronising agent that forces a common temporal structure on the entire body. In this chapter the authors review the evidence for this hypothesis, consider possible mechanisms, and address whether chronotherapeutics should be used more widely as a therapeutic approach to assist in the treatment of neuropsychiatric disorders.

Chapter III - Knowledge is recently building regarding the factors that affect the development of an optimal sleep-wake cycle. Fetuses near term experience arousal fluctuations and sleep episodes already in the darkness of the womb. Yet, the factors that allow for a smooth emergence of an optimal sleep wake cycle are the conditions that place newborn at higher risk to develop sleep disorders are not yet fully understood. The current chapter proposes an integrative model for the development of circadian sleep-wake rhythms in preterm infants. This model is based on a comprehensive review of the basic science literature as well as clinical work with at-risk preterm infants. This integrative model proposed a three source risk source: 1) infant dependent neurobiological vulnerability risk, such as brainstem mediated functional disability; 2) a familial source that entails genetic predispositions, parental style and support resources; and 3) an environmental stimulation source, such as excessive NICU related stimulation and handling. These three realms act as main effects and interact with the infants' prenatal and postnatal age and with the infants' self-regulation mechanisms. The manner in which these sources operate and interact lead to directions of interventions that ameliorate the development of circadian sleep-wake rhythm deficit in infants born at-risk for sleep disorders. The chapter details each component of the model and illuminates the complexity of the interactions involved in it to deepen the understanding the mechanisms involved in early organization of sleep-wake rhythms in infants.

Chapter IV - Research of influence pinealectomy and epithalamin injection to pinealectomized animals on circadian rhythm proliferation epithelium in jejunum crypts and the spermatogenesis rhythm in albino rats were studied. Epithalamin is melatoninless extract of pineal oligopeptides. Animals were kept under 12/12h day/night regimen. Researches were carried out on day 40-41 after pinealectomy.

Epithalamin (in dose 2.5 mg/kg) was injected before beginning night period of day since 26 till 40–th day after pinealectomy.

Circadian rhythm of mitotic index epithelium in jejunum crypts, which disappeared after pinealectomy and restored after epithalamin injection to pinealectomized animals, was observed in intact animals. Daily dynamic of the spermatogonia mitotic index and stage frequency of spermatogenic cycle, which was cyclic change of cell combination spermatogenic coat in process of spermatogenesis, were carried out for studied of spermatogenesis rhythmicity. Circadian rhythm of mitotic index type B spermatogonia, the duration of reproduction period and differentiation intermediate type spermatogonia and also meiosis was observed, that it is shown on spermatogenesis circadian rhythm existence in general. Pinealectomy caused disappearance the circadian rhythms of spermatogenesis and epithalamin injection restored circadian rhythms.

According to the authors results the pineal gland is the necessary part in formation of circadian rhythm of various tissue proliferations. The involvement of pineal gland peptides in regulation of proliferation circadian rhythm shows the necessity of extraction their in clean form, testing of biological activity and usage ability in desynchronizes correction, accompanying various diseases.

Chapter V - In mammals, hibernators have a unique ability to adapt organisms to an extremely decreased body temperature (Tb) and survive for several months by consuming much less energy. Many years of research it has shown that during hibernation, animals are protected from various lethal diseases and events, including bacterial infections, tumorigenesis, hypothermia and ischemia, suggesting that hibernation is an adaptive state for maintaining health by protecting organisms. If its control mechanisms were understood, hibernation would offer myriad applications in medicine and health sciences. Therefore, discovery of an endogenous factor that controls hibernation has long been dreamed of. The authors early studies on cardiac muscles of chipmunks, a rodent hibernator, demonstrated that Ca^{2+} is intracellularly recycled in the hibernating state by circannually modulated Ca^{2+} source for excitation-contraction coupling and suggested that this modulation may prevent intracellular Ca^{2+} overload at a decreased Tb. Based on these findings, the authors have isolated a novel hibernation-specific protein (HP) complex in the blood, which is produced in the liver and downregulated during hibernation in association with the circannual cardiac modulation. Recently, this complex was identified as a promising candidate hormone that is essential for circannual control of hibernation in the brain. In this review, a significant role of cirannually modulated cardiac Ca^{2+} regulation and HP complex in the brain will be discussed in maintaining health and life, and hibernation may be redefined as a hormonally regulated state independent of Tb changes.

Chapter VI - In recent years many studies have been published trying to establish the differences among individuals depending on their circadian typology (morning-, neither-, and evening-type). This is an individual difference underlying differences in the rhythmic expression of both biological and behavioural parameters, especially the circadian period. In general, morning-type subjects show phase advances with respect to evening-type subjects, which vary from 2 to 12 hours, depending on the variable studied. The neither-type group is usually in an intermediate position regarding the extreme groups (morning- and evening-type), although many studies do not contemplate this group despite the fact that it is predominant in the general population (around 60 %). The authors study exposes the optimal moments of efficiency according to the type of cognitive ability that healthy subjects must perform depending on their circadian typology. This aspect is of vital importance for the subjects' performance at school and work, and also when the environmental conditions of the light-dark cycle (jet-lag and shift-work) are modified. The authors also comment on the observed differences in eating habits, consumption of psychoactive substances, and preferred spare-time activities among chronotypes, which in general are healthier in the morning-type than in the evening-type. Finally, the authors present data on the associations found between other personality characteristics, such as extraversion or impulsivity, and with the vulnerability to suffer certain psychopathologies such as Seasonal Affective Disorder (SAD), bulimia nervosa or Attention Deficit Hyperactivity Disorder. The relevance of this individual difference makes it advisable that research which is not directly interested in the study of this variable, but which evaluates the parameters involved in the circadian typology, should have an adequate control of it.

Chapter VII - In a neutral and controlled environment, physical performance during a short-duration exercise fluctuates during the day with an acrophase close to 18:00 h. The diurnal increase in muscle power is mainly attributed to a diurnal increase in muscle force. This diurnal increase in muscle force is mainly attributed to a diurnal increase in muscle contractile properties. Today, the physiological basis of this variation in muscle contractile function remains unclear and represents a challenging topic in the future because the circadian variation in muscle force is generally in phase with the circadian variation in body core temperature. It is hypothesized that the diurnal increase in core temperature could have a beneficial passive warm-up effect.

It is possible to shift the acrophase of the muscle power variation by a couple of hours following a few days by adapting sleep and meal schedules. Moreover, the use of another source of passive warm-up (e.g., a warm environment) seems to blunt the diurnal variation in muscle power. With long duration exercise, the thermal stress induced by physical exercise in a warm (and humid) environment should be higher in the evening, when the initial body temperatures are the highest.

Physical activities have to be considered for their effect on health. For example, cardiac events occur mainly between 06:00 - 12:00 h and a vigorous physical activity could be a trigger for them. Clinicians should advise at risk patients to avoid exercise at this time-of-day. Physical activity can also have some beneficial effects. For example, physical activity increases the amplitude of some circadian variations in elderly people allowing for the improvement both nocturnal sleep quality (i.e., less) and diurnal vigilance.

Chapter VIII - The suprachiasmatic nuclei (SCN) of the mammalian hypothalamus coordinate a wide range of circadian rhythms in behavior and physiology. Recent technical advances allow the recording of firing rate rhythms from individual SCN neurons and the measurement of clock gene expression rhythms from various cell types over several days. These techniques provide an opportunity to study the generation and synchronization of cellular rhythms *in vitro*. Studies of transgenic mice lacking vasoactive intestinal polypeptide (VIP) or its receptor $VPAC_2$ show that impaired behavioral rhythms in these mice stem from dysfunction at the level of individual SCN neurons. In both mutants, the proportion of SCN neurons with circadian rhythms in firing is reduced, while circadian synchrony between the remaining rhythmic neurons is absent. Studies mimicking or blocking GABAergic signaling between SCN neurons have shown that while daily GABA treatment can synchronize SCN neurons *in vitro*, endogenous GABA signaling is not required for SCN timekeeping or synchrony. Pharmacological manipulations of wild-type SCN neurons suggest that daily rhythms in G-protein mediated signaling pathways regulating adenylyl cyclase are required to maintain rhythms and synchrony within the SCN. It is hoped that such studies will clarify the signaling pathways involved in circadian rhythm coordination *in vivo*, leading to novel, targeted interventions for sleep disturbances and health problems stemming from circadian dysfunction.

Chapter IX - Circadian rhythmicity in mammals is chiefly governed by a master pacemaker located in the suprachiasmatic nucleus of the hypothalamus. The functioning of this pacemaker is known to be sensitive to a number of centrally and peripherally derived signals, such as neuropeptides and circulating hormones. Recently attention has been drawn as to how the immune system may impinge on circadian rhythms and alter the endogenous timekeeping system under basal conditions and immune challenge.

The brain is now realised to be far from the immune privileged site it was once thought to be, and as such the roles of immune modulators such as cytokines have been much studied in the central nervous system, with regard to both neuropathological processes and normal neurophysiology. Recent evidence has implied that circadian rhythms may be modulated by peripheral immune challenge, by infections such as trypanosmiasis, as well as by treatment with specific cytokines (e.g. interferon-γ, tumour necrosis factor-α). There is also the suggestion that age-related changes in brain cytokine levels may contribute to alterations in clock function with senescence. Furthermore, actions on the circadian clock of neuro-immune modulators may contribute to the syndrome of "sickness behaviour" by producing alterations in sleep, feeding, attention and memory, and the roles of cytokines in the psychopathology of affective illness may also be partly mediated via circadian rhythm disturbance. This chapter will further examine these interactions and look forward to exciting avenues of research that are about to be opened in the field.

Chapter X - Morningness-Eveningness has been theorized as a continuum along which subjects can be arranged or classified into coarsely determined categories: owls (or Evening-types), intermediates and larks (or Morning-types). This trait is thought to be universal chiefly because it is biologically and genetically determined. The number of categories changes to 5 after adding the "Definitively Evening" and "Definitively Morning" extreme types. This procedure is justified for epidemiological, clinical and research purposes, and it is common in psychology, but it also raises a number of concerns the authors shall discuss in this chapter.

To proceed, one first needs some valid instrument: the Morningness-Eveningness Questionnaire is the most widely used scale, though others exist that deserve attention. Reliability is a necessary but insufficient property on the road toward validity of a scale's score. Next, beyond the fact that the commonly used methods for categorisation are not theoretically sound or equivocal (i.e. for what reasons should we use extreme deciles or extreme quartiles? how are the cut-off scores determined?), the most important criticism is that demographical variables such as age and gender are not taken into consideration. Some recent data emphasize that evening-types are not rare in middle-aged, an erroneous conclusion derived from the rigid categorization. To say it simpler, norms for Morningness scores have not been determined. Moreover, we, like others, have claimed that cross-cultural validity is a desirable property for scales intended to measure personality traits including Morningness-Eveningness. This property might be extended to the factors extracted in the studied cultures leading to the problem of factorial invariance.

In this chapter, the authors review the existing scales and their general properties in different cultures before focusing on the problem of categorization with examples showing its limits (e.g. lack of coherence between types determined on different scales). The authors will eventually propose some remedies. Finally, the authors will address the issue of cross-gender and cross-cultural factorial invariance of these scales. The authors conclude that the way is long that will lead to valid, between-group comparisons in terms of Morningness-Eveningness scores and/or frequencies of the corresponding types.

Chapter XI - In this brief review on the cortisol awakening response (CAR), the authors provide a description of this useful and easy to measure marker of hypothalamic-pituitary-adrenal (HPA) axis activity. The CAR is a discrete entity superimposing the circadian cycle. To date, the measurement of the CAR by collecting saliva samples in the first hour after awakening proofed to be useful in psychosomatic medicine and psychoneuroendocrinological research. In particular, the CAR technique allows the assessment of basal HPA axis functioning in ambulatory settings. Beside technical issues, known sources of inter and intra individual variability are discussed. Finally, the existing literature on associations with psychological states, chronic stress, and psychopathology is presented.

In: Progress in Circadian Rhythm Research
Editor: Anne-Laure Léglise, pp. 1-34
ISBN: 978-1-60021-876-7

Chapter I

Reinforcing Biological Rhythms with Hypnosis: A New Paradigm for Immunotherapy in Oncology?

Fabrice Kwiatkowski[1], Nancy Uhrhammer[2] and Alain Blanchet[3]
[1] Bio-statistician, Centre-Jean Perrin and University Paris-8
[2] Biologist, Centre Jean Perrin, Clermont-Ferrand, France
[3] Depaertment of Psychology, University Paris-8 (Saint-Denis), France

Abstract

Oncology is a domain of particular interest in chronobiology research, focusing the efforts of a wide range of scientists. A great deal of data concerning the association between biological rhythms and cancer is now available, associating genetics, cell biology, animal experimentation, epidemiology, and clinical research. Despite these efforts, little has been done to associate biological rhythmicity and therapeutics, although some teams have shown the advantage of such an approach on survival and toxicity control. There have been two major approaches: chronochemotherapy and melatonin supplementation. Preliminary data suggest that improving circadian rhythms may improve global cancer prognosis.

Several groups have studied psychosocial aspects in relation to cancer, initially intending to improve overall survival, with equivocal results. Presently such surveys focus on the quality of life and nursing management, although improved study design may show an impact on survival after all.

We present here the rationale for a new type of investigation, based partially on Ernest Rossi's suggested impact of hypnosis on immunity and gene expression, and then an original protocol being conducted among palliative care cancer patients. This is prospective randomized trial testing the impact of Ericksonian hypnosis on immunity and biological rhythms. Biological parameters and activity/rest rhythms are measured before inclusion in the trial and after two months; hypnosis intervention is planned once a week for 8 weeks. In the second arm, patients receive a training CD for relaxation that they are told to use *ad libitum*. Using suggestions aimed at behavioral changes and personnal insights, we hope to obtain an immunitary rebound thanks to the amplification of ultradian and circadian rest/activity rhythms. The principal measurements include rhythms of salivary melatonin, cortisol and leptin, rest/activity cycle measured by actimetry, temperature. Levels of interleukins IL-1, IL-2, IL-6, IL-15, INF-γ and TNF-α and lympocytes counts are included. Circulating expression of clock genes are measured in parallel with the one of the interleukin receptor genes. Other genes concerning hypnosability and some brain memory managment functions will also be studied. Overall survival and quality of life are secondary goals. A multicenter inclusion is planned; we intend to compare the effect of different therapists, and if significant differences are observed, to search for ways to improve hypnosis managment.

Last inclusions are forecast for 2009, for a total of 100 patients per arm, while an interim analysis should give us a good estimate of the final results after 30 patients are accrued in each arm.

Introduction

In ancient times, the Greeks and Romans never started an important project, such as the construction of a new temple, a long princely voyage, or the decision to engage in battle, without considering the augurs, sacrificing a beast to read in its bowels or asking the oracle what the omens said, or the astrologer to search for the time favorable to success. Centuries have passed and today's physicians do not call up the intervention of Asklepios or the medical knowledge of Chiro the centaur. The names that refer to particular illnesses (Parkinson, Hodgkin...) or specialized practices (Patey, Chiropractic...) have changed. Although the hours of the day are no longer governed by genies, they can remain favorable or unfavorable to specific therapies or remedies. If the former beliefs have faded away and this latter science has become well established, grounded on a vast body of scientific evidence, this is quite recent: in the 1970s, when a researcher at the CNRS[1] started to investigate in the domain of chronobiology and had to present his topic to his national supervisors, he was accused of flirting with some sort of witchcraft or esotericism [Reinberg, 1997].

Research into chronotherapy still draws the same kind of criticism from some scientists, despite the positive results obtained these last decades. New hypothesis need to be tested to keep up with the advances made in fundamental research, especially in oncology where therapeutic progress seems rather slow and no efficient treatment exists for several tumor types (kidney, pancreas, lung, brain...), and the prognosis for metastatic disease remains very poor. These new chronotherapy hypotheses take into account results from such diverse scientific disciplines as biology, medicine, epidemiology and psychology, together demonstrating the importance of rhythms in oncology. The underlying idea: *doesn't disruption of the global rhythmic harmony of the body contribute to cancer (and in a more general way to most pathologies)?*

The first part of this work will be devoted to rhythmicity and cancer. Then we shall try to assess the respective roles of the two major clock-hormones (melatonin and cortisol). The importance of the immune system in oncology and the mediating role it plays between cognitive functions and metabolism will be discussed, although cancer immunotherapies have

[1] CNRS : French National Center of Scientific Research

thus far been disappointingly inefficient. The fourth part will address the ubiquity of rhythms in the organism. Based on these considerations, we will propose a global chronotherapeutic approach. Finally, since hypnosis may be a method of choice to modulate immune function, a fifth section will propose a new type of psychological intervention against advanced cancer. The integration of this intervention in clinical practice will be described with the design of a trial that will start in 2008. The methods used to address these questions are the same as those found in other clinical trials.

1. Chronobiology, Psychosomatic and Cancer

1.1. Biological Rhythms Disruptions and Cancer

The first scientific references to a chronobiologic cause for carcinogenesis are quite recent and can mainly be found in Pukkala [1995], Davis [2001], Hansen [2001] and Schernhammer [2001], concerning the impact of frequent jet-lag, nights of insomnia, night-work or shift-work on breast cancer incidence, and also in Pukkala [2002] regarding a significant increase in melanoma (relative risk RR > 2) and prostate cancer among airline pilots: for this latter population, responsibility that could be attributed to exposure to cosmic radiation was rejected [Pukkala, 2003] although this was the authors' initial hypothesis. In 2005, a meta-analysis concluded that disturbed life rhythms increased breast cancer risk by 50% [Megdal, 2005].

In a study of the relationship between hypnotics consumption (number of nights with sleeping pills) and mortality in an American population of 1.1 million people, Kripke [1998] showed a significant association between the number of nights with hypnotics and the risk of death. Despite a very significant p value, the authors were right not to conclude that hypnotic drugs had a straightforward and dramatic impact on mortality. Rather, the constant use of hypnotics (i.e. sleep disturbance) was considered a strong indicator of a lethal risk (RR = 1.35 for men and 1.22 for women). Among breast or colon cancer patients, an unfavorable prognosis associated with use of hypnotics was also significant and similar to what was found with other causes of death.

Sleep disturbances are one indication of alterations in biologic rhythms. The prognostic value of biological rhythm disruption was corroborated by several studies: the first originates from F. Levi's team [Mormont, 2000] which showed differences in overall survival among metastatic colorectal cancer patients with good performance status: poor prognosis was correlated to a narrower amplitude of the rest-activity cycle. Others studies investigated the role of physical activity upon the prognosis of cancer: for example a side study of the CAL GB 89803 trial [Meyerhardt, 2006] about the impact of physical activity on recurrence and survival among stage III colon cancer patients. Physical activity appeared to reduce by two the risk of recurrence. Similar conclusions are suggested in breast cancer, where the primary aim of physical activity is to prevent weight gain during and after treatment, a gain which has proven to favor relapses. The discussion of this study focused on subjects other than biological rhythms (diet, weight, quality of life) [Rock, 2002], but on that point one could argue that physical activity modifies the rest-activity cycle and is very likely to favor a good sleep as well as cellular immunity.

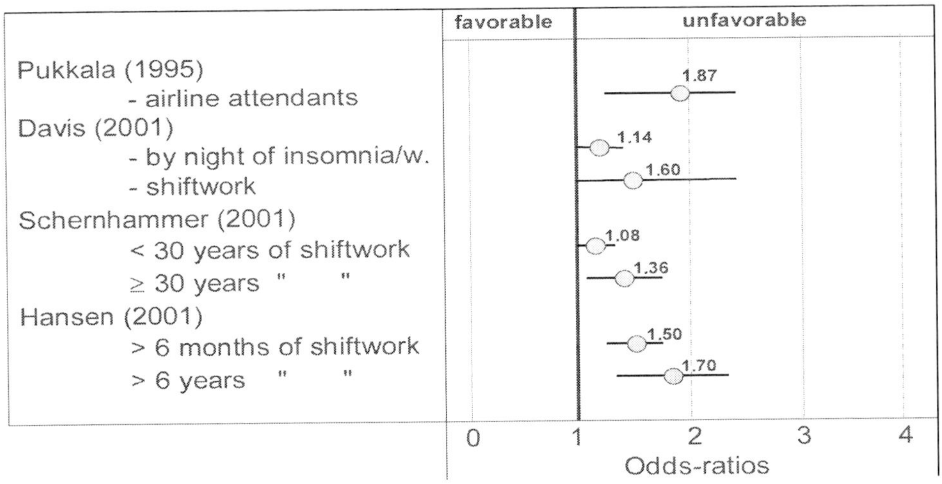

Figure 1. Odds-ratios[2] for breast cancer risk in studies investigating circadian rhythm disturbances (segments show 95 %-confidence intervals).

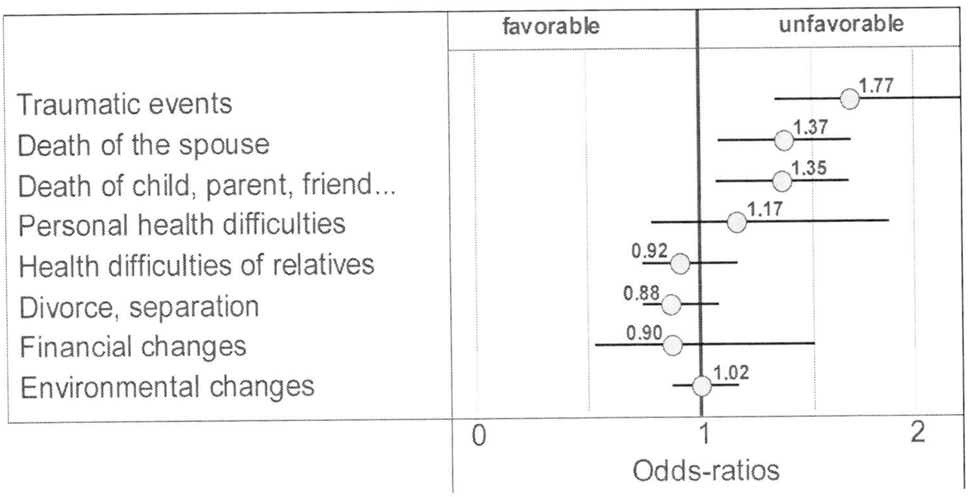

Figure 2. Odds-ratios associated with breast cancer risk in relation to life events happening not more than six months before diagnosis [Duijts, 2003].

[2] Relative-risk (RR) is the ratio of the frequency of a particular sign or disease in a group over the same frequency in a reference group. Odds-ratio (OR) applies to case-control studies, and differs from RR in the way it is calculated: if the occurrence of the sign or the disease is less than 20 %, these figures are comparable. Hazard-ratio (HR) is similar to RR except it applies to survival and is calculated using Cox's regression model.

1.2. Life Events and Cancer

Another way to estimate the impact of sleep troubles on cancerogenesis may be to consider stressful life events, since sleep difficulties are reported very frequently in both depression and post-traumatic stress disorder [Empereur, 2003]. Duijts [2003] investigated this point during the 6-month period preceding the diagnosis of breast cancer in a meta-analysis of 27 studies (figure 2).

The author concluded that, once heterogeneity had been taken into account, only one weak association with breast cancer risk was confirmed: that of the death of a spouse. But six months seems a very short period to investigate a link between stressful life events and breast cancer, which is thought to usually need several years before it is clinically detectable. Lillberg [2003] approached this topic with stronger means, considering a longer period (5 years) before the diagnosis of breast cancer and, instead of referring to mixed sources, she prospectively studies 10,808 women belonging to the Finnish Twin Cohort (figure 3).

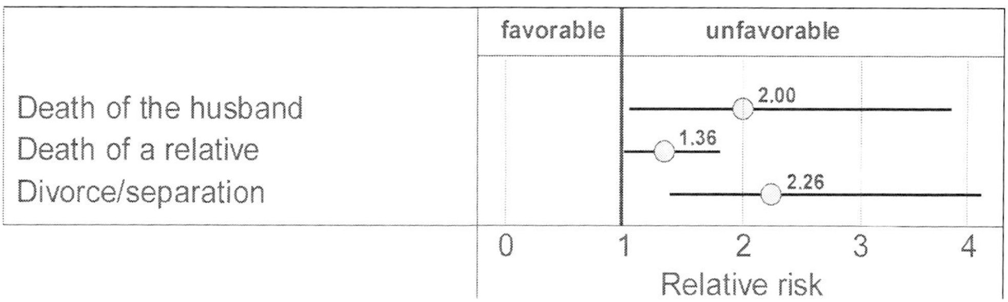

Figure 3. Association between life events and breast cancer risk in the following 5 years [Lillberg, 2003].

The results of her survey reinforce the tendencies that emerged from Duijts' works, with a doubling of risk of breast cancer for the twin who had lost her husband through either death or divorce. She concludes also that each stressful event induced a 7 % rise in breast cancer risk and a 35 % increase for each serious event (death of a husband, divorce/separation, personal illness or injury, death of close relative or friend, loss of a job). These associations were independent of the stress of daily activities, which appeared unrelated to breast cancer risk [Lillberg, 2002]. However, the reasons evoked for such relationships were not clearly defined by authors, citing "either hormonal or other mechanisms".

Two other case-control studies analysing adult life events should be cited. The first [Kvisktad, 1996] compared 14,700 mothers having lost a child to 29,750 who had not. No association between child loss and the risk of breast cancer (OR = 0.96 [0.87-1.07]), or overall survival (HR = 1.08 [0.92-1.26]) was found. Lambe [2004] in an even larger survey (27,570 cases and 141,800 controls), obtained the same globally negative statistics (OR = 1.05 [0.96-1.15]), but some differences arose within subgroups: OR = 1.27 [0.98-1.64] if the death concerned the first child and OR = 2.65 [1.06-6.60] if the lost child was less than 4 years old.

Other authors report early historical factors that could increase the risk of cancer in adults. For example, Jacobs [2000] found that the loss of the mother during childhood multiplied by 2.5 the adult risk of breast cancer. Tacon [2002] explored another direction :

Bowlby's attachment theory [Bowlby, 1973][3]. In his review, he reports the pejorative impact of an insecure childhood attachment on immunity and illness (including cancer) that could develop through a now well-described process: the hypothalamic-pituitary-adrenocortical axis responsible for the secretion of cortisol (one of the major circadian hormones). The additional risk of breast cancer in the sub-group of Lambe's study experiencing the death of a child less than 4 years old suggests that Bowlby's attachment theory may concern both the child and the mother during this sensitive period when interactions are crucial for development. Exploring a different kind of stress, Elias [2004b] found an increased risk of breast cancer (hazard ratio = 1.48 [1.09-2.01]) for adults who underwent the 1944-1945 Dutch famine, especially for those who were less than 10 years old during that war period. No significant relationship was found with any other cancer location [Elias, 2004a], raising the question of whether stress-related (whether physical or psychological stress or a combination) cancers might tend to be hormonally-driven malignancies.

One last way to correlate sleep loss and cancer may be to consider depression, a condition where sleep disturbances are a major clinical sign. This subject was reviewed by Dalton [2002], who showed a moderate relationship between depression and cancer (figure 4).

The global odds-ratio is around 1.2 but with a trend toward a greater risk when depression is major or chronic: for instance, in Penninx [1998a], the odds-ratio is close to 1.9, almost as high as the impact of a husband's death for breast cancer. In contrast and to compare with the other main cause of mortality in developed countries, chronic depression was less correlated to cardiovascular disease than new depression in the elderly [Penninx, 1998b].

Although many reasons have been proposed for these relationships between psychosocial factors and the risk of cancer, no definitive conclusions on the biological processes involved can be made. A few comments, however, may help untangle the skein: if these links are grounded, oncogenesis appears to be a very slow process that may gain strength through time with the accumulation of risk factors, and may manifest itself rather suddenly after severe life difficulties. This pathological process is likely to concern biological rhythms, in keeping with the data on shift-work or frequent jet-lag. The pathway seems more indirect for adverse life events, although sleep loss is a common element. And for famine, aside from attachment difficulties that may very well be more frequent in time of war and that have proven to weaken immunity in mammals [Coe, 1993], we should keep in mind the contribution of the associated nutritional and physical stress.

---

[3] Attachment theory is a psychological theory concerning the evolved adaptive tendency of very young primates to maintain proximity to an attachment figure, mainly a caregiver. Works of Harry Harlow (1958) are complementary : he showed that infant rhesus monkeys preferred emotional attachment over food. The attachment tie between infant and mother seems particularly important in humans where infants are for a long time very vulnerable. Insecure attachments have proven to induce among primates abnormal social behaviors.

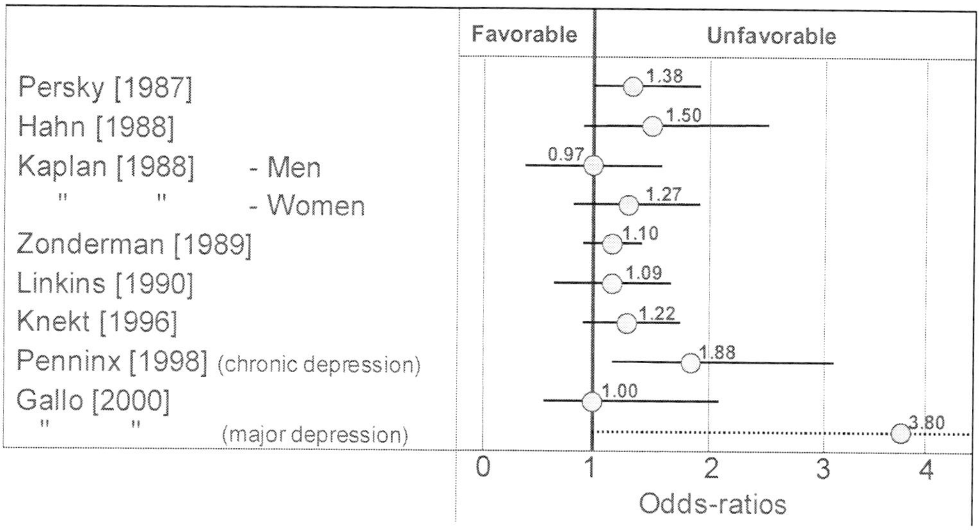

Figure 4. Association between depression and cancer risk [Dalton, 2002].

2. Cortisol or Melatonin?

The hypothalamic-pituitary-adrenal axis is often cited as a major component of the psychosomatic pathway. It constitutes a direct link between stress and physical/emotional reactions, and further, via the secretion of corticoids, with the immune system where it helps mobilize defensive reactions against attacks from external agents (UV rays, cosmic radiation, chemical agents...). Corticoids are secreted in equilibrium with melatonin and enable the metabolism to switch between diurnal catabolic functions (protection against external aggressions) and nocturnal anabolic ones (growth, repair, consolidation, cellular immunity...). Experimentation on animals has shown that stress can induce large changes in both cortisol and melatonin production and a general drop in immunity [Dhabhar, 1994 ; Campmany, 1996]. To us, cortisol and melatonin do not play the same role in carcinogenesis. The first seems not to directly intervene in the process except to perturb the production of melatonin, which may modulate immune surveillance for abnormal cells as well as oxidative stress in general.

2.1. Cortisol First?

One of the suspected general pathways for adult cancer susceptibility starts, to our mind [Kwiatkowski, 2005], with in-utero brain development. If this development is abnormal, it can favor the installation of ineffective coping strategies or deleterious life habits (in particular addictions) which in turn expose cells to chronic aggressions (alcohol...), increase the risk of mutations (via exposure to carcinogens in tobacco smoke, for example), and weaken the immune system.

Many animal studies have concluded that overproduction of corticoids provoked by stress in the mother influences the intrauterine growth of embryos, and particularly effects brain neurogenesis (for example delayed myelination of neurons [Huang, 2002]), which may in turn induce various behavioral alterations in the newborn individuals [McCrabe, 2001; Bardi, 2006 ; Owen, 2007; De Vries, 2007]. Although no similar experimentation can be conducted in humans, there is no reason to believe that this should not apply to our own species. In fact, the side effects of certain medical treatments may extend these conclusions, since the prescription of antenatal corticoid to prevent preterm birth has been associated with hyperactivity later in childhood [French, 2003]. Since corticoids pass through the placental barrier, maternal stress is directly transmitted to the embryo through the umbilical cord. The consequences of such exposure may be more serious than in adults because it occurs in a particularly important phase: that of brain formation. Because of this, maternal stress may effect the whole future of the individual. Several cerebral targets of hypercortisolemia during neurogenesis have been described [Gallarda, 2003], including reduced hippocampal volume (memory deficiencies and alteration of spatial functions) and increased amygdala volume (exaggerated emotional responses). The alteration of emotional control may also result from reduced numbers of corticoid receptors on neurons in the amygdala.

2.2. Finally Melatonin?

Nevertheless, no author yet evokes any direct relationship between this kind of alteration and future cancer. Indeed, an intervening stage needs to be addressed: life habits and life events modulated by coping strategies. As we learned from French [2003] that hypercortisolemia during pregnancy seemed to favor hyperactivity of the child years later, it is very likely that once adult, the consequences manifest themselves through poor coping attitudes adopted during youth, and more frequent addictive tendencies [Empereur, 2003]. We have already discussed life events that we suspect to favor cancer through disturbance of sleep and biological rhythms. Deleterious life habits appear to play a similar role, especially on melatonin secretion. In particular, alcohol consumption inhibits melatonin production in a dose-dependent manner [Ekman, 1993], as do some classes of hypnotics (benzodiazepines) [Rosenstein, 1990]. In animals, nicotine delays the peak of melatonin secretion and lowers global nocturnal production [Ferguson, 1999] as efficiently as exposure to constant light [Van den Heiligenberg, 1999]. Light during nocturnal sleep has been shown to favor breast cancer in humans [Glickman, 2002].

Many addictions alter the production profile of melatonin in the same way as old age, by diminishing the global amplitude of its production cycle [Bartsch, 2001]. The same change precedes menopause by 3 or 4 years [Vakkuri, 1996] and a trial of melatonin supplementation in 18 peri-menopausal women resulted in a 30 % recurrence of menstruation [Bellipanni, 2001]. Is exposure to estrogens then sufficient to explain that the peak incidence of breast cancer occurs an average of ten years after menopause [Menegoz, 1999]? Ageing is correlated to an increased incidence of cancer [Hill, 2003]. Should we add the diminishing rhythmicity of the pineal gland to the accumulation of risk factors responsible for oncogenesis?

Fortunately, good life habits may play a role on the other pan of the balance. For example, diurnal physical activity is now recommended for breast cancer patients in order to prevent weight gain, but it may very well produce other effects thanks to a different mechanism: physical activity favors the production of daily serotonin, an indispensable hormone without which no melatonin will be secreted at the beginning of the night (serotonin is a precursor of melatonin). And since melatonin possesses hypnotic properties, sleep onset is faster and easier.

On a biological level, many *in-vitro* studies have examined the effect of melatonin on tumor cells. It has been shown to exert anticancer activity through several mechanisms [Lissoni, 2007]:

- direct cytostatic or cytotoxic action by inducing apoptosis in cancer cells expressing melatonin receptors;
- stimulation of the antitumor immune response by promoting IL-2 release from TH1 lymphocytes expressing melatonin receptor, and by enhancing dendritic cell IL-12 production in response to IL-2;
- modulation of oncogene expression, with differentiating activity on cancer cells mainly through endocrine receptor expression, with consequently reduced biological malignancy [Regelson, 1987];
- melatonin may counteract the activation of EGF-receptor [Siu, 2002], whose expression plays an important role in determining the biological malignancy of cancer cells and their metastatic capacity.
- some anti-angiogenic activity has been reported by Pawlikowski [2002]
- melatonin is one of the most potent antioxidant agents [Reiter, 2002]

The last point may be the most important for cancer onset, since oxidizing agents frequently provoke mutagenic DNA damage in the nucleus and mitochondria. This damage have shown to be highly carcinogenic [Yamamoto, 2001 ; Athar, 2002 ; Carew, 2002].

For therapy against advanced cancer, prescription of melatonin (one or two hours before sleep, 1 to 20 mg orally every day for several months) has been tested either alone or in combination with conventional chemotherapy. Significant positive results have been obtained for disease control and survival [Neri, 1998 ; Lissoni, 1999 ; Lissoni, 2007]. Another advantage of melatonin supplementation during chemotherapy or radiotherapy: it significantly reduces treatment-related toxicities (thrombocytopenia, neurotoxicity, asthenia and cachexia [Lissoni, 2007]), mainly thanks to its antioxidant power. But melatonin is not a miracle medicine. Supplementation is not effective in all patients, nor in all types of cancer, nor does it prevent relapse among the patients who respond. In these instances, perhaps melatonin itself is not at fault, and one should question the quality and number of melatonin receptors on lymphocytes or wonder about other elements that could interfere with the normal cellular immune process.

At this point an interim conclusion seems useful: we have considered how disruption of biological rhythms could favor cancer. Thanks to epidemiology, we may hypothesize that life difficulties contribute to a higher risk of cancer, especially when these difficulties induce deep and durable perturbations and likely rhythm alterations. Then we examined how and at what stage both major clock hormones could intervene in a global carcinogenesis process. The first hypothesis is, for the moment, rather daring: it features corticoids in a prominent role in the stress cascade. The critical period starts *in-utero* and covers the entire neurogenesis phase (i.e. mainly until 3 years of age). During this period, excess corticoids may disrupt normal brain development, and favor long term behaviors deleterious to health. Related behaviors can lead to other types of life threats (for example dangerous acting, risky sports, violence...) and impeach life stability, and usually induce distress. Unfortunately, today the main social solution to containing distress remains addiction (to medicines or other agents). Here starts the second level of the intervention of melatonin. This nocturnal clock hormone has proven its importance in global body homeostasis, partly thanks to its role as a free radical scavenger and also because of its links to the immune system, a point we will consider later. Its production profile is easily disrupted by sleep troubles, light at night, changing life habits, chronic disease and drug consumption. Its reduced secretion is suspected to favor carcinogenesis; it has also proved to be an interesting complement to cancer treatment. But to introduce a new therapeutic approach, additional steps need to be treated: first the essential role that immunity plays in cancer, then the ubiquity of rhythms from cells to global human behavior. Finally, this will enable us to develop new strategies combining psychosocial and medical interventions, chronobiologic and immunologic dosages and gene expression measures.

3. The Essential Role of Immunity and the Melatonin Connection

In oncology, research concerning immunity is relatively recent, especially concerning the complex communication that enables cells to act harmoniously toward homeostasis. It is very likely that new elements and even new means of intracellular communication are still to be discovered. An example of such an unexpected discovery in another domain is that of small RNA fragments that were thought to be waste but are now known to be involved in RNA interference (RNAi), regulating gene expression, RNA turnover, and (at least in plants) immunity to viral infection [Carrington, 2003]. In this section, we will show that strategies based on immunity have for the most part tried either to use specific immune defenses against malignant cells (adding more of…), or to compensate for deficiencies of the whole system (replacement…). It is disappointing that in spite of valid hypotheses and well-designed protocols, the results were not as postive as expected.

But the immune system does not act independently. It belongs to an organism and exchanges information with the different components, generally in agreement with these components. The mind appears to be one of those components, although exactly how it interacts with the rest of metabolism we are only beginning to investigate. Much work needs to be completed before a clear understanding of the mechanisms bi-directionally relating mind and body is reached. For the moment, one thread can be drawn between various functions; its witness and representative is melatonin. As we have seen, this thread vibrates, echoes to our joys and pains, to our behavior and life style… to our remaining time.

3.1. Immunity and Cancer: Disappointing Results?

One of the characteristics of cancer is that possibly oncogenic mutations occur all the time. Current thinking is that about seven complementary genetic events may be necessary for the malignant phenotype to firmly establish itself. Even at that point, cancer usually cannot develop, since the immune system can eliminate the abnormal cells. Some of the mechanisms involved in this immune control have been discovered, and many hopes have been placed on those biologic substances responsible for intercellular communication and able to trigger actions at distance in other cells, hence their name: cytokines.

Several vaccine therapies, with the goal of stimulating the whole immune system and thus triggering an anticancer activity, have been tested, but demonstrated no significant effect. It is hoped that with the development of micro-array based gene profiling, new concepts will emerge and be developed into efficient therapies in the next decade [Mocellin, 2007]. Immunotherapies (notably the delivery of IFN-γ, IL-1, IL-2 or IL-6) have also focused much energy in the past decades with mixed results. A major target today is melanoma, where spontaneous recoveries have been described [Wagner, 1998; Curiel-Lewandrowski, 2002]. Renal carcinoma is a target for IL-2 therapy, but with mixed results suggesting that selection of patients is necessary [McDermott, 2007]. Only the recent success of monoclonal antibodies has crowned years of difficult research (for example Herceptin against breast cancer cells expressing Her2-Neu receptor). Unfortunately, these latter tools seem to present a real but narrow interest, mainly since cells in a given tumor are often heterogenious and thus are difficult to target with a single antibody. In contrast, if the antibodies are not specific to tumor cells, their use will induce drastic toxicities.

A completely different strategy has been used to compensate for cellular immune deficiency: allogenic bone marrow grafting is an approach with encouraging success in hematologic and solid tumors. Horowitz [1990] explained the activity of such a strategy through the "graft-versus-host effect". The host's malignant cells can be recognized and eliminated by a foreign immune system, a task the original system was not able to perform. For the implanted immune system to perform its cytotoxic role efficiently, it must be sensitive to the biological environment of the host, and in particular the receiver's levels of melatonin and cortisol. Although this therapeutic approach sometimes results in good disease control, relapses usually occur eventually. Does the oncostatic function of the new immune system gradually fade? Do the cancer cells become selected for evasion or resistance to lysis by the donor's lymphocytes? Or is it the biological environment that appears deleterious to the new immune system as happened with the original one? If this last hypothesis is correct, research needs to be done to describe what represents a good biological environment. Once this description is done, some new therapeutic means could be searched in order to promote in metabolism an "environmentally" sufficient level capable to keep the immune system efficient.

The hypothesis of tumor-immunity may be reinforced by the observation that many small tumors (thyroid, prostate) show no apparent evolution for years without any treatment and are sometimes discovered only after autopsy or by accident during surgery for another purpose. In Norway and Sweden, the very large study of Zahl [2004a] revealed a one-third increased incidence of breast cancer in women concerned by nationwide screening, in comparison to women living in regions not covered by the screening. The authors concluded that the main reason for this discrepancy was the over-diagnosis of low grade tumors that would never have progressed (2/3 of the screened tumors and about 1/3 if *in-situ* tumors are put aside). But should we exclude that in some cases, immunity may very well have played its tumor suppressor role against cells that it had been unable to control some months before when it was depressed? We mean that if malignant cells appear, it is likely that the immune cells have temporarily been either unable to recognize them or not in a state that permitted their elimination. But after a more or less long period, immunity seems to recover its abilities and performs its tumor-suppressor role in an efficient manner. Of course, one could hypothesize that a threshold exists beyond which immune cells are insufficient. In fact such an hypothesis does not solve the question : it just delays the problem.

This could happen more often than believed [Zahl, 2004b ; Zahl, 2006] since some types of micro-calcifications seen on mammography are thought to be scars of old tumors and also because histopathological examination after surgery often uncovers, aside from the invasive tumor, areas of fibrous tissues considered to have replaced the original intraductal cells [Horii, 2005]. The Japanese authors explain that this kind of immune reaction, principally mediated by lymphocytes, can eventually eliminate all malignant cells, "the ductal structure being changed into a round-to-oval scar consisting of concentric layers of collagen and elastic fiber". Since these phenomena occurred in 7% of the cases included in Horii's survey, it would be interesting to fully understand the immune process involved in such healing and then, based on this understanding, to elaborate the best treatment or management able to boost such mechanisms.

3.2. Mind and Immunity

Immunity is another self! It acts like an autonomous system that inherited a full set of skills and knowledge, is able to react, to learn and memorize. It can also forget and one has to train it if one wants to keep the security level sufficiently high. It seems to possess its own character traits: some are slow, others are fast. Some appear so reactive they can provoke general inflammations and others seem lazy, leaving the host in a constant sickly state. Some seem stupid and attack their own host perhaps because it is bored (well-vaccinated, for years it had no real enemy) or because it intends to protect the body from a misleading threat.

Although like another self, the immune system is not isolated: it entertains close dependencies with both the general metabolism and the brain. For example, battlefield surgeons discovered that wounded soldiers recovered faster when their army won than when it lost. Closer to common experience, herpes and shingles crises happen more frequently during personal difficulties. On the other hand, when a person is sick, he feels tired, loses his appetite and only wants to sleep. One of the reasons seems to be the action of proinflammatory cytokines (IL-1, Tnf-β, IL-6) on the central nervous system [Wood, 2006], that probably provoke fatigue so that more energy can be directed to rescue (or in the other sense, the redirection of energy resulting in fatigue). A normal switch between secretion of IL-2 versus IL-4 and IL-10 related to the rest/activity cycle was found in healthy individuals, where IL-2 seemed to enhance sleep [Marshall, 2002]. Interactions between brain and immunity are in fact continuous and diverse; mediated by many different types of messengers that have been categorized in three main groups: hormones, cytokines and neurotransmitters. These messengers also inter-communicate and some authors wonder if a new, integrated model considering these components as parts of a unique bi-directional interaction system would not be more satisfying [Petrovsky, 2001].

The best example of the narrow link between mind and immunity is found in the placebo effect. Commonly this effect accounts for 20 % to 30 % of recoveries. In cancer no such effect is thought to exist, but some examples have been reported, suggesting this effect should not be set aside in oncology. For instance, Rossi [1993] reported the case of M. Wright, who presented with metastatic lymphosarcoma. Despite his dramatic physical condition, the man had not lost hope for rescue and, awkwardly, he was sure that a brand new medicine he had heard about would cure him. Despite this drug not being indicated for his case, he persuaded the physician to include him in the laboratory trial, and against all expectations suddenly obtained a remission with disappearance of the main tumor sites. He was rapidly able to return to his activities. A few months later, he learned bad news in the medical press: preliminary results of the trial showed little effect of the medicine. He then experienced a rapid recurrence that drove him back to the hospital in very poor condition. But the doctors decided to lie to him since the placebo effect had worked so well once. They persuaded M. Wright that the first samples were not purified enough and that they awaited another batch, really more efficient. After a couple of days during which the hope of the patient culminated, they started new infusions, this time with purified water. And a new remission occurred, even more unexpected than the previous one. The patient died after several months, soon after definitive results were published in the medical press announcing the medicine had no effect at all.

But one case, even among rare others, does not make a significant statistic nor an argument for evidence-based medicine. It only suggests a direction for research until more scientific knowledge is available. In addition to the placebo effect having a limited range of action in normal conditions, it has another limitation, the "packaging". It is beyond our reach in many cases to make an appropriate suggestion and to present it in such a way as to be accepted by a particular individual. Also, the distance from suggestions to modified metabolism is a long and little-known path that cognitive and experimental sciences have only begun to investigate.

To sum up, it appears that the mind is one of the environmental components of the immune system, perhaps a major one. Therefore, psychological tools may exist that could modify immunity, that we could benefit from in a global strategy against cancer. Perhaps also, no immune rebound may be expected, with melatonin or cytokine therapies, if the mental environment continues to give deleterious orders.

3.3. Importance of Melatonin for Immunity

The importance of the immune system's functional shift between day and night, between catabolic and anabolic functions, has been very elegantly highlighted by Blask [2005] with his experimentation on rats xenografted with human breast cancer. The experimental treatment consisted of samples of human blood. While blood from premenopausal women collected during a normal night suppressed the proliferative activity of tumors, blood collected during the day did not. Moreover, nocturnal blood from women who had been exposed to light during their sleep was as inefficient as blood collected during the day, possibly because of reduced melatonin levels. This model also confirmed a previously developed point that corticoids do not seem to play any role against established cancer.

Several studies have investigated the role of the small lipophilic melatonin molecule, essentially secreted by pineal gland but also by lymphocytes and some skin cells [Carillo-Vico, 2004]. Humans express three types of membrane receptors (MT1, MT2, MT3), and two types of nuclear receptors (ROR and RZR). The large variety of tissues expressing these receptors attests to the various functions melatonin may exert in metabolism. MT1 receptors are mainly localized to the suprachiasmatic nuclei (SCN), the pars tuberalis of pituitary gland, the cortex, hippocampus, blood vessels and some immune cells. MT2 can be found in retina, SCN, and central nervous system. MT3 receptors are located in the hippocampus, cortex and hypothalamus. Nuclear receptors are present mainly in the pars tuberalis of the pituitary gland [Witt-Enderby, 2003]. The high concentration of melatonin receptors in the pars tuberalis of the adenohypophyse explains the influence of seasons (annual light cycle) on mammalian sexual activity and reproduction. This point is indirectly confirmed by the flattening of the melatonin circadian rhythm that anticipates menopause. The number of melatonin receptors on SCN indicates that this pineal hormone may exert retro-control on the major clock regulator. This function is necessary to recover from jet-lag or other shifts when the organism has to adapt to changing external conditions.

During the night, melatonin stimulates cell immune functions, in particular T lymphocytes and monocytes, by increasing their production of the cytokines IFN-γ, IL-1, IL-2 and IL-6. IFN-γ activates macrophages, increases the formation and activity of natural killer cells (NK). Interleukines –1 –2 and -6 are involved in the production and maturation of B and T lymphocytes. IL-2 stimulates NK activity and IL-6 intervenes in platelet formation. Finally, melatonin antagonizes the immunosuppressive effects of cortisol and thus favors the fight against viruses and bacteria [Von Gall, 2002]. It also plays a growth factor role for granulocytes in the bone marrow. Melatonin can induce apoptosis of tumor cells by increasing the expression of the tumor suppressor gene P53 [Winczyk, 2001; Blask, 2002], while preventing this response in normal cells [Currier, 2000].

4. Rhythms, Rhythms, Rhythms!

Rhythms of melatonin and cortisol, our two major circadian clock hormones, show how metabolism needs regulation in order to adapt itself to its multi-phasic environment without running out of energy. It reserves certain periods of time for growth and repair, while others are devoted to various activities: feeding, housing, reproduction, social relations, etc. The predominant rhythm is the rest/activity cycle based on the light/dark cycle, regulated by the impact of light on the retina, which sends signals to SCN. These signals are transmitted through intermediates to reach the pineal gland [Reppert, 2002]. Other rhythms can be found at all levels of life, from complex behavior to individual cells. In this section we will explore these different levels in the field of cancer treatment. The primary interest of this complex exploration will be summed up by identifying different entry points where chronotherapy could intervene for better efficiency and less toxicity.

4.1. Advantages of Chronotherapy

The problem with radiation and chemotherapy is that they intend to kill malignant cells, but they also kill healthy cells. The primary goal of chronotherapy was to help remedy this problem. Haus [1983] in his extensive review of chronobiology discusses some work on chronotherapy from the early seventies. The main principles were set with the early discovery in murine models of the relationship between biological rhythms and treatment toxicity or death. The essential interest of chronotherapy is to increase the therapeutic index of standard chemotherapy thanks to a sinusoidal delivery of the drugs. Delivery outflows are programmed to culminate when the drugs appear least toxic for the organism. If drug combinations have mixed toxicity nadir times, portable multichannel pumps can be used to deliver each product in its own schedule (figure 5).

Using this strategy, the standard maximum tolerated dose can be exceeded, and cytotoxicity against cancer cells is increased, often resulting in better tumor response [Levi, 2006]. In Europe, several thousand patients have taken advantage of chronochemotherapy, mostly for colorectal cancer. Although the pharmaceutical industry pays no real attention to this type of administration in clinical trials, such research should be sponsored, particularly for drugs that seemed to present real anticancer efficacy but with excessive toxicity. It is possible that certain abandoned drugs could put into use thanks to chronomodulated delivery.

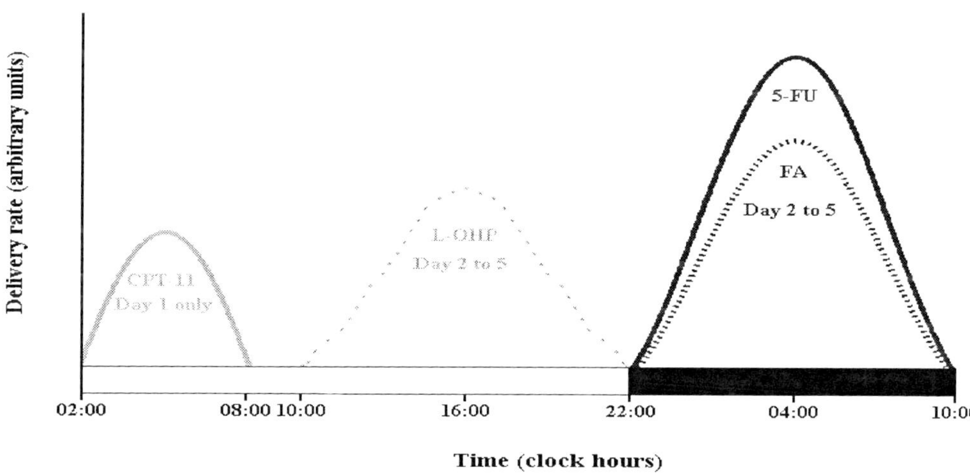

Figure 5. Example of a chronomodulated quadritherapy schedule for metastatic colorectal cancer: delivery of 5-FU (5-Fluorouracile), FA (folinic acide), L-OHP (oxalyplatin) and CPT-11 (irinotecan) [Focan, 2007].

4.2. Rhythms at the Behavioral Level

Long rhythms exist in mammals; seasonal periods are the most obvious, but there are also weekly and monthly periods [Haus, 1983]. Seasons regulate mammalian reproduction cycles through the rise in melatonin secretion during the winter, slowing down the production of sexual hormones after a hormonal cascade, and affecting sexual activity. Very short cycles also exist; ultradian behavioral rhythms have been described in humans. These were discovered by researchers investigating the causes of "human errors" responsible for so many accidents [Brown, 1982]. They found that attention varied in a diurnal cycle of about 90 minutes and, that after each period, the attention was lowered and some rest had to be taken, either by stopping the activity for a while, or by doing something that did not demand concentration (short coffee break, etc.). If this rhythm was respected, creativity and efficiency was increased. Klein [1979] discovered alternating sequences of the same length between the two brain hemispheres, letting us change cognitive style. During the night, a similar period separates REM sleep from deep sleep (slow wave sleep). Some authors think that the reinforcement of this ultradian periodicity may favor a stronger circadian activity/rest cycle {Rossi, 1993; Moser, 2006].

4.3. Rhythms and "Zeitgebers"

Zeitgeber means time-giver in German. It was defined by Aschoff [1965] as "cues which help the organism to establish and stabilize its biologic rhythms". We have talked indirectly about the major *zeitgeber* for humans: morning daylight and evening darkness. Because of its periodicity, it forces organisms to regulate their activity according to light/dark phases. Naturally, the sense of this regulation depends on life type, and nocturnal mammals (mice for example) have inverted cycles in comparison to humans and hence biological rhythms that are different although still circadian. Other *zeitgebers* are meals. In our modern society, television and artificial light probably alter the nightfall *zeitgeber*. In the past 50 years, members of developed societies have lost an average of one hour of nightly sleep, mainly because of night programmes. Finally, daily exercises may contribute to the realignment of circadian phase for shift-workers and transmeridian travelers [Barger, 2004]. As Moser explained [2006], these represent levers that could be used in a global approach to chronotherapy all together, but with certain flexibility since biologic rhythms comprise adjustment capacities and humans are not all alike.

4.4. A Succession of Biological Functions

Many biological parameters show circadian profiles, with peaks (acrophase) and nadirs, and while much variety can be found between individuals, on average and when pathological states are excluded, humans present similar rhythms. Some discrepancies can be found between reports exploring the circadian rhythms of these parameters, one of the possible causes being the time scale expressed sometimes in local time and sometimes not. Also, a single parameter may present more than one peak per day (ex. TNF-α, IL6). Nevertheless, it is possible to arrange a simple time-table indicating peak production. Factors other than normal endogenous periodicity may influence the secretion of certain hormones, such as meals [Haus, 1983] and sleep shifts [Simon, 1998] for insulin. For example, morning carbohydrate-rich feeding has been shown to induce significant alterations in melatonin production in humans [Krauchi, 2002]. Moreover, a peak should not be considered to have a very narrow slope or sharp summit; it is rather a point around which secretion culminates and often the major period of production lasts a few hours.

Table 1. Circadian production peaks of different biological parameters in peripheral blood

Peak times	Blood cell count	Cytokines	Hormones	Other
1:00	Lymphocytes T lymphocytes Reticulocytes		Growth hormone	
1:30			Prolactin	
2:00		IL-1β		
2:30	Eosinophils		Melatonin TSH	
3:00		IL-12	Leptin	
4:00		IL-1 IL-6		
5:00	B lymphocytes			
6:00		TNF-α		
6:30				Endorphin β
7:00			ACTH	
7:30		TNF-γ		
8:00			Cortisol	
9:00			Insulin	
10:00		TNF-α		NK cytolytic activity
11:00		IL-4 IL-10		
12:00	Redblood cells	IFN-γ		
13:00				Cell bone marrow proliferation
13:30		IL-2 TNF-γ		
14:00			Noradrenaline, adrenaline	Immune complement
16:00			Insulin	Immunoglobulin G-H-A
19:00	Neutrophils Platelets	IL-10		
20:00		IL-6		
21:00	Monocytes			
22:00		TNF-α		
23:00	Leukocytes			

Many other parameters could be added to this table, but it is sufficient to show the chain of processes that follow each other, with of course some overlapping since peaks do not indicate cycle duration. Under normal conditions, each function seems to come into play at a particular time, either because it requires some previous task to be completed before it can happen, or because too many functions at once would overwhelm the metabolic resources.

For instance, leukocytes are mainly produced during the night, while red cells have a production peak at noon [Haus, 1983], a moment when physical activity is maximal. Interesting too is the observation of NK cytolytic activity with a peak in the morning rather distant from the melatonin peak. Although global melatonin production and NK activity are correlated, this occurs through a long chain of intermediates, reported in paragraph 3.3. A shorter time separates the adreno-corticotrophic hormone (ACTH) and cortisol peaks, which seems logical since ACTH secretion by the pituitary gland produces a direct effect on adrenals.

Measures of rhythms have also been used to compare circadian relationships between peripheral blood variables and bone marrow proliferative activity in healthy young men [Smaaland, 1995]. The same kind of periodicity has been shown in bone marrow, but with phase shifts of 4 to 12 hours.

4.5. The Clock-Genes

The lowest level of rhythm was shown at the cellular level, with the transcription of genes, in the late 1990s [Tosini, 1996; Balsalobre, 1998], and this discovery brought biologists to speak of peripheral clocks. At least a dozen of clock-genes have since been described in mammals (*Per-1*, *-2*, and *–3*; *Cry-1*, and *–2*; *Tim*, *Bmal1*, *Clock*, *Rev-erb*, *CKIε*, *Dec-1* and *-2*), most of which are also present in Drosophila. In nature, when a solution appears to work, it is kept.

These peripheral clocks do not work independently from the master biological clock ruled by the SCN. Thanks to regulation of pituitary and pineal secretions by the hypothalamus, the master clock coordinates the local tasks of the peripheral clocks in tissues. For example, Motzkus [2002] showed that in humans, the peripheral clock gene *Per1* was induced by IL-6, an interleukin mainly secreted by immune cells (T lymphocytes, monocytes and fibroblasts) that are stimulated by melatonin. In whole blood, clock genes in normal subjects show circadian expression patterns, in particular *Per –1*, *–2* and *-3* (peak around 6 a.m.), *Bmal1* and *Clock* (peak around 2 a.m.) [Takimoto 2005]. Interestingly, these peaks occur during the sleep phase, thus during the nocturnal period of melatonin secretion. In the same survey, a patient with sleep disorder presented a phase shift that could be modulated by treatments such as light therapy, exercise and medicinal therapy.

Most important for oncogenesis is the discovery of the key role of clock genes in DNA damage-responsiveness [Fu, 2002]. After gamma irradiation, *Per2* mutated mice showed a marked increase in tumor development, in particular because of deregulated transcription of the oncogene *c-Myc*. Thus the loss of periodic functions at the cellular level may interfere with many different functions via this multifunctional transcription factor, since Myc is involved in the regulation of cellular proliferation, apoptosis, cycle checkpoints, and DNA repair [Dang, 2006].

The multiplicity of clock-genes may probably be explained by their different roles in complement to their clock function [Numano 2006]. *Per* genes for example intervene in chromosomal unfolding in preparation for transcription, and in the regulation of leptin during bone formation [Fu, 2005]. Bmal1 possesses specific functions which help regulate physiological integration of tissues [McDearmond EL, 2006]. In Dorsophila, *Per* genes play a key role in the formation of long-term memory [Hamet, 2006]. Clearly, if no rhythm occurs or is dischronic, either no function is performed or functions are performed that are not appropriate and may be deleterious. To finish this short evocation of clock-gene interactions with different factors, Spanagel [2005] showed that addiction to alcohol could alter clock-gene expression.

4.6. Synthetic View for a Global Chronotherapy Approach

The extensive bibliography of works and discoveries in the field of chronobiology allows us to identify different levels where therapeutic efforts may produce an effect. Of course, from a public health perspective, it would be interesting to promote social changes that may contribute to the reduction of risk factors in general. For example, as some countries encourage people to walk or to eat so many fruits or vegetables per day, so government or associations could promote healthy rhythms of life: limit shift-work, holidays in winter, early bedtime (except for people with natural 'night owl' habits), education about rhythms, better health care for pregnant women and infants to reduce stress, limitation of addiction, school and work schedules adapted to ultradian rhythms, etc. But in our consumer society, to expect such political measures is day-dreaming. As usual then, measures will have to be taken in emergency, once the disease is diagnosed.

We propose that a global, innovative immuno-chronotherapy should include three major components (figure 6). Interventions should be performed at each level in order to prevent any blocking effect from one level over the other.

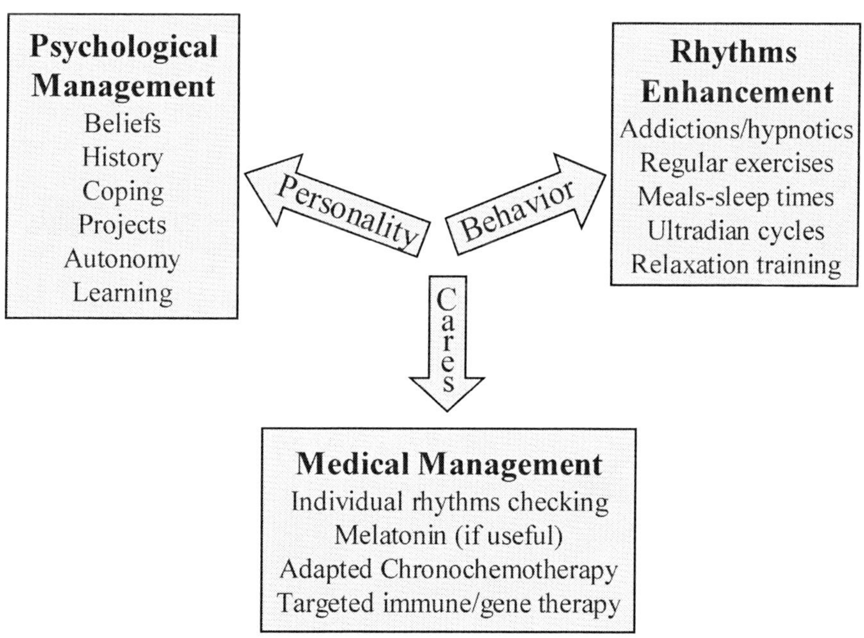

Figure 6. Three-component global approach targeting rhythms in cancer patient management.

Rhythm enhancement (top right) cannot be expected without important behavioral changes from patients themselves. Addictions and use of hypnotics should be suppressed if possible because of their direct impact on melatonin secretion. These may also interfere with and minimize the benefit of psychosocial interventions. Environmental rhythms (sleep and meal times) seem to be the most accessible to regulation, although in order for meals to remain effective *zeitgebers*, snacking should be proscribed. Advice concerning food quality is beyond the scope of this paper, and could occupy a book by itself. Ultradian rhythms (the approximately 90-minute cycle) deserve special mention. For example, although most people are informed about regular physical exercise (walking, jogging, swimming, stretching), few people know how to relax or breathe correctly, even for a short while. It is very unusual to advise people to isolate themselves in order to find a quiet moment and place, where they can just do nothing at all, have a few minutes meditation or save a moment for introspection. Indeed, such practices may be unbearable to persons who try to escape from their inner world by involving themselves in continuous activity, occupying their attention with perpetual distractions, fearing silence over all (ex. People who turn on the television as soon as they arrive home, although they do not look at it). For these people in particular, art therapy may be an excellent indirect approach to introspection.

Besides cultural[4], there might be some reason for maladaptive behaviors: for instance, alcohol acts as an antidepressor, and so does tobacco. This is where a psychological or psychosocial intervention may play a major induction role, by helping the background problems to be solved or lightened. As detailed elsewhere [Kwiatkowski, 2005], long-term interventions such as psychoanalysis cannot be recommended for cancer pathologies for which survival is likely to be short. Therefore brief interventions are preferable. We will address the use of Ericksonian hypnosis in the next section, a practice that seems to be a psychological lever of choice for several reasons. But other kinds of psychotherapy or personal development training may very well be proposed.

Of primary concern is the control of the patient's biological rhythms. As Moser said [2006]: "Optimal therapy should incorporate flexible and individualized rhythms, instead of metronomic periods, "*Kairos*" instead of *Chronos* (right timing instead of arbitrary set timing)." This principle has been studied in a clinical setting in an EORTC multinational prospective trial including 564 metastatic colorectal cancer patients comparing a 2 day flat regimen to a 4 day chronomodulated one [EORTC, 2006]. The overall results were disappointing, but further investigation showed that while the arbitrary timing of the chronomodulated schedule was correct for men (and improved their survival), it was not for women and they did not benefit from it. Undoubtably, personal adaptation of the schedule could be easily achieved using salivary dosages of melatonin and cortisol, temperature and actimetry. Data collected and analysed before the onset of chemotherapy could help predict the individual moment of best tolerance. Some adaptation could be performed during the chemotherapy cycle, with pharmacokinetic follow-up of 2 or 3 blood samples during drug delivery, as is already done sometimes to anticipate toxicity. If a delay appears between

[4] For example, in our society, to share a glass of alcohol sometimes appears as a convivial attitude

observed and expected kinetics, it is possible to correct the delivery schedule for the following cycles.

5. Hypnosis, Immunity and Cancer

5.1. Why Hypnosis?

In French the same word means packaging and conditioning. We spoke of packaging in section 3.2 about the way a suggestion should be made to the best effect. Ader's finding that a conditioned immune response could be obtained in animals [1993] particularly concerns our topic. The experiment was rather simple, consisting of the combination of an unconditional immune stimulus (ex. the delivery of a dose of Cyclosporin) with a neutral stimulus (ex. a colored light, a bell or some food) and the repetition of their presentation together. Each time after the infusion, blood was sampled and some immune factors measured, the results showed an immune drop as expected. After a certain number of iterations, only the neutral stimulus was given, but surprisingly, there was a similar drop in immune markers without the use of any drug. Immunity can thus be conditioned.

One could object that there is a great distance from Pavlovian conditioning to hypnosis, and we do not propose to lump the two together. What Ader's finding shows is that a non biological intervention can modify an immune response. In other words, immunity can be trained, even if this experimentation does not prove it can be trained in a positive way. Of course, with vaccination, it is already possible to train the immune system. The difference with Ader's experiment concerns the means used for the training. If conditioning worked with his dogs, this means first, that neuronal activity can have a direct and rapid effect on immunity. Second, these neural interactions with immunity can be developed, trained and stimulated by a environmental intervention. In everyday life, this can consist of some repetitive stress or perturbation. For our research, it can be psychological interventions and in particular hypnosis. Hypnosis, more than other approaches, has shown a real ability to produce a psychosomatic effect. A well-known example concerns hemorrhage [Enqvist, 1995]. In a trial, a hypnosis tape containing pre-operative suggestions was associated with a 26 % reduction in blood loss during maxillofacial surgery. Suggestions also permitted a significant decrease in blood pressure. Banks [1985] also reported the interest of suggestion in controling gastro-intestinal hemorrhage.

Several trials have observed positive effects of hypnosis on immune-based diseases. For example, Hackman [2000] reviewed the literature associating hypnosis and asthma and concluded in favor of the efficacy of this practice in this pathology, while Marshall [2000] reported several clues showing associations between stress, immune regulation and asthma. Langewitz [2005] attested similar advantages in a randomized prospective trial of patients suffering from hay-fever: patients trained for self-hypnosis showed a 29 % reduction of pollinosis symptoms during the following year and 25 % the year after. The improvement in well-being was similar and also statistically significant. Finally, in cancer, Spiegel [1989] showed an impact of psychosocial intervention on survival and quality of life for metastatic breast cancer patients, with a doubling of median survival. The intervention comprised training in self-hypnosis by patients in the test arm. Although the methodology of this latter study has been criticized, it is the only trial testing hypnosis on survival among cancer patients, and one of the few trials that obtained a significant result for this end-point. It does represent an interesting starting point.

The management of chronic pain is another domain where hypnosis has shown major successes. Even pain related to terminal phase cancer can be partially or totally suppressed [Erickson, 1959]. To bridge the gap between conditioning and hypnosis, Erickson explained that hypnosis uses the resources of our past conditionings in a positive fashion. We have all learned to forget about a terrible pain when a very threatening event happens, and for this, an immediate action is required. Furthermore, we can repeat the performance each time it is necessary, without even being aware of how we do it. Why not use this ability when we suffer physically or mentally? The main and usual problem, argued Erickson, is that we are simply ignorant of the fantastic possibilities of our mind.

One last argument in favor of hypnotherapy that could encourage its more generalized use as a complement to standard cancer treatments, is that it is well accepted by oncologists [Newell, 2000], and considered to be harmless : *primum non nocere* !

5.2. What to Investigate with Hypnosis?

As shown in figure 6, two main axes can be addressed during hypnosis sessions, the first concerning personality and the second behavior. A third axis could be evoked: it concerns the hypnotist himself, who can also provide data.

5.2.1. Personality Axis

Psychological investigations should concern in priority all that the cancer has changed in a patient's life and thoughts. According to Ronson [2004, 2005], cancer itself causes a kind of post-traumatic stress disorder (PTSD), that starts soon after the announcement of the diagnosis and that may again impair the immune system. Some psychological "repair" may be needed to deal with this shocking revelation and its consequences. The same care could be necessary for other eventually related material, like older life events that have penalized the patient's life for years and sometimes since infancy, even if such a belief may be imaginative. Very practical tools exist in hypnosis for the treatment of life history.

Of course, as Erickson remarked, the disease may offer some secondary benefits (more attention from a spouse or relatives, social consideration, end of an unpleasant job, health insurance that pays for lost income...) and some sessions may be assigned to the discovery of alternative solutions by the patient.

It seems particularly important for patients with no hope of recovery to reintroduce the ability to build projects for the future, even if this future seems short. For example, in the very special situation of such a severe disease, the need for new insights and/or beliefs concerning essential questions may be answered quite easily with hypnosis.

Confidence in the powerful resources of the human subconscious may emerge from such an investigation in trance. Since nothing can arise without motivation, everything that favors this force should be encouraged.

Another topic that therapists will have to face is approaching death. Fortunately, hypnosis is a good context to represent this major event, and since palliative cancer patients are in a situation where this question will necessarily arise, hypnotists should prepare in advance some processes adapted to such a demand.

5.2.2. Behavioral Axis

Once new beliefs are introduced (if needed), coping attitudes may change and old addictions become useless. Since addictions contribute greatly to immune weakness, such an evolution is encouraged. The autonomy of patients can also be increased thanks to self-hypnosis training that can be integrated into the global psychological management of patients.

The recommendation to respect ultradian rhythms may be devoted to a short trance, and such a practice may be reinforced by post-hypnotic suggestion. A similar approach can be proposed to reinforce circadian rhythms and sleep quality. We have not discussed breathing, which with the pulse, is one of our most evident rhythms. A deeper and wider awareness of these functions may very well serve as starting point for indirect suggestions concerning immune functions, since direct suggestions have proven to be inefficient (ex. fights between white warriors – immune cells - against dark enemies - cancer cells). Since breathing has a direct impact on the sympathetic and parasympathetic nervous systems, and thus on immune function, this domain deserves some special investigation. Unfortunately, such a topic is very large and beyond the scope of the present work.

Physical training may need too to be emphasized from the patient's point of view. Different strategies may be useful to encourage the patient to engage himself in more practice. Erickson's regression in the past appears a very good tool for that. Other strategies depend on data given during the patient's interview. Of course, the same encouragement can be given for art skills that combine both pleasure and project.

5.2.3. About Hypnotists and Sessions

How, at the end of experimentation, will we be able to know what works and what does not? The patients do not seem to be appropriate witnesses to discriminate between the different processes they will have undergone or asked for. Of course, the statistical analysis may show us a therapist-effect and leave it unexplained, arguing this might arise from some kind of personal quality or knack. As another extreme, we could describe a series of processes that should be given one after the other to all patients independently of their will. This latter solution should not occur for two reasons:

- First, with Ericksonian hypnosis, the patient is the director of the process and the hypnotist only a backing point.
- Second, it is possible that some hypnotists will obtain results while others do not, due to different approaches to therapy. If this is so, the obligation to follow fixed processes would impede the introduction of new favorable practices and limit our chances of progress.

If we knew what worked, such an attitude would be relevant. But we do not at present, and the choice that remains to us is to ask hypnotists to document the events of each session.

An estimate of the depth of the trance should be included since a direct impact of hypnosis on the soma may be discussed. According to Roustang [2003], it is possible that the surrender induced by hypnosis is in itself sufficient to provoke deep changes (not only in the personality of an individual, but also in his body).

In the same order, the quality of affects (positive or negative) between patient and hypnotist need some estimate. Like trance depth, it can be noted on a visual analog scale after each session. From a statistical point of view, these data should be corrected before analysis, in order to erase large differences between therapists' appreciation: for example by dividing each score by the therapist's average notation.

The therapist's judgement of the importance of the patients' insights can be reported. Some factual data can also be noted: reminders during or between sessions, age and type of traumas selected, recurrence/disappearance of old pains or fears, patient's new attitudes/beliefs concerning his disease, and other elements that he considers important.

Introspection may lead the patient to develop new values. Our western cultures usually favor activity, movement and social involvement. Sickness may be a good opportunity to center the patient on his real desires, to encourage the expression of his deep will and inclinations. He may discover his whole life has been directed by others, or that his beliefs do not properly belong to him. He may also realize that his working life has been a soporific burden and become depressive: such an insight must be handled very carefully by the hypnotist. Such psychological management should only be given to skilled professionals. Therefore, the hypnotist should also be either a psychologist or a trained physician.

Finally, it is advised that several therapists be involved together in this kind of investigation. Large discussions are recommended before the trial starts and several meetings all along accrual should be reserved to share impressions and discoveries. In our domain, progress will only arise from objective confrontation and be validated by assessed methodology.

5.3. Time to Act: Design of a Trial

The right methodology for this kind of research is a randomized prospective trial. Of course this can only be an open-label trial, where both patients and therapists know who gets what. Therefore, we have designed a multicentre trial that will start in the summer 2007 and last two years. Its purpose is to measure the impact of hypnosis on immunity among cancer patients in palliative care [Kwiatkowski, 2007]. It includes most of the points developed in figure 6, excepting chronomodulated chemotherapy since patients receive a palliative treatment. Another regrettable missing element is the use of melatonin. In France this molecule is not considered a medicine by AFSAPPS[5], nor may it be utilized as an dietary supplement. With special temporary authorization, the role of melatonin will be addressed in

[5] AFSAPPS : the French equivalent of the US Food and Drug Administration

the future, since we expect to build a complementary trial when this one ends. Patients will be randomized to two arms: in the test arm, patients will receive 6 to 8 personal hypnosis sessions. Session will be spaced once a week or every ten days, spread over a total of about 2 months. Patients in the control arm will either receive a relaxation CD they will be asked to use *ad libitum*, or they will be placed on a "waiting list" and receive hypnotherapy 2 months after inclusion. We may prefer this second solution since it appears more ethical.

Four cancer locations will be included: breast, prostate, lung and colorectal cancer. An OMS performance status of less than 3 is required, and life expectancy superior to 6 months. brain metastasis represent an exclusion criterion. Signed informed consent will be requested before inclusion.

The primary end-point for this trial concerns immune parameters and rhythms. Concerning immunity, we have focused our attention on cytokines (IL-1, IL-2, IL-6, IL-15, INF-γ and TNF-α) and blood lymphocytes counts (CD-4 T-helper, CD-8 T-suppressor, CD-56 NK). The count of B lymphocytes (differentiation class 19) more related to specific immunity is also included. Because of the periodicity of leukocyte production, samples will be taken before breakfast at 8:00. Salivary kinetics of melatonin, cortisol and leptin (this hormone has shown some NK stimulation properties [Zhao, 2003]) will be assessed every 3 hours between 8:00 and 23:00. Body temperature and actimetry over 2 days complete these kinetics. These measures will take place for all patients at the time of inclusion and after 2 months, in order to be able to compare the evolution of immune parameters between the two arms. Secondary end-points include quality of life, pain, sleep quality, reduction of addictions, and overall survival.

A gene expression study will complete the analysis. We will quantitate the expression of genes related either to circadian rhythms or to immunity, and that have proven to be expressed in whole blood. The following genes are included:

- Clock and Per-3 (but not Per-1 or Per-2 since all three Period gene expressions are correlated) [Takimoto, 2005].
- MLT-1, MLT-2, MTRN1 a-b, ROR a-b (melatonin receptors)
- rIL-1, rIL-2, rIL-6, rINF-γ and rTNF-α (receptors of the corresponding cytokines)
- ESR1 (oestrogenic receptors)
- NR3C1 (NK glucocorticoid receptors)
- Ob-r_l, Ob-r_s, LEPR (leptin receptors; the two first appear to promote NK cytolytic activity [Zhao, 2003])
- COMT (Catechol-O-Methyltransferase) (this gene is thought to characterize hypnosability [Rossi, 2005])

Two hundred patients are to be accrued, one hundred per arm. Considering Neri's trial [1998] showing a 40 % increase in interleukins after a melatonin-based treatment, we anticipate a 20 % rise in interleukins and/or in the amplitude of circadian rhythms. But taking into account the subjects needed to observe a 0 % versus 20 % increase (standard deviation = 20 %), risks α = 5 % and β = 20 %, a bilateral situation and Bonferroni adjustment for multiple testing (10 parameters), 35 patients appear sufficient in each arm. Stratification will concern centers (or hypnotists), location of cancer and sex.

An interim analysis is programmed after the accrual of 35 patients.

6. Perspectives

The unifying thread of this work can be summed up as follows: we first described the importance of biological rhythms in the cancer process. Second, several arguments have highlighted the possible impact of distress, trauma and major life events on rhythm disturbances, a connection that could account for the related increase of cancer incidence. Then we focused on the two master clock hormones, cortisol and melatonin, evoking the idea that chronic stress, via hypercortisolemia, could impair brain neural development and function, especially if these deleterious conditions happened during early childhood, during the main phase of neurogenesis. Later, especially if childhood has been perturbed but also if the culture encourages them, stressful life difficulties may be faced with poor coping attitudes, inducing sleep troubles and favoring addictions. The second master clock hormone, melatonin, is then disrupted, its rhythms flattened, its production lowered.

Melatonin rhythm and production level are indispensable for efficient cellular immunity. This hormone synchronizes many biological functions, including gene expression (intermediate messengers may intervene). Melatonin also eliminates free-radicals, promotes NK activity and apoptosis of mutant cells, and helps DNA repair in non-malignant cells. Both frequent time shifts and strong personal difficulties converge toward the same deleterious effect on melatonin production and on an increased risk of cancer, indicating a long-term "psychosomatic" cancer pathway. We then pointed out that spontaneous healing of cancer had been observed and that the most consistent hypothesis for such recoveries involves an immune rebound. The first condition for this rebound appears to be the enhancement of biological rhythms. Moreover, several clues suggest that a possible cause of immunotherapy failure could be partly due to an "environmental" shutdown of the immune system. Because of the numerous interactions between mind and immunity, we finally suggested a way to favor rhythm reinforcement: cognitive approaches on different levels offer to our mind the best perspectives, and hypnosis appears to be an especially efficient biological lever. Such an approach, to take advantage of all the data of chronobiology, should be completed with complementary patient management aiming for a global rhythm-reinforcement strategy: regular physical exercise and meal times, pauses every 90 minutes to relax or escape for a while, strong *zeitgebers*, and eventually melatonin prescription at the beginning of the night. Of course, cancer chemotherapies adapted to individual rhythms will complete patient management in order to minimize toxicities. Co-treatments impairing the immune system (corticotherapy…) should be questioned.

Clinical trials testing some of the management proposed here are easily feasible and do not require large populations. Of course, they should include follow-up of immune parameters and specific gene expression measurements. Such data are today necessary for real understanding of the mechanisms governing the mind-immunity interaction in the future. Most care management strategies respectful of chronobiological conclusions do not cost much. No new drug needs to be discovered hence no huge financial profit may be expected. Awkwardly, this may represent an obstacle to chronotherapy. The only indispensable matter for the present research is time and sweat. It is therefore very likely that only patient associations, personal sponsoring and government funds will support such research. On the other hand, psychological questioning may echo more general social problems and the global cultural loss of reference. Are we ready for such challenges?

Back to clinical considerations, there is a gap between scientific methodological concerns and rapid optimization of therapeutics. For example, it may be tempting to install this new therapeutic approach and bring modifications to actual cancer treatments so they fit the new paradigm. Even if such a strategy represents a real improvement for patient health and quality of life, it would probably delay a real understanding of the very biological and psychological mechanisms, efficient and useful for therapy. Probably too, such an attitude would induce a great amount of useless practices that already burden the daily medical workload. In 2002 with Francis Levi, the chronotherapy group of the EORTC[6] [Coudert, 2002] declared: "it is time for chronotherapy!" It still is, but at present many more scientific contributions converge toward this yesterday certainty. What differs from them: this certainty has filled out with parts coming from all domains of science, including neuroscience, and biological tools have progressed so that individual adaptations appear possible. We are probably allowed to think that treatments, either pathological or psychological, still need considerable efforts to meet the desired results against the disease.

Today, we know that our consciousness oscillates at a rhythm programmed by our genes, shaped by our past and regulated by environmental conditions. Let us be positive and bet that tomorrow, it will be possible to counter-balance the erratic growth rhythms of cancer cells with well-suited medical treatments and cognitive individual interventions... delivered at the right moments of the day.

References

Ader R, Cohen N (1993) Psychoneuroimmunology : conditioning and stress. *Annu. Rev. Psycho*; 44 : 53-85

Arjona A, Sarkar DK (2006) Evidence supporting a circadian control of natural killer cell function. *Brain Behavior and Immunity;* 10: 469-76

Aschoff J (1965) Circadian rhythms in man. *Science*; 148: 1427-32

Athar M. (2002) Oxidative stress and experimental carcinogenesis. *Indian J. Exp. Biol*; 40(6):656–67.

Balsalobre A, Damiola F, Schibler U. (1998) A serum shock induces circadian gene expression in mammalian tissue culture cells. *Cell*; 93: 929–37.

Banks W (1985) Hypnotic suggestion for the control of bleeding in the angiography suite; *Ericksonian Monographs*, I: 76-88

Bardi M, Huffman MA (2006) Maternal behavior and maternal stress are associated with infant behavioral development in macaques. *Dev. Psychobiol*; 48(1): 1-9

[6] European Organisation for Research and Treatment of Cancer

Barger LK, Wright KP, Huges RJ, Ceizler CA (2004) Daily exercise facilitate phase delays of circadian melatonin rhythm in very dim light. *Am. J. Physiol. Regul. Comp. Physiol*; 286(6): R1077-84

Bartsch C, Bartsch H, Blask DE, Cardinali DP, Hrusheski WJM, Mecke D (eds) (2001) The pineal gland and cancer: Neuroendocrine Mechanisms in Malignancy, Berlin : Springer

Bellipanni G, Bianchi P, Pierpaoli W, Bulian D, Ilya E (2001) Effects of melatonin in perimenopausal women: randomized and placebo controlled study. *Exp. Gerontol*; 36(2): 297-310

Blask DE, Sauer LA, Dauchy RT (2002) Melatonin as a chronobiotic/anticancer agent : cellular, biochemical, and molecular mechanisms of action and their implications for circadian-based cancer therapy. *Curr. Top. Med. Chem*; 2(2): 113–32.

Blask DE, Brainard GC, Dauchy RT, Hanifin JP, Davidson LK, Krause JA, Sauer LA, Rivera-Bermudez MA, Dubocovich ML, Jasser SA, Lynch DT, Rollag MD, Zalatant F (2005) Melatonin-depleted blood from premenopausal women exposed to light at night stimulates growth of human breast cancer xenografts in nude rats. *Cancer Res*; 65(23): 11174-84

Bowlby J (1973) *Attachment and loss*. Vol. 2. New-York, NY: Basic books

Brown F, Graeber R (Eds), (1982) *Rhythmic aspects of behavior*. Hillsdale, NY, Erlbaum

Campmany L, Pol O, Armario A (1996) The effect of two chronic intermittent stressors on brain monoamines. *Pharmacol. Biochem. Behav*; 53(3):517–23.

Carew JF, Huang P (2002) Mitochondrial defects in cancer. *Mol. Cancer*;1(1):9–20.

Carrillo-Vico A, Calvo J, Abreu P, Lardone P, Garcia-Mauriño S, Reiter R, Guerrero JM (2004) Evidence of melatonin synthesis by human lymphocytes and its physiological significance: possible role as intracrine, autocrine, and/or paracrine substance. *FASEB J*; 18(3): 537-9

Carrington J, Ambros V (2003). Role of microRNAs in plant and animal development. *Science;* 301 (5631): 336-8

Coe C (1993) Psychosocial factors and immunity in non human primates: a review. *Psychosom. Med*; 55: 298-308

Coudert B, Focan C, Donato di Paola E, Lévi F for the EORTC *Chronotherapy Group*. (CTG), (2002) It's time for chronotherapy! *Eur. J. Cancer*; 38(Suppl 4): 50–3.

Curiel-Lewandrowski C, Atkins MB (2001) Immunotherapeutic approaches for the treatment of malignant melanoma. *Curr. Opin. Investig. Drug;* 2(11): 1553-63

Currier NL, Sun LZY, Miller SC. (2000) Exogenous melatonin: quantitative enhancement in vivo of cells mediating non-specific immunity. *Journal of Neuroimmunology;* 104(2): 101–8

Dalton SO, Boesen EH, Ross L, Schapiro IR, Johansen C (2002) Mind and Cancer: do psychological factors cause cancer ? *European journal of cancer*. 38 : 1313-1323

Dang CV, O'Donnell KA, Zeller KI, Nguyen T, Osthus RC, Li F. (2006) The c-Myc target gene network. *Semin. Cancer Biol*; 16(4):253-64

Davis S, Mirick DK, Stevens RG (2001) Night shift work, light at night and the risk of breast cancer. *J. Natl. Cancer Inst*; 93(20):1557–62.

De Vries A, Holmes MC, Heijnis A, Seier JV, Heerden J, Louw J, Wolfe-Coote S, Meaney MJ, Levitt NS, Seckl JR (2007) Prenatal Dexamethasone induces changes in nonhuman primate offspring cardiometabolic and hypothalamic-pituitary-adrenal axis function. *J. Clin. Invest*; 111(4): 1058-67

Dhabhar FS, Miller AH, Stein M, McEwen BS, Spencer RL. (1994) Diurnal and acute stress-induced changes in distribution of peripheral blood leukocyte subpopulation. *Brain Behav. Immun*; 8:66–79.

Duijts S, Zeegers M, Borne V (2003) The association between stressful life events and breast cancer risk: a meta-analysis. *Int. J. Cancer*. 107 : 1023-9

Ekman AC, Leppaluoto J, Huttunen P, Aranko K, Vakkuri O. (1993) Ethanol inhibits melatonin secretion in healthy volunteers in a dose-dependent randomized blind cross-over study. *J. Clin. Endocrinol. Metab*; 77(3):780–3.

Elias SG, Peeters PH, Grobbee DE, van Noord PA (2004a) The 1944-1945 Dutch famine and subsequent overall cancer incidence. *Cancer Eoidemiol. Biomarkers Prev*; 14(8): 1981-5

Elias SG, Peeters PH, Grobbee DE, van Noord PA (2004b) Breast cancer risk after caloric restriction during the 1944-1945 Dutch famine. *J. Natl. Cancer Inst*; 96(7): 539-46

Empereur F, Baumann M, Alla F, Briancon S (2003) Factors associated with the consumption of psychotropic drugs in a cohort of men and women aged 50 and over. *J. Clin. Pharm. Ther*. 28(1) : 61-8

Enqvist B, Von Konow L, Bystedt H (1995) Pre- and perioperative suggestion in maxillofacial surgery: effects on blood loss and recovery. *Int. J. clin. Exp. Hypn*; 43(3): 284-94

EORTC chronotherapy group: Giacchetti S, Bjarnason G, Garufi C, Genet D, Iacobelli S, Tampellini M, Smaaland R, Focan C, Coudert B, Humblet Y, Canon JL, Adenis A, Lo Re G, Carvalho C, Schueller J, Anciaux N, Lentz MA, Baron B, Gorlia T, Levi F. (2006) Phase III trial comparing 4-day chronomodulated therapy versus 2-day conventional delivery of fluorouracil, leucovorin, and oxaliplatin as first-line chemotherapy of metastatic colorectal cancer: the European Organisation for Research and Treatment of Cancer Chronotherapy Group. *J. Clin. Oncol*; 24(22):3562-9

Erickson MH (1959) Hypnosis in painly diseases in terminal phase. *J. Clin. Hypnosis*; 1: 117-121

Ferguson SA, Kennaway DJ, Moyer RW. (1999) Nicotine phase shifts the 6-sulphatoxymelatonin rhythm and induces c-Fos in the SNC of rats. *Brain research bulletin*; 48(5): 527–38.

Focan C, Kreutz F, Longree L, Graas MP, Moeneclaey N, Demolin G, Focan-Henrard D (2007) Interest of chronotherapy in multidisciplinary management of oesophageal and gastric cancers. *Pathol. Biol.* (Epub ahead of print)

French NP, Hagan R, Evans SF, Mullan A, Newnham J (2004) Repeated antenatal corticosteroids: effect on cerebral palsy and childhood behavior. *Am. J. obst. Gyn*; 190: 588-95

Fu L, Pelicano H, Liu J, Huang P, Lee C (2002), the circadian gene Period2 plays an important role in tumor suppression and DNA damage response in-vivo. *Cell*; 111: 41-50

Fu L, Patel MS, Bradley A, Wagner EF, Karsenty G. (2005) The molecular clock mediates leptin-regulated bone formation. *Cell*;122(5): 803-15.

Gallarda T (2003) Stress, depression and cognitive troubles. In Thurin JM, Baumann N. *Stress, pathologie et immunité*. Chapt. 12. Médecine-Sciences Flamarion – Paris: 125-33

Gallo JJ, Armenian HK, Ford DE, Eaton WW, Khachaturian AS (2000) Major depression and cancer: the 13-year follow-up of the Baltimore epidemiologic catchment area sample. *Cancer Causes Control*; 11(8): 751-8

Glickman G, Levin R, Brainard GC (2002) Ocular input for human melatonin regulation : relevance to breast cancer, *Neuro Endocrinol. Lett.*; 23(2) : 17-22

Hackman RM, Stern JS, Gershwin ME (2000) Hypnosis and asthma: a crirical review. *J. Asthma*; 37(1) : 1-15

Hahn RC, Petitti DB (1988) Minnesota Multiphasic Personality Inventory-rated depression and the incidence of breast cancer. *Cancer*; 61(4): 845-8

Hamet P, Tremblay J. (2006) Genetics of the sleep-wake cycle and its disorders. *Metabolism*; 55(10 Suppl 2): S7-12

Hansen J (2001) Increased breast cancer risk among women who work predominantly at night. *Epidemiology*;12:74–7.

Haus E, Laktua DJ, Swoyer J, Sackett-Lundeen L (1983) Chronobiology in Hematology and Immunology. *Am. J. Anat*; 168: 467-517

Hill C, Doyon F (2003) Fréquence des cancers en France, *bull cancer*; 90(3) : 207-13

Horii R, Akiyama F, Kasumi F, Koike M, Sakamoto G (2005) spontaneous "healing" of breast cancer. *Breast Cancer*; 12(2): 140-4

Horowitz MM, Gale RP, Sondel PN, Goldman JM, Kersey J, Kolb HJ, Rimm AA, Ringden O, Rozman C, Speck B (1990) Graft versus leukemia reactions after bone marrow transplantation, *blood;* 75 : 555-62

Huang WL, Harper CG, Evans SF, Newnham JP, Dunlop SA (2001) Repeated prenatal corticosteroid administration delays myelination of the corpus callosum in fetal sheep. *Int. J. Dev. Neurosci*; 19: 415-25

Jacobs JR, Bovasso GB (2000) Early chronic stress and their relation to breast cancer. *Psychol. Med.* 30(3) : 669-78

Kaplan GA, Reynolds P (1988) Depression and cancer mortality and morbidity: prospective evidence from the Alameda County Study. *J. Behav. Med*; 11(1): 1-13

Klein R, Armitage R (1979) Rhythms in human performance : one-and-a-half-hour oscillation in cognitive style. *Science*; 2004: 1236-8

Knekt P, Raitasalo R, Heliovaara M, Lehtinen V, Pukkala E, Teppo L, Maatela J, Aromaa A (1996) Elevated lung cancer risk among persons with depressed mood. *Am. J. Epidemiol.* 144 : 1096-1103

Krauchi K, Cajochen C, Werth E, Wirz-justice A (2002) Alteration of internal circadian phase relationships after morning versus evening carbohidrate-rich meals in human. *J Bio Rhythms* 17(4): 364-76

Kripke DF, Klauber MR, Wingard DL, Fell RL, Assmus JD, Garfinkel L (1998) Mortality hazard associated with prescription hypnotics. *Biol. Psychiatry*. 43(9) : 687-93

Kvikstad A, Vatten IJ (1996) Risk and prognosis of cancer in middle-aged women who have experienced the death of a child. *Int. J. Cancer*; 67(2): 165-9

Kwiatkowski F, Abrial C, Gachon F, Chevrier R, Curé H, Chollet P (2005a) Stress, cancer and circadian rhythm of melatonin, *Pathol. Biol*; 53 : 269-272

Kwiatkowski F, Dalle C, Auvray H (2005b) Cancer independent from the psyche: a lack of creative interpretation? *Rev. Francoph Psycho-Oncologie*; 2 : 105-16

Kwiatkowski F, Bignon YJ (2007) To test the impact of hypnotherapy upon immunity and circadian rhythms among palliative cancer patients: a promising goal ? *Pathol. Biol.* (Epub ahead of print)

Lambe M, Cerrato R, Askling J, Hsieh CC (2004) Maternal breast cancer risk after the death of a child, *Int. J. Cancer* 110 : 763-6

Langewitz W, Izakovic J, Wyler J, Schindler C, Kiss A, Bircher AJ. (2005) Effect of self-hypnosis on hay fever – a randomised controlled intervention study. *Psychother. Psychosom*; 74(3) : 165-72

Levi F (2006) Chronotherapeutics: the relevance of timing in cancer therapy. *Cancer Causes Control*; 17(4): 611-21

Lillberg K, Verkasalo PK, Kaprio J, Teppo L, Helenius H, Koskenvuo M (2003) Stress of daily activities and risk of breast: a prospective cohort study in Finland. *Eur. J. Cancer*; 91: 888-93

Lillberg K, Verkasalo PK, Kaprio J, Teppo L, Helenius H, Koskenvuo M (2003) Stressful life events and risk of breast cancer in 10 808 women: a cohort study. *Am. J. Epidemiol*; 157(5): 415-23

Lissoni P, Barni S, Mandala M, Ardizzoia A, Paolorossi F, Vaghi M, et al. (1999) Decreased toxicity and increased efficacy of cancer chemo-therapy using the pineal hormone melatonin in metastatic solid tumour patients with poor clinical status. *Eur. J. Cancer*; 35(12): 1688–92

Lissoni P (2007) Biochemotherapy with standard chemotherapy plus the pineal hormone melatonin in the treatment of advanced solid neoplasms. *Path. Bio*; (under press)

Linkins RW, Comstock GW (1990) Depressed mood and development of cancer. *Am. J. Epidemiol.* 132 : 962-72

Marshall GD, Agarwal SK (2000) Stress, immune regulation, and immunity: applications for asthma. *Allergy Asthma Proc.*; 21(4): 241-6

Marshall L, Born J (2002) Brain immune interactions in sleep. *Int. Rev. Neurobiol*; 52: 93-131

McCrabe L, Marash D, Li A, Matthews SG (2001) Repeated antenatal glucocorticoid treatment decreases hypothalamic corticotropin releasing hormone mRNA but not corticosteroid receptor mRNA expression in the fetal guinea-pig brain. *J. Neuroendocrinol*; 13(5): 425-31

McDearmon EL, Patel KN, Ko CH, Walisser JA, Schook AC, Chong JL, Wilsbacher D, Song EJ, Hong HK, Bradfield CA, Takahashi JS. (2006) Dissecting the functions of the mammalian clock protein BMAL1 by tissue-specific rescue in mice. *Science*; 314(5803): 1304-8.

McDermott DF (2007) Update of the application of interleukin-2 in the treatment of renal cell carcinoma. *Clin. Caner Res*; 13: 716s-20s

Megdal SP, Kroenke CH, Laden F, Pukkala E, Schernhammer ES (2005) Night work and breast cancer risk: a systematic review and meta-analysis. *Eur. J. Cancer*; 41: 2023-32

Menegoz F, Cherie-Challine (1999) Le cancer en France : incidence et mortalité. Réseau FRANCIM - La Documentation Française; Paris

Meyerhardt JA, Heseltine D, Niedzwiecki D, Hollis D, Saltz LB, Mayer RJ, Thomas J, Nelson H, Whittom R, Hantel A, Schilsky RL, Fuchs CZ (2006) Impact of physical activity on cancer recurrence and survival in patients with stage III colon cancer : findings from CAL GB 89803. *J. Clin. Oncol*; 24(22): 3535-41

Mocellin S, Lise M, Nitti D (2007) Tumor immunology. *Adv. Exp. Med. Biol*; 593: 147-56

Motzkus D, Albrecht U, Maronde E (2002) The human PER1 gene is inducible by interleukin-6. *J. Mol. Neurosci*; 18(1-2): 105-9

Mormont MC, Waterhouse J, Bleuzen P, Giachetti S, Jami A, Bogdan A, Lellouch J, Misset JL, Touitou Y, Levi F (2000) Marked 24-h rest/activity rhythms are associated with better quality of life, better response and longer survival in patients with metastatic colorectal cancer and good performance status. *Clin. Cancer Res*; 6(8): 3038-45

Moser M, Schaumberger K, Schernhammer E, Stevens RG (2006) Cancer and rhythm. *Cancer Causes Control*; 17: 483-7

Neri B, de Leonardo V, Gemelli MT, di Loro F, Mottola A, Ponchietti R, Rauqei A, Cini G (1998) Melatonin as biological response modifier in cancer patients. *Anticancer Res*; 18(2B) : 1329-32

Newell S, Sanson-Fisher RW (2000) Australian oncologists' self-reported knowledge and attitudes about non-traditional therapies used by cancer patients. *Med. J. Aust*; 172(3): 110-3

Numano R, Yamazaki S, Umeda N, Samura T, Sujino M, Takahashi R, Ueda M, Mori A, Yamada K, Sakaki Y, Inouye ST, Menaker M, Tei H. (2006) Constitutive expression of the Period1 gene impairs behavioral and molecular circadian rhythms. *Proc. Natl. Acad. Sci*; 103(10): 3716-21.

Owen D, Matthews SG (2007) Prenatal glucocorticoid exposure alters hypothalamic-pituitary-adrenal function in juvenile guinea pigs. *J. Neuroendocrinol*; 19(3): 172-80

Pawlikowski M, Winczyk K, Karasek M (2002) Oncostatic action of melatonin : facts and question marks, *Neuro endocrinol. Lett*; 23(1), 24-9

Penninx BW, Guralnik JM, Pahor M, Ferrucci L, Cerhan JR, Wallace RB, Havlik RJ (1998a) Chronically depressed mood and cancer in older persons. *J. Natl. Cancer Inst*; 90(24): 1888-93

Penninx BW, Guralnik JM, Mendes de Leon CF, Pahor M, Visser M, Corti MC, Wallace RB, (1998b) Cardiovascular events and mortality in newly and chronically depressed persons > 70 years of age. *Am. J. Cardiol*; 81(8): 988-94

Persky VW, Kempthome-Rawson J, Shekelle RB (1987) Personality and risk of cancer. 20-year follow-up of the Western Electric Study. *Psychosom. Med*; 49(5): 435-49

Petrovsky N (2001) Toward a unified model of neuroendocrine-immune interaction. *Immunol. Cell Biol*; 79(4): 350-7

Pukkala E, Aspholm R, Wahlberg G (1995) Incidence of cancer among Finnish airline attendants, 1967-92. *BMJ;* 311: 649-52

Pukkala E, Aspholm R, Auvinen A, Eliasch H, Gundestrup M, Haldorsen T, Hammar N, Hrafnkelsson J, Kyyronen P, Linnersjö A, Rafnsson V, Storm H, Tveten U (2002) incidence of cancer among Nordic airline pilots over five decades: occupational cohort study. *BMJ.* 325 : 567-9.

Pukkala E, Aspholm R, Auvinen A, Eliasch H, Gundestrup M, Haldorsen T, Hammar N, Hrafnkelsson J, Kyyronen P, Linnersjö A, Rafnsson V, Storm H, Tveten U (2003)

Cancer incidence among 10.211 airline pilots: a Nordic study. *Aviat. Space Environ. Med*; 74(7):699–706.

Regelson W, Pierpaoli W. (1987) Melatonin: a rediscovered antitumor hormone ? *Cancer Invest*; 5 : 379-85

Reinberg A, (1997) les rythmes biologiques : mode d'emploi, 2nd edition, Paris, Flamarion: 5-6

Reiter RJ, Tan DX, Sainsz RM, Mayo JC, Lopez-Burillo S (2002) Melatonin: reducing the toxicity and increasing the efficacy of drugs. *J. Pharm. Pharmocol*; 54 : 1299-321

Reppert SM, Weaver DR (2002) Coordination of circadian timing in mammals. *Nature*; 418(6901): 935–41.

Rock CL, Denmark-Wahnefried W (2002) Can lifestyle modification increase survival in women diagnosed with breast cancer ? *J. nutr*; 132(11 suppl.): 3504s-7s

Ronson A. (2004) Psychiatric disorders in oncology: recent therapeutic advances and new conceptual framework. *Current opinion in oncology*; 16 : 318-323

Ronson A. (2005) Le trouble de l'adaptation en oncologie : un cadre conceptuel à préciser. *L'Encéphale*; 31(2) : 118-26

Rosenstein RE, Chuluyan HE, Diaz MC, Cardinali DP (1990) GABA as a presumptive paracrine signal in the pineal gland. Evidence on an intrapineal GABAergic system. *Brain Res. Bull*; 25: 339-344

Rossi EL (2002) Psychobiologie de la guérison. Edition "Le souffle d'or", Barret-sur-Méouge, France. Original edition (1993) WW Norton and Company Inc. New-york

Rossi EL (2005) Prospects for exploring the molecular-Genomic Foundation of therapeutic hypnosis with DNA microarrays. *Am. J. Clin. Hypn*: 165-182

Roustang F (1993) Il suffit d'un geste. Odile Jacob editions, Paris

Schernhammer ES, Laden F, Speizer FE, Willet WC, Hunter DJ, Kawachi I, et al. (2001) Rotating night shifts and risk of breast cancer in women participating in the nurses' health study. *J. Natl. Cancer Inst*; 93(20):1563–8

Simon C, Gronfier C, Schlienger JL, Brandenberger G (1998) Circadian and ultradian variations of leptin in normal man under continuous enteral nutrition: relationship to sleep and body temperature. *J. Clin. Endocrinol. Metab*; 83(6): 1893-9

Siu SW, Lau KW, Tam PC, Shiu SY (2002) Melatonin and prostate cancer cell proliferation: interplay with castration, epidermal growth factor, and androgen sensitivity. *Prostate*; 52 : 106-22

Smaaland R, Sothern RB, Lote K, Sandberg S, Aakvaag A, Laerum OD (1995) Circadian phase relationships between peripheral blood variables and bone marrow proliferative activity in clinical health. *In vivo*; 9(4): 379-89

Spanagel R, Rosenwasser AM, Schumann G, Sarkar DK. (2005) Alcohol consumption and the body's biological clock. *Alcohol Clin. Exp. Res*; 29(8): 1550-7.

Spiegel D, Bloom JR, Kraemer HC, Gottheil E (1989)Effect of psychosocial treatment on survival of patients with metastatic breast cancer. *Lancet* 2: 211-22

Takimoto M, Hamada A, Tomoda A, Ohdo S, OhmuraT, Sakato H, Kawatani J, Jodoi T, Nakagawa H, Terazono H, Koyanagi S, Higuchi Sn Kimura M, Tukikawa H, Irie S, Saito H, Miike T (2005) Daily expression of clock genes in whole blood cells in healthy subjects and a patient with circadian rhythm sleep disorder. *Am. J. Physiol. Regul Integr. Comp. Physiol*; 289: R-1273-79

Tosini G, Menaker M. (1996) Circadian rhythms in cultured mammalian retina. *Science*; 272: 419–21.

Vakkuri O, Kivela A, Leppaluoto J, Valtonen M, Kauppila A (1996) Decrease in melatonin precedes follicle-stimulating hormone increase during perimenopause. *Eur. J. Endocrinol*, 135(2) : 188-92

Van den Heiligenberg S, Depres-Brummer P, Barbason H, Claustrat B, Reynes M, Levi F (1999) The tumor promoting effect of constant light exposure on diethylnitrosamine-induced hepatocarcinogenesis in rats. *Life Sci*;64:2523–34.

Von Gall C, Stehle JH, Weaver DR (2002) Mammalian melatonin receptors: molecular biology and signal transduction. *Cell Tissue Res*; 309:151–62.

Wagner SN, Schultewolter T, Wagner C, Briedigkeit L, Becker JC, Kwasnicka HM, Goos M (1998) Immune response against human primary malignant melanoma: a distinct cytokine mRNA profile associated with spontaneous regression. *Lab. Invest*; 78(5): 541-50

Winczyk K, Pawlikowski M, Karasek M. (2001)Melatonin and RZR/ROR receptor ligand CGP 52608 induce apoptosis in the murine colonic cancer. *J. Pineal Res*; 31(2): 179–82.

Witt-Enderby PA, Bennett J, Jarzynka MJ, Firestine S, Melan MA. (2003) Melatonin receptors and their regulation: biochemical and structural mechanisms. *Life Sciences*; 72(20): 2183–98

Wood LJ, Nail LM, Gilster A, Winters KA, Elsea CR (2006) Cancer chemotherapy-related symptoms: evidence to suggest a role of proinflammatory cytokines. *Oncol. Nurs. Forum;* 33(3): 535-42

Yamamoto H, Mohanan P. (2001) Preventive effect of melatonin against DNA damage induced by cyanide, kainate, glutathione/Fe3+/O2, or H2O2/Fe2+. *J. Pineal Res*;31:314–9.

Zahl PH, Strand BH, Maehlen J. (2004a) : Incidence of breast cancer in Norway and Sweden during introduction of nationwide screening : prospective cohort study. *Br. Med. J*; 328: 921-4

Zahl PH, Andersen JM, Maehlen J (2004b) Spontaneous regression of cancerous tumors detected by mammography screening. *JAMA*; 292(21): 2579-80

Zahl PH, Maehlen J. (2006) Do model suggest spontaneous regression of breast cancer ? *Int. J. Cancer*; 118(10): 2647

Zhao Y, Sun R, You L, Gao C, Tian Z. (2003) Expression of leptin receptors and response to leptin stimulation of human natural killer cell lines. *Biochem. Biophys. Res. Comm*; 300: 247-252

Zonderman AB, Costa PT, McCrae RR (1989) Depression as a risk for cancer morbidity and mortality in a nationally representative sample. *JAMA*. 262 : 1191-5

In: Progress in Circadian Rhythm Research
Editor: Anne-Laure Léglise, pp. 35-65
ISBN: 978-1-60021-876-7

Chapter II

The Interaction of Light and the Circadian Clock Network: A New Therapeutic Approach for the Treatment of Neuropsychiatric Disorders?

Henrik Oster* and Russell G. Foster
Circadian and Visual Neuroscience Group; Wellcome Trust Centre for Human Genetics; University of Oxford; Roosevelt Drive; Oxford OX3 7BN; United Kingdom

Abstract

The 24h cycle of day and night is the most powerful synchroniser (*Zeitgeber*) of the human circadian system thereby regulating a wide range of behavioural, physiological and metabolic rhythms. A common feature of varied forms of mental health conditions are circadian rhythm disruption of hormonal and sleep/wake cycles. Significantly, various neuropsychiatric disorders ranging from schizophrenia and depression to age-related medical conditions like Alzheimer's and dementia can be improved with specific light treatments that are designed to act as *Zeitgeber* for the circadian clock. In recent years our understanding of the circadian timing system has changed profoundly. Less than a decade ago is was thought that the hypothalamic suprachiasmatic nuclei (SCN) imposed a 24h temporal structure on the entire body, but the discovery of tissue and organ specific circadian oscillators has radically changed this view. It now appears that the SCN does not drive circadian rhythms in the periphery but acts to coordinate and perhaps sustain their activity. This has led to the hypothesis that pathologies may arise from or are exacerbated by sustained internal desynchronisation – the inappropriate alignment of physiological rhythms in different organ systems over the 24h day. If this is

* Henrik Oster: Email: henrik.oster@well.ox.ac.uk; Phone: +44-1865-287773; Fax: +44-1865-287664

indeed the case, then this might explain why bright light given at the appropriate time can alleviate some of the symptoms associated with mental health conditions. Light might be acting as a strong synchronising agent that forces a common temporal structure on the entire body. In this chapter we review the evidence for this hypothesis, consider possible mechanisms, and address whether chronotherapeutics should be used more widely as a therapeutic approach to assist in the treatment of neuropsychiatric disorders.

Introduction

Our life is shaped by multiple environmental cycles. One of the most influential is the succession of day and night. Most species – from unicellular cyanobacteria to humans – have evolved internal timekeepers (so called *circadian clocks*) to anticipate these daily events and fine-tune physiology and behaviour to the varying demands of activity and rest [178]. The circadian control of our body is pervasive and encompasses all aspects of biology from behaviour down to cell cycle control and chromatin modulation [40,51]. A master clock in the *suprachiasmatic nucleus* (SCN) is synchronised to geophysical time via a non-visual photoreceptive system located in the ganglion cells of the inner retina [174]. From here, external time information is transmitted to numerous peripheral clocks throughout the body. Perturbations of the circadian system have a profound impact on health and wellbeing. The physiological and psychological disturbances following trans-meridian travel (*jet lag*) are caused by a transient state of internal de-synchronisation while the body's clocks struggle to adapt to a rapidly shifted external light/dark cycle [86]. Similar and more severe symptoms including suppressed immunity and an elevated susceptibility to cancer are the consequences of the sustained disruption of the circadian clock in shift workers [103].

Sleep and circadian rhythm disturbances affect more than one third of the Western population [49], Both frequently accompany neuropsychiatric disorders like depression, schizophrenia and Alzheimer's [13,58,240], suggesting a tight interplay between the circadian system and higher neurological functions. In this chapter we summarise new findings on the molecular organisation of the circadian system and the rhythmic aspects of these diseases. Combining physiological and clinical findings we propose new mechanisms of how the circadian system may influence pathological processes in the brain and how this pathway can be employed in the treatment of neuropsychiatric disorders.

The Internal Clockwork

In mammals, the paired suprachiasmatic nuclei (SCN) located in the ventral part of the hypothalamus function as the master pacemaker of an endogenous circadian timekeeping system [102]. The SCN receive photic input from the retina via direct and indirect pathways, thus forming the prime relay between external and internal times [190]. Neuronal firing patterns within the SCN show robust 24h rhythms which are self-sustaining under constant environmental conditions [136] and even in explant cultures [207]. SCN transplants from short-period mutant donor animals superimpose their rhythm onto the behaviour and physiology of the host [185]. Furthermore, SCN lesions have been shown to abolish behavioural and physiological circadian rhythms [149,217] thereby demonstrating the prime role of this locus in the organisation of the mammalian circadian timing system.

Structurally, the SCN can be roughly divided into two parts. The ventral region (or *core*), directly adjacent to the optic chiasm, receives neuronal input directly from the retina via the retinohypothalamic tract and is believed to mediate the synchronisation of the internal clock to the external light/dark cycle [114]. The dorsomedial region – the *shell* – is innervated from various adjacent hypothalamic nuclei (e.g. the dorsomedial nucleus of the hypothalamus; DMH), the basal forebrain, the limbic cortex, and the brainstem [147] suggesting a role in the so called *non-photic* regulation of the clock [150]. Of note, despite this compartmentalisation all SCN neurons show sustained 24h firing rhythms both in vivo and in vitro and therefore are equally capable of acting as endogenous circadian pacemakers [205].

Even though pharmacological and genetic data suggest that coupling between SCN neurons is critical for sustained rhythmicity in the absence of a *Zeitgeber* signal [142], the molecular clockwork itself is based on an intracellular mechanism [246]. In SCN neurons and – as we believe now – most other cells of the body (see below) a set of so called *clock genes* is organised in a system of interlaced transcriptional/translational feedback loops (TTLs) that drive the rhythmic expression of thousands of *clock controlled genes* (CCGs), thereby transmitting time information to the physiological status of the cell [104,190]. In the core loop of the TTL the two *basic helix-loop-helix* (*bHLH*) transcription factors CLOCK and BMAL1 (ARNTL; MOP3) control the rhythmic expression of three *Period* (*Per1*, *Per2*, and *Per3*) and two *Cryptochrome* genes (*Cry1* and *Cry2*) by binding to E-box elements in the corresponding promoters [68]. PER and CRY proteins form hetero-multimers that translocate into the nucleus where they interfere with CLOCK/BMAL1 and inhibit their own transcription [162,227]. Complexation, translocation, and post-translational modifications of PER and CRY proteins substantially delay this transcriptional feedback, resulting in a stable 24h rhythm of E-box activation [66]. Additional feedback loops have been identified that stabilise and regulate the resetting properties of the core TTL. These include the E-box-controlled nuclear orphan receptor genes *RevErbα* (*Nr1d1*) and *Rorα* (*Rora*) that regulate rhythmic transcription of *Bmal1* [181,204], the PAR *bZip* transcription factor *Dbp* modulating *Per* expression [253], and the *bHLH* PAS proteins *Dec1* and *Dec2* (*Bhlhb2/3*; *Sharp2/1*) that have also been implicated in the repression of CLOCK/BMAL1 activity in the SCN [90]. Clock gene variants – like the *Clock* homologue *Npas2* – have been detected in other tissues where they may compensate for the absence of some of the *canonical* clock genes [135,188].

Recent data from diverse model organisms – including rodents – have challenged the TTL model by indicating that rhythmic transcription of clock genes does not necessarily lie at the very heart of the circadian timing system but may rather constitute just another output mechanism of the true pacemaker [110,141,154,232]. It was suggested that either metabolic circuits or oscillations in post-translational protein modifications could continue to measure time in the absence of transcriptional rhythms. Bacterial clocks can even be re-built *in vitro* and in the absence of any cellular host system [155]. However, so far no appropriate experimental approach has been proposed to demonstrate circadian rhythmicity without a functional TTL in multicellular organisms.

Initial experiments by Rae Silver and co-workers using SCN transplants encapsulated in a semi-permeable membrane suggested that the SCN synchronises behaviour and physiology by means of secreted molecules [208]. Several candidates for such soluble factors have been suggested ranging from peptides [42,107,108,211] to nucleotide analogues [258]. The action of these substances, however, is largely restricted to behavioural rhythms while for endocrine control neuronal connections seem to be more relevant [38]. The SCN projects to several intra- and proxy-hypothalamic targets such as the preoptic area, an important regulator of the sleep/wake cycle (figure 1) [203], the paraventricular nucleus of the hypothalamus (PVN), an integrative centre concerned with neuroendocrine, autonomic and behavioural processes [224], and the dorsomedial nucleus of the hypothalamus (DMH) that links the clock to the arousal system of the brainstem [12]. The DMH has also been implicated in controlling rhythmic behaviour in response to metabolic timing signals [74]. Lesion experiments have shown that the autonomic nervous system is essential for circadian regulation of e.g. the pineal [226]. The physiology of other organs, however, like the liver [39] or the adrenal gland [92,164] seems only in part to depend on functional autonomic innervation from the SCN but instead relies on endocrine or metabolic signals [46,164].

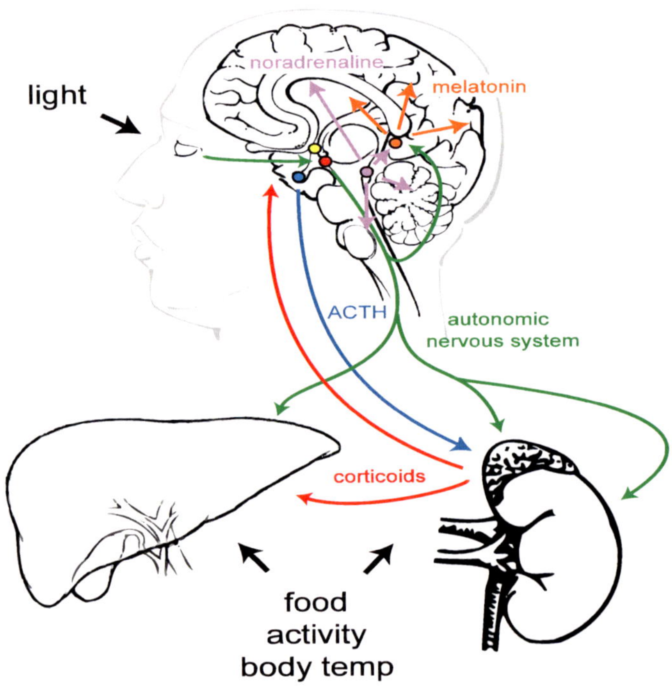

Figure 1. Synchronisation Pathways of the Circadian System. A network of circadian clocks is synchronised with external time via humoral and neuronal pathways. Light regulates clock gene expression in the SCN (red circle). From there, sleep (VLPO – yellow circle) and arousal centres (locus coeruleus (LC) – purple circle), endocrine tissues like the pituitary (blue circle) and the adrenal, and other peripheral organs (shown are the kidneys and the liver) are synchronised via the autonomic nervous system. Noradrenaline from the LC regulates clock gene rhythms in the corticolimbic system and the pineal gland (orange circle). Other rhythmically released hormones – ACTH from the pituitary, melatonin from the pineal and corticoids from the adrenal – further contribute to synchronisation of peripheral clocks and feedback to the CNS. Other Zeitgeber like food availability and systemic cues further strengthen internal coupling by regulating clock phases in the periphery [161].

The hunt for a common synchroniser of peripheral rhythms remains undiscovered. Mounting evidence, however, suggests that there may not be a single mechanism but instead different organs may use different clues to stay in tune with the SCN and external time.

Resetting the Pacemaker

The major Zeitgeber (from German time giver) of the circadian system in mammals is light. The SCN clock stays tuned to the external light/dark (LD) cycle via a special, non-visual photoreceptive system located in the retinal ganglion cells of the inner retina (see below). Short light exposures during the subjective night – the activity phase in nocturnal and the rest phase in diurnal animals – can reset activity and physiological rhythms. The timing of the light exposure determines the direction of this shift while light wavelength, duration and intensity define the magnitude of the phase advances or delays.

Other – so called non-photic – synchronisers have been shown to affect behavioural rhythms in rodents and humans. They include various chemical substances like benzodiazepine and opiates [129,236], hormones like melatonin and several neuropeptides [11,176,177], temperature changes [189] as well as behaviourally arousing stimuli like forced exercise, novelty-induced wheel-running, sleep deprivation and social interactions [143-146,150,151]. Like light, the effectiveness of non-photic Zeitgeber is dependent on the time of their application. However, the sensitivity of the circadian clock to these cues – unlike light – seems largely restricted to the (subjective) day [150].

Light resets the SCN clock via direct and indirect innervation by the retina [102]. In mammals the eye is the only light-perceiving tissue capable of synchronising the circadian timing system [250]. However, the visual photoreceptive system of the rods and cones does not or only in part contribute to circadian phototransduction. Experiments in mice lacking both rods and cones show that these animals are perfectly capable of synchronising their activity to external light/dark (LD) cycles [60]. Other light responses like the suppression of melatonin synthesis in the pineal gland and pupil constriction are also preserved in these animals [125,126] indicating that there exists a third photoreceptor system in the mammalian eye. Intense studies from several labs finally resulted in the discovery of melanopsin (OPN4), an atypical opsin form expressed in a subgroup of ganglion cells of the innermost layer of the retina [183]. OPN4-positive cells are intrinsically photosensitive [82]. They innervate the SCN as well as other targets of the brain implicated in photic responses [81]. Opn4-deficiency causes attenuated illumination responses including clock resetting, pupil constriction and melatonin suppression [127,170,199]. When Opn4-deficiency is combined with an ablation of rods and cones in mice, the activity rhythms of these animals no longer respond to light but instead free-run with their internal period [83]. These data clearly show that the OPN4 and the visual system together entrain the circadian clock and that no other – so far undiscovered – photosystem contributes to this mechanism.

Nocturnal light exposure releases the neurotransmitters glutamate and pituitary adenylate cyclase activating polypeptide (PACAP) at the synapses of SCN-projecting photosensitive retinal ganglion cells (pRGCs) [78,79]. These activate several intracellular signalling cascades including nitric oxide (NO), PKA, PKG and MAP kinase pathways all of which converge on the transcriptional activator CREB (cAMP-response element binding protein)[137]. Activated CREB binds to cAMP-response elements (CREs) at the promoter sites of the clock genes Per1, Per2 and Dec1 [3,90]. All three genes are relatively inactive during the night when the circadian clock is most responsive to light. CREB activation and the subsequent de novo synthesis of the corresponding clock proteins results in prolonged suppression of CLOCK/BMAL1 activity by PER, CRY and DEC protein complexes and thus either phase-advances or phase-delays the TTL depending on the timing of light exposure. Light during the early part of the night leads to sustained CLOCK/BMAL1 suppression in the early morning and hence causes a phase delay. In the later part of the dark phase light stimulates premature synthesis of daily PER and DEC protein levels, resulting in an accelerated CLOCK/BMAL1 suppression and phase advances on the following cycle [41,71].

Post-translational events have been shown to contribute to clock-resetting in invertebrates [59,113]. Though likely, it remains to be shown whether similar mechanisms also play a role in the mammalian circadian system.

The Circadian "Clock Shop"

The expression of clock genes is not restricted to the SCN. In fact, the molecular components believed to make up the circadian clockwork are found in most if not all tissues of the mammalian body [15]. Moreover, many extra-SCN tissues display circadian expression of clock genes and this rhythm is preserved even without SCN control, e.g. when the tissue is cultured outside the body [254,255]. This apparent contradiction of the finding that all circadian rhythms stop when the SCN pacemaker is removed (see above) is resolved if one assumes that the SCN does not drive rhythms in the periphery but merely synchronises the different peripheral clocks with external time and with each other, thus coordinating the different physiological and behavioural rhythms and aligning them with external time [47]. When the pacemaker is lost the peripheral clocks become desynchronised resulting in overall arrhythmicity [221].

There remains the possibility that the cellular circadian timekeeping is not based on a transcriptional timekeeping mechanism (see above). However, transcriptional rhythms surely comprise the most significant molecular manifestation of circadian clocks. By means of transcriptional control of *clock controlled genes* (ccgs) peripheral oscillators transmit time information onto the physiological state of the cell [190]. Microarray analyses have shown that in mammals in any rhythmic organ about five to ten percent of all expressed genes are transcribed in a circadian fashion [54,124,163]. The nature of these genes varies significantly between different organs and often key regulators of organ-specific pathways are under clock control hence offering a simple mechanism by which the clock can organise a tissue's physiological functions in respect to other tissues and to external time [169,220]. Not all of these rhythmic genes, however, are directly controlled by the peripheral clock. Instead they respond to rhythmic signals originating from outside the tissue like hormones [159], activity-driven changes in body temperature [106], metabolic cues [91,99], and light [92].

The SCN synchronises peripheral clocks via both humoral and neuronal pathways (figure 1) [38]. Early cell culture based work suggested endocrine pathways as prime regulators of peripheral circadian rhythms. Dexamethasone, a synthetic glucocorticoid analogue, is capable of synchronising circadian rhythms in cultured cells and causes time-dependent phase shifts in clock gene expression in the liver when injected in mice [16]. However, glucocorticoid receptor deficiency does not abolish clock gene synchronisation in the liver [112]. Several other drugs can elicit clock gene rhythms in cell culture [18,130] and temperature cycles help to sustain synchrony among these cells [34]. Noradrenaline induces clock gene expression in the liver [228], the pineal gland [209], and in the cerebral cortex [243], and light has been shown to activate *Per* transcription in the adrenal via the autonomic nervous system [92]. In the latter two cases clock gene rhythms critically depend on functional innervation, highlighting the importance of neuronal pathways in the synchronisation of at least some peripheral clocks. In summary, it seems unlikely that a single agent will be discovered that is capable of synchronising the whole periphery with the SCN. Instead, specific combinations of timing signals seem necessary to align the different tissues with the master pacemaker and keep the organism in tune (figure 1).

Asynchrony amongst different organ systems underlies the various symptoms of jet lag [244]. While the different organs receive inconsistent timing cues from the SCN – which is itself in the process of resetting its circadian phase – and from the environment (e.g. by irregular sleep/activity schedules, light exposures and food intake) the normal regulation and coordination of different physiological processes is lost [8]. After a few days synchrony is again established, and the symptoms retreat. Jet lag not only affects our physiology. It also severely impacts on our cognitive system and psyche [94]. The implications of such clock desynchrony on neuropsychiatric disorders are considered below.

Rhythmic Aspects of Neuropsychiatric Disorders

A long established correlation exists between psychiatric dysfunction and components of sleep. The 19th century English neurologist John Hughlings Jackson firmly believed that the key to understanding psychiatric disease lies in our dreams [45] and even Sigmund Freud made dream interpretation one of the key elements of his psychoanalytic theory [61].

Mental health problems are almost always accompanied by substantial sleep disturbances. Subjective sleep complaints have been reported in more than 80% of patients with either depression or schizophrenia [52,191]. Sleep abnormalities are similarly common in both Alzheimer's (AD) and Parkinson's disease (PD) patients, with more severely affected patients showing both more nocturnal sleep fragmentation, alterations in sleep structure, and worsening sleep/wake rhythms [2,241]. Not surprisingly, sleep problems were repeatedly suggested as diagnostic markers for psychiatric disorders in general or as a first sign of impeding aggravating episodes [157,173,194,216]. However, the diversity of alterations in the various sleep parameters in different patients – even within the same group of patients, age and gender dependencies, the high impact of most psychoactive medications on sleep/wake structure, plus the total absence of these symptoms in some of the patients have made it difficult to assign a clear strategy for using defined sleep parameters as diagnostic markers in psychiatric praxis [9,10].

Sleep is regulated by two different mechanisms, a homeostatic one (also called *process s*) that accumulates a need for sleep depending on the duration of staying awake, and a rhythmic one controlled by the circadian clock in the SCN (hence called *process c*) that under normal conditions confines sleep to the night in humans (or to the day in nocturnal animals like mice) [29]. Unlike that of the circadian clock, the anatomic locus controlling the homeostatic sleep drive remains elusive [140,218]. Several studies have suggested that an accumulation of the ATP breakdown product adenosine in the basal forebrain and more specifically in the ventrolateral preoptic nucleus (VLPO) suppresses the arousal system of the upper brain stem thereby promoting sleep [202]. The VLPO is, however, also a key player in the regulation of process c. The SCN sends projections to the VLPO via the dorsomedial nucleus of the hypothalamus (DMH) which also integrates other rhythmic signals from the periphery/from peripheral clocks [203]. Interestingly, clock genes are expressed in many of the areas implicated in the regulation of sleep [115] and it will be interesting to dissect how the SCN and peripheral clocks in the brain coordinate their activities to regulate process c and to what extent they impact on sleep drive (process s).

In depression, sleep architecture – based on sleep electro encephalogram (EEG) analyses – has proven a good marker to distinguish depressal from non-depressal subjects. The first group regularly shows increased and phase-advanced bouts of REM sleep which can be explained by a diminished accumulation of sleep pressure during wakefulness (*process s deficiency*) and, hence, a reduced amount of NREM sleep in the beginning of the night [30]. Other parameters are less clearly defined. On one hand, insomnia (or hyposomnia) is the most pervasive self-reported sleep disturbance in depression. However, 15-35% of patients complain of hypersomnia, and difficulty getting up in the morning; this feature being most prevalent in bipolar as well as in younger patients [171]. Insomnia and sleep fragmentation are also regularly reported in schizophrenic patients, often accompanied by increased stage 2 NREM sleep latency and reduced overall slow wave sleep. These findings, however, are frequently disturbed by antipsychotic medication, medication history, and the sheer heterogeneity of this disease [148].

The impact of psychiatric disease on sleep regulation by process c is less obvious, most likely because period and phase alterations of sleep timing are easily masked by strong changes in sleep structure as described above. In general, moderate phase-advances of sleep have been observed in all classes of mood, psychotic, and neurodegenerative disorders suggesting either altered *Zeitgeber* sensitivity or free-running period lengths [120,245]. However, closely controlled studies using defined *Zeitgeber* and medication-free conditions are needed to resolve this question.

Alterations in endocrine rhythms – both in the brain and in the periphery – have been frequently reported in many neuropsychiatric disorders [132,134,230]. They include both amplitude and phase changes of hormonal rhythms. Some of these findings are summarised in table 1. While the most commonly observed symptom is a decrease in amplitude of the hormonal rhythm, often accompanied by a phase advance of the rhythm with regard to external time, the opposite was also reported for some parameters (e.g. elevated cortisol secretion in schizophrenic patients [65]). Despite technical caveats (like poor temporal synchronisation amongst individuals, heterogeneous cohorts, and small group sizes) collectively these findings suggest a close link between the circadian system and the (neuro-)

physiological changes accompanying these disorders. It was hypothesized that – like jet lag – depression represents a state of internal desynchronisation with altered phase angles between different rhythmic physiological processes [62,94,233]. With the discovery of peripheral circadian clocks a new facet has been added to this phenomenon: Weakened synchronisation strengths between tissue clocks or between the periphery and the SCN could indeed underlie the phase angle alterations observed in hormonal and behavioural rhythms in psychiatric patients. Moreover, disruptions of peripheral oscillators in mice can result in deregulated responses of endocrine tissues to external stimuli and, hence, affect the diurnal rhythm of hormone secretion as has been shown for corticoid secretion from the adrenal gland [163,164]. Very similar effects on sensitivity have been observed in the adrenal, the pituitary, and the thyroid in various neuropsychiatric diseases [28,73]. On the other hand, several psychoactive drugs like lithium, fluoxetine, or rubidium have the ability to phase-shift the circadian system [84,192], indicating a common regulatory mechanism (see also the therapeutic use of circadian *Zeitgeber* below).

Corticoid Feedback

In the context of neuropsychiatric disease circadian oscillations of the adrenal glands deserve particular attention. The adrenals (also known as suprarenal glands) are two triangle-shaped endocrine glands located above the kidneys. The outer part of the adrenal, the cortex, produces glucocorticoids, mineral corticoids and sex steroids. The inner part, the neuronal crest derived adrenal medulla, secretes the catecholamines adrenaline and noradrenaline. Hormonal secretion from both cortex and medulla follows a robust circadian pattern [93]. At the same time, acute, stress-related regulation can rapidly overrun the resting production of these hormones, preparing the body for action by increasing blood flow, oxygen uptake, and releasing glucose from storage [210]. Clock genes are expressed throughout both adrenal cortex and medulla but rhythmic expression is mostly confined to the two outer layers of the cortex, the *zona fasciculata* and the *zona glomerulosa*, where both classes of corticoids are synthesized [27,164]. The transcriptional rhythm in the adrenal cortex is regulated by the SCN via the splanchnic nerve [92,164]. Circulating melatonin seems also important, as clock protein rhythms are strongly attenuated in mouse strains with melatonin deficiencies [231]. Interestingly, ocular light exposure is capable of entraining the adrenocortical clock even in the absence of a functional SCN pacemaker [164] and glucocorticoid secretion from the adrenal can be acutely induced by nocturnal light exposure [92]. Together with the fact that glucocorticoids can reset clock gene rhythms in other peripheral tissues [16] the adrenal seems to play an important role in the coordination of peripheral circadian rhythms and their alignment with the external LD cycle (figure 1).

Table 1. Modulation of Endocrine Rhythms in Neuropsychiatric Diseases

disease	major depression		bipolar disorder		schizophrenia		Alzheimer's		Parkinson's		references
hormone	level	phase	level	phase	level	phase	level	phase	level	phase	
ACTH	normal	advance	normal	advance	normal	normal	lower	normal	lower	normal	[22,25,121,152, 235]
cortisol	higher	advance	higher	advance	higher	delay	higher	-	higher	-.	[22,65,80,180,2 47]
CRH	higher	-	higher	-	higher	normal	lower	-	normal	-	[19,22,222]
GH	lower/ higher	advance	higher	advance	normal	normal	normal	normal	lower	normal	[25,87,139,201, 235]
melatonin	lower	advance/ delay	lower	advance /delay	lower	advance	lower	advance	lower	advance	[22,31,44,122,1 58,187,251]
prolactin	higher	-	higher	-	normal/ higher	normal	normal	normal	higher/ lower/ normal	normal	[14,24,97,153,1 66,235,247]
T3 / T4	normal/ higher	-	normal	-	lower	normal	normal	normal	lower	-	[20,33,43,187,2 25,234]
testosterone	lower	-	lower	-	higher	-	lower	-	lower	-	[89,160,168]
TSH	lower	normal	lower	normal	lower	advance	higher	normal	normal	normal	[43,100,166,18 6,187,234]
AVP	higher	normal	higher	-	normal	normal	lower	-	normal	normal	[23,116,156,19 6,212]

- : no data available.

Adrenal corticoids are distributed via the bloodstream and initiate a broad range of actions that regulate the function of multiple organ systems, including the liver, muscle, the immune system, the pancreas, adipose tissue, as well as the brain [55]. Both mineral and glucocorticoid receptors are expressed throughout the CNS [96]. The highest abundance of both receptors is found in neurons of the corticolimbic system, including the hippocampus, the amygdala, the septum, and the dorsal thalamus. Furthermore, strong expression is observed in the olfactory bulb, the cerebellar cortex, and the arousal centres of the brain stem [138]. Via these structures corticoid hormones have a strong impact on brain development, sexual differentiation, cognitive and memory processes, and behavioural control. Specifically in the hippocampus, glucocorticoids modulate fast excitatory amino acid mediated synaptic transmission and synaptic plasticity, which are considered to underlie learning and memory processes [32,98]. By contrast, stressful events and elevated corticosteroid hormones, as frequently observed in the development of neuropsychiatric diseases (see table 1), suppress synaptic potentiation when stress precedes the learning stimulus [67].

There is evidence that the immunosuppressive activity of glucocorticoids can also trigger inflammatory changes in the brain that are pathological features of both depression and dementia. A reduction in the synthesis of neurotrophic factors caused by a rise in brain glucocorticoids, may play a role in the pathology of these disorders and predispose chronically depressed patients to dementia [117,219].

Hypercortisolaemia, possibly resulting from a breakdown in glucocorticoid-receptor-mediated negative feedback mechanisms within the hypothalamus/pituitary/adrenal (HPA) axis, may be central to the pathogenesis of major and bipolar depression and the neurocognitive deficits which characterise this disorder [256,257] (see also below). Both preclinical and clinical studies suggest that lowering circulating cortisol levels or blocking the effects of elevated cortisol with antagonists (e.g. mifepristone/RU-486) has therapeutic benefits in terms of improvements in both depressive and manic symptoms and some domains of neurocognitive function [256]. Elevated cortisol levels are also regularly reported in schizophrenic patients, but current treatments focus on the modulation of 5-HT- and glutamate-mediated synaptic signalling – without great success [36,215]. In Parkinson's and Alzheimer's, negative (hippocampal depletion) and positive (neuroprotection) effects of modulating HPA axis regulation have to be carefully balanced to achieve positive symptomatic net effects [179,182]. A chronotherapeutic approach – e.g. using short-acting glucocorticoid agonists in the morning and antagonists in the late afternoon – could potentially improve the results by stabilising rhythmic adrenal cortisol production and re-establishing normal feedback regulation at the level of the pituitary [35].

Zeitgeber Therapy in Neuropsychiatric Diseases

The robust rhythmic secretion of melatonin is directly controlled by the circadian clock of the SCN and by the external light/dark cycle (see above and figure 1) [131]. It has the ability to phase shift both peripheral and central circadian rhythms [175] and has – because of the strong regulation of its secretion profile by day length – been implicated in adapting human behaviour and physiology to seasonal changes [7,165]. Melatonin levels are regularly altered in neurological disorders (see table 1) and melatonin has been described as a 'trait marker' for depression [21]. Melatonin strongly improves sleep in depressive subjects, its direct impact on mood, however, remains a matter of debate [53]. Good results have been achieved with agomelatine, a synthetic melatonin receptor agonist with additional serotonin receptor blocking potential [48,197].

No conclusive data exists on the use of melatonin in bipolar and schizophrenic disorders. It has been shown to improve sleep and prevent cell loss in Alzheimer and Parkinson patients [213]. In both cases melatonin effectively restores sleep structure in these patients, its proposed antioxdative effect, however, may have been at least as important as its chronobiotic capabilities for the progression of the disease itself [123,133].

The most powerful synchroniser of the mammalian circadian system is light (see above). It is believed to act by direct regulation of *Period* gene expression in the SCN and in few other peripheral oscillators like the pineal and the adrenal [3,63,92,164]. From these relays the other peripheral clocks are subsequently coordinated via hormonal secretion, the autonomous nervous system, and indirectly via the regulation of the rest/activity cycle, and with it, body temperature and metabolic activity [38,221]. Interestingly, light – or phototherapy – is also a commonly used as a treatment for psychiatric disorders. This is especially true for one form of major depression, so called seasonal affective disorder (SAD) [128]. SAD is characterised by long episodes of major depression that tend to recur during specific times of the year, usually in autumn and winter. SAD is more prevalent in countries of higher latitude (e.g. in Scandinavia) correlating with the extremely shortened daily light period during the winter season. SAD can be successfully treated with melatonin or by exposure to bright light, both of which also result in a marked phase-advance of behavioural and endocrine rhythms in treated patients [119,214].

Phototherapy is also more and more frequently used in non-seasonal depression and bipolar disorders [109,229,248], exclusively or together with other conventional treatments. Only few data exist on the efficacy of light exposure on schizophrenia [85]. However, irregular natural light exposure correlates with disease severity [69] and controlled light treatment, especially in the blue wavelength range, can re-establish normal endocrine functionality in schizophrenic patients [237].

Bright light has also successfully been applied to alleviate some of the associated symptoms of Alzheimer's disease like disrupted sleep/wake patterns [167]. A direct effect on the pathogenesis has not been proven, however it was shown that light treatment can reactivate atrophic neurons during early Alzheimer stages [223]. The effect of light on Parkinson's remains unclear.

It was originally thought that light acts primarily via the regulation of melatonin secretion from the pineal (s.a.) [214]. With the discovery of peripheral clocks it may, however, also become relevant for the – direct or indirect – synchronisation of non-SCN oscillators. Interestingly, light has been shown to via the eye regulate corticoid secretion from the adrenal [92]. Corticosteroids have a strong feedback effect on neurophysiological functions in the CNS [257]. Together with the alterations of glucocorticoid levels in the CNS of psychiatric patients (see below) these findings suggest an alternative pathway for light action in neuropsychiatric disorders.

Arguably, the most potent treatment in depression is just another strong *Zeitgeber* of the circadian system: sleep derivation [5,145]. Sleep deprivation was introduced by the German psychiatrist Walter Schulte in the mid-1960s [206]. One night of total sleep deprivation improves depressive symptoms in 40–60% of patients [70]. It has played a certain although modest role in the therapy of unipolar depression since then - modest because the therapeutic effect is variable and mostly confined to one or a few days. Moreover, sleep deprivation or sleep reduction can trigger a manic or hypomanic episode in people with this kind of disposition. Consequently, its use in bipolar depression is limited to the depressive episodes of the disease [193]. While it seems clear that sleep deprivation affects the homeostatic control of sleep rather than process c, the relapse of the depressive symptoms could also be explained by a gradual desynchronisation of different, weakly coupled oscillators after a single strong *Zeitgeber* signal. A similar effect can be observed when the circadian clocks of fibroblasts in culture are temporally synchronised by serum shock or the glucocorticoid analogue dexametasone [17,18]. Interestingly, the rhythm of cortisol relase from the adrenal can be temporally stabilised on the day after SD in depressive patients. No such effect is seen in the control group [252], indicating that the timing signal elicited by the prevention of sleep does not – or only to a lesser extend – affect internal synchronisation with good health. In mice deficient for the clock gene *Bmal1* recovery after sleep regulation is affected, highlighting the close interaction between both homeostatic and circadian control in the response to SD [111]. Of note, this effect is not seen in *Per1/2* double mutant mice which show normal homeostatic sleep drive [105]. No conclusive data exist on the effects of SD in schizophrenia and Alzheimer's disease while positive effects in Parkinson's patients have been reported [4,26,118].

The Social Zeitgeber Theory

Social *Zeitgeber* are a subgroup of non-photic synchronisers of the circadian system. Social stimuli – intra- or interspecies interaction, arousal, aggression etc. – may affect the circadian system, either directly or indirectly via temporally regulating the access of other *Zeitgeber*, e.g. by influencing daily patterns of light exposure, or by associative learning processes that utilize circadian time as a discriminative or conditioned stimulus [146]. In rodents, social contact between adult individuals or between mother and child have been shown to synchronise internal circadian rhythms [184,238,239]. Entrainment by aural and olfactory signals has been reported in rodents [75,101]. Meal timing – which can be classified as a *socially controlled Zeitgeber* – is a strong stimulus for the synchronisation of peripheral rhythms in animals and humans [6,46,72,95]. Alterations of the natural sleep/wake cycle, e.g. by shift work or by ongoing sleep deprivation (so called *social jet lag* [249]), do also belong in this category.

In our modern 24h society with its constant access to electricity (and hence, artificial light) more and more of our activities are transferred into times that would naturally be used for sleep [195]. This *social jet lag*, combined with a persistent cacophony of misleading *Zeitgeber* signals puts our circadian system under permanent stress and renders it hypersensitive to further destabilising events. It has been postulated that *stressful life events*

(SLEs) [50] have the potential to act as strong social *Zeitgeber* by changing our normal behavioural, physiological, and metabolic rhythms. SLEs could include e.g. the death of a close person, a traumatic incident, abuse, finding a new love, or losing it etc. It is easily imaginable how such incidents may change our normal lifestyle, affecting our endogenous rhythms, and ultimately affect the causal chain that leads to depressive or psychotic episodes [57]. SLEs have been shown frequently to precede depressive but also manic episodes [56,172] and influence the development of psychotic disorders [242].

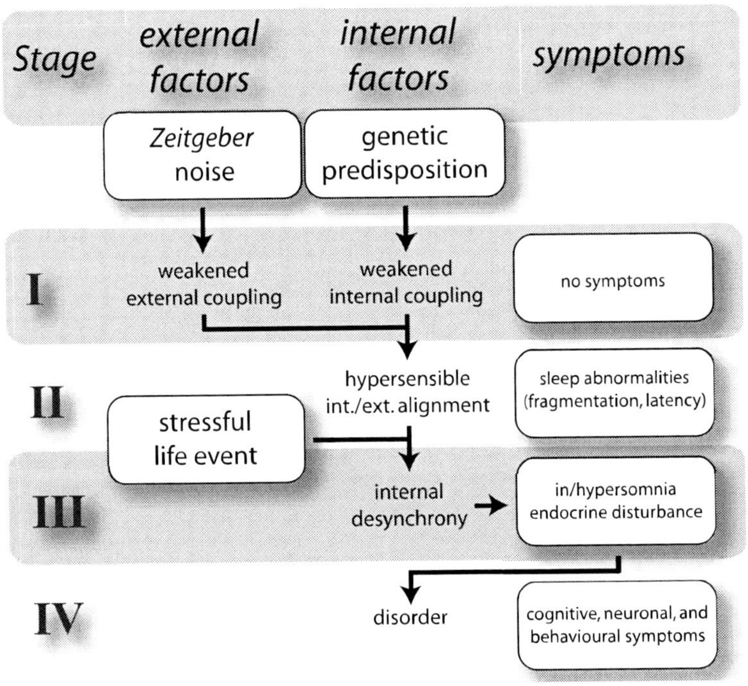

Figure 2. A 4 Stage Desynchrony Model for the Pathogenesis of Neuropsychiatric Disorders For comments see text.

Several other theories have been developed explaining the influence of stress on the pathogenesis of neuropsychiatric disorders (see e.g. [77,200]). Our increasing knowledge about the circadian organisation of physiology and behaviour, however, combined with the accumulating experimental evidence linking the circadian system to neurophysiological function as outlined above, provides compelling support for the so called *Social Zeitgeber Theory* of neuropsychiatric disorder development [76]. One can summarise this concept in a four step model of neuropsychopathogenesis as depicted in figure 2. During the first stage, genetic disposition – general neurotransmission defects or circadian clock related polymorphisms – or strong artificial external *Zeitgeber* input can weaken both external and internal coupling within the body's circadian system. At this time, no or only very subtle symptoms may be observed. When both factors come together circadian alignment becomes hypersensitive to further external disturbances (stage II). One of the first processes affected by such hypersensitivity is the sleep/wake rhythm with increased sleep fragmentation and subtle changes in sleep architecture (e.g. REM sleep latency). External disturbances in the form of an SLE then trigger stage III: SLE-induced social *Zeitgeber* conflicts significantly alter the entrainment parameters of the internal timing system, resulting in internal desynchrony. At this stage, strong symptoms like in- or hypersomnia and endocrine dysfunctions appear. Over time these symptoms become increasingly distressing and affect motivation and social behaviour. This increasing stress level has strong impact on mood and cognitive performance and further enforces the pathological progress, finally leading to the full-blown depressive, manic, or psychotic episode (stage IV).

From a therapeutic perspective it is remarkable that changes in sleep structure are the very earliest signs of impeding neuropsychiatric disease. Given the rising prevalence of these disorders in our modern society [1], the close control of sleep by the circadian clock, and the relative ease (e.g. light exposure or sleep timing) by which the circadian system can be regulated, sleep monitoring and *Zeitgeber* hygiene – that is the controlled exposure to stimuli with clock-shifting abilities – should be amongst the first steps in the prevention and treatment of neuropsychiatric disease.

Conclusion

We are just beginning to understand the complex interactions of the component parts of the circadian system that coordinates the physiological status of our body, and even its contribution to higher brain functions in health and disease. Strong feedback circuits link circadian clocks and physiological oscillations in the periphery and the CNS with the central pacemaker of the SCN [37,64,161]. By contrast, multiple inputs exist into the circadian network allowing for modulation by environmental events to impinge on the balance of this system [88,221]. Clock regulated processes like the sleep/wake cycle and many endocrine rhythms are amongst the first symptoms in the pathogenesis of neuropsychiatric disease. Their potential in the early detection and medication of these highly detrimental disorders is still widely underestimated. We possess powerful tools to influence the circadian system, like timed light exposure, sleep restriction, or melatonin. With the advances in molecular genetics and our increasing knowledge of the synchronisation pathways between the different circadian oscillators within the body, new means will be discovered to more efficiently use these tools, whilst new ones will be developed to more specifically attack different aspects of the systemic circadian regulatory network.

Mood, psychotic, and neurodegenerative disorders display many characteristics of internal circadian desynchronisation and the strengthening of coupling within the circadian system has been shown to be beneficial in such cases. Whether clock defects can actually cause affective, psychotic, or even neurodegenerative diseases, remains to be demonstrated. Intriguingly, a very recent study reports that a mutation in the *Clock* gene can induce mania-like behaviour in mice [198]. Comparable studies – in animal models as well as in humans – are needed to show by which mechanism circadian mutations and defects influence the development of depression, schizophrenia and neurodegenerative disorders, how the synchronisation of different peripheral clocks contributes to this phenomenon, and in which way we can use our knowledge about the circadian signalling pathways to slow down or even prevent the progression of neuropsychiatric pathologies.

References

[1] Mental health: facing the challenges, building solutions, *Report from the WHO European Ministerial Conference.* WHO Europe, Copenhagen, 2005.

[2] C.H. Adler and M.J. Thorpy, Sleep issues in Parkinson's disease, *Neurology. 64* (2005) S12-20.

[3] U. Albrecht, Z.S. Sun, G. Eichele and C.C. Lee, A differential response of two putative mammalian circadian regulators, mper1 and mper2, to light, *Cell.* 91 (1997) 1055-1064.

[4] L.A. Andrade, J.G. Lima, S. Tufik, P.H. Bertolucci and E.A. Carlini, Rem sleep deprivation in an experimental model of Parkinson's disease, *Arq. Neuropsiquiatr.* 45 (1987) 217-223.

[5] M.C. Antle and R.E. Mistlberger, Circadian clock resetting by sleep deprivation without exercise in the Syrian hamster, *J. Neurosci.* 20 (2000) 9326-9332.

[6] M. Apfelbaum, A. Reinberg and D. Lacatis, Effects of meal timing on circadian rhythms in 9 physiologic variables of young healthy but obese women on a calorie restricted diet, *Int. J. Chronobiol.* 4 (1976) 29-37.

[7] J. Arendt, Melatonin and human rhythms, *Chronobiol. Int.* 23 (2006) 21-37.

[8] J. Arendt and V. Marks, Physiological changes underlying jet lag, *Br. Med. J.* (Clin. Res. Ed.) 284 (1982) 144-146.

[9] R. Armitage, Sleep and circadian rhythms in mood disorders, *Acta Psychiatr. Scand. Suppl.* (2007) 104-115.

[10] R. Armitage and R.F. Hoffmann, Sleep EEG, depression and gender, *Sleep Med. Rev.* 5 (2001) 237-246.

[11] S.M. Armstrong, V.M. Cassone, M.J. Chesworth, J.R. Redman and R.V. Short, Synchronization of mammalian circadian rhythms by melatonin, *J. Neural. Transm. Suppl.* 21 (1986) 375-394.

[12] G. Aston-Jones, Brain structures and receptors involved in alertness, *Sleep Med.* 6 Suppl. 1 (2005) S3-7.

[13] D.S. Baldwin and G.I. Papakostas, Symptoms of fatigue and sleepiness in major depressive disorder, *J. Clin. Psychiatry.* 67 Suppl. 6 (2006) 9-15.

[14] J. Balldin, C.G. Gottfries, I. Karlsson, G. Lindstedt, G. Langstrom and J. Walinder, Dexamethasone suppression test and serum prolactin in dementia disorders, *Br. J. Psychiatry.* 143 (1983) 277-281.

[15] A. Balsalobre, Clock genes in mammalian peripheral tissues, *Cell Tissue Res.* 309 (2002) 193-199.

[16] A. Balsalobre, S.A. Brown, L. Marcacci, F. Tronche, C. Kellendonk, H.M. Reichardt, G. Schutz and U. Schibler, Resetting of circadian time in peripheral tissues by glucocorticoid signaling, *Science.* 289 (2000) 2344-2347.

[17] A. Balsalobre, F. Damiola and U. Schibler, A serum shock induces circadian gene expression in mammalian tissue culture cells, *Cell.* 93 (1998) 929-937.

[18] A. Balsalobre, L. Marcacci and U. Schibler, Multiple signaling pathways elicit circadian gene expression in cultured Rat-1 fibroblasts, Curr Biol 10 (2000) 1291-1294.

[19] C.M. Banki, L. Karmacsi, G. Bissette and C.B. Nemeroff, CSF corticotropin releasing hormone, somatostatin, and thyrotropin releasing hormone in schizophrenia, *Psychiatry Res.* 43 (1992) 13-21.

[20] A. Baumgartner, K.J. Graf, I. Kurten and H. Meinhold, Thyrotropin (TSH) and thyroid hormone concentrations during partial sleep deprivation in patients with major depressive disorder, *J. Psychiatr. Res.* 24 (1990) 281-292.

[21] J. Beck-Friis, B.F. Kjellman, B. Aperia, F. Unden, D. von Rosen, J.G. Ljunggren and L. Wetterberg, Serum melatonin in relation to clinical variables in patients with major depressive disorder and a hypothesis of a low melatonin syndrome, *Acta Psychiatr. Scand.* 71 (1985) 319-330.

[22] J. Beck-Friis, J.G. Ljunggren, M. Thoren, D. von Rosen, B.F. Kjellman and L. Wetterberg, Melatonin, cortisol and ACTH in patients with major depressive disorder and healthy humans with special reference to the outcome of the dexamethasone suppression test, *Psychoneuroendocrinology.* 10 (1985) 173-186.

[23] H. Beckmann, R.E. Lang and W.F. Gattaz, Vasopressin--oxytocin in cerebrospinal fluid of schizophrenic patients and normal controls, *Psychoneuroendocrinology.* 10 (1985) 187-191.

[24] R.D. Bell, A. Carruth, R.N. Rosenberg and R.M. Boyar, Effect of dopamine agonist (Lergotrile mesylate) therapy on twenty-four hour secretion of prolactin in treated Parkinson's disease, *J. Clin. Endocrinol. Metab.* 47 (1978) 807-811.

[25] G. Bellomo, L. Santambrogio, M. Fiacconi, A.M. Scarponi and G. Ciuffetti, Plasma profiles of adrenocorticotropic hormone, cortisol, growth hormone and prolactin in patients with untreated Parkinson's disease, *J. Neurol.* 238 (1991) 19-22.

[26] P.H. Bertolucci, L.A. Andrade, J.G. Lima and E.A. Carlini, Total sleep deprivation and Parkinson disease, *Arq. Neuropsiquiatr.* 45 (1987) 224-230.

[27] E.L. Bittman, L. Doherty, L. Huang and A. Paroskie, Period gene expression in mouse endocrine tissues, *Am. J. Physiol. Regul. Integr. Comp. Physiol.* 285 (2003) R561-569.

[28] K. Boelaert and J.A. Franklyn, Thyroid hormone in health and disease, *J. Endocrinol.* 187 (2005) 1-15.

[29] A.A. Borbely, A two process model of sleep regulation, *Hum. Neurobiol.* 1 (1982) 195-204.

[30] A.A. Borbely and A. Wirz-Justice, Sleep, sleep deprivation and depression. A hypothesis derived from a model of sleep regulation, *Hum. Neurobiol.* 1 (1982) 205-210.

[31] R. Bordet, D. Devos, S. Brique, Y. Touitou, J.D. Guieu, C. Libersa and A. Destee, Study of circadian melatonin secretion pattern at different stages of Parkinson's disease, *Clin. Neuropharmacol.* 26 (2003) 65-72.

[32] J. Born and U. Wagner, Memory consolidation during sleep: role of cortisol feedback, *Ann. N Y Acad. Sci.* 1032 (2004) 198-201.

[33] J.P. Brouwer, B.C. Appelhof, W.J. Hoogendijk, J. Huyser, E. Endert, C. Zuketto, A.H. Schene, J.G. Tijssen, R. Van Dyck, W.M. Wiersinga and E. Fliers, Thyroid and adrenal axis in major depression: a controlled study in outpatients, *Eur. J. Endocrinol.* 152 (2005) 185-191.

[34] S.A. Brown, G. Zumbrunn, F. Fleury-Olela, N. Preitner and U. Schibler, Rhythms of mammalian body temperature can sustain peripheral circadian clocks, *Curr. Biol.* 12 (2002) 1574-1583.

[35] B. Bruguerolle and N. Simon, Biologic rhythms and Parkinson's disease: a chronopharmacologic approach to considering fluctuations in function, *Clin. Neuropharmacol.* 25 (2002) 194-201.

[36] P.F. Buckley and S.M. Stahl, Pharmacological treatment of negative symptoms of schizophrenia: therapeutic opportunity or cul-de-sac?, *Acta Psychiatr. Scand.* 115 (2007) 93-100.

[37] R.M. Buijs, F.A. Scheer, F. Kreier, C. Yi, N. Bos, V.D. Goncharuk and A. Kalsbeek, Organization of circadian functions: interaction with the body, *Prog. Brain Res.* 153 (2006) 341-360.

[38] R.M. Buijs, C.G. van Eden, V.D. Goncharuk and A. Kalsbeek, The biological clock tunes the organs of the body: timing by hormones and the autonomic nervous system, *J. Endocrinol.* 177 (2003) 17-26.

[39] C. Cailotto, S.E. La Fleur, C. Van Heijningen, J. Wortel, A. Kalsbeek, M. Feenstra, P. Pevet and R.M. Buijs, The suprachiasmatic nucleus controls the daily variation of plasma glucose via the autonomic output to the liver: are the clock genes involved?, *Eur. J. Neurosci.* 22 (2005) 2531-2540.

[40] L. Canaple, T. Kakizawa and V. Laudet, The days and nights of cancer cells, *Cancer Res.* 63 (2003) 7545-7552.

[41] E. Challet, I. Caldelas, C. Graff and P. Pevet, Synchronization of the molecular clockwork by light- and food-related cues in mammals, *Biol. Chem.* 384 (2003) 711-719.

[42] M.Y. Cheng, C.M. Bullock, C. Li, A.G. Lee, J.C. Bermak, J. Belluzzi, D.R. Weaver, F.M. Leslie and Q.Y. Zhou, Prokineticin 2 transmits the behavioural circadian rhythm of the suprachiasmatic nucleus, *Nature.* 417 (2002) 405-410.

[43] J.E. Christie, L.J. Whalley, J. Bennie, H. Dick, I.M. Blackburn, D.H. Blackwood and G. Fink, Characteristic plasma hormone changes in Alzheimer's disease, *Br. J. Psychiatry.* 150 (1987) 674-681.

[44] M. Crasson, S. Kjiri, A. Colin, K. Kjiri, M. L'Hermite-Baleriaux, M. Ansseau and J.J. Legros, Serum melatonin and urinary 6-sulfatoxymelatonin in major depression, *Psychoneuroendocrinology.* 29 (2004) 1-12.

[45] M. Critchley and E.A. Critchley, John Hughlings Jackson: *Father of English Neurology,* 1st edn., Oxford University Press, USA, New York, 1998, 228 pp.

[46] F. Damiola, N. Le Minh, N. Preitner, B. Kornmann, F. Fleury-Olela and U. Schibler, Restricted feeding uncouples circadian oscillators in peripheral tissues from the central pacemaker in the suprachiasmatic nucleus, *Genes Dev.* 14 (2000) 2950-2961.

[47] A.J. Davidson, S. Yamazaki and M. Menaker, SCN: ringmaster of the circadian circus or conductor of the circadian orchestra?, *Novartis Found Symp.* 253 (2003) 110-121; discussion 121-115, 281-114.

[48] J.A. den Boer, F.J. Bosker and Y. Meesters, Clinical efficacy of agomelatine in depression: the evidence, *Int. Clin. Psychopharmacol.* 21 Suppl 1 (2006) S21-24.

[49] P.P. Doghramji, Recognizing sleep disorders in a primary care setting, *J. Clin. Psychiatry.* 65 Suppl 16 (2004) 23-26.
[50] B.P. Dohrenwend and G. Egri, Recent stressful life events and episodes of schizophrenia, *Schizophr. Bull.* 7 (1981) 12-23.
[51] M. Doi, J. Hirayama and P. Sassone-Corsi, Circadian regulator CLOCK is a histone acetyltransferase, *Cell.* 125 (2006) 497-508.
[52] Y. Doi, M. Minowa, M. Uchiyama, M. Okawa, K. Kim, K. Shibui and Y. Kamei, Psychometric assessment of subjective sleep quality using the Japanese version of the Pittsburgh Sleep Quality Index (PSQI-J) in psychiatric disordered and control subjects, *Psychiatry Res.* 97 (2000) 165-172.
[53] O.T. Dolberg, S. Hirschmann and L. Grunhaus, Melatonin for the treatment of sleep disturbances in major depressive disorder, *Am. J. Psychiatry.* 155 (1998) 1119-1121.
[54] G.E. Duffield, DNA microarray analyses of circadian timing: the genomic basis of biological time, *J. Neuroendocrinol.* 15 (2003) 991-1002.
[55] D. Duma, C.M. Jewell and J.A. Cidlowski, Multiple glucocorticoid receptor isoforms and mechanisms of post-translational modification, *J. Steroid Biochem. Mol. Biol.* 102 (2006) 11-21.
[56] D.L. Dunner, V. Patrick and R.R. Fieve, Life events at the onset of bipolar affective illness, *Am. J. Psychiatry.* 136 (1979) 508-511.
[57] C.L. Ehlers, E. Frank and D.J. Kupfer, Social zeitgebers and biological rhythms. A unified approach to understanding the etiology of depression, *Arch. Gen Psychiatry.* 45 (1988) 948-952.
[58] R.G. Foster and K. Wulff, The rhythm of rest and excess, *Nat. Rev. Neurosci.* 6 (2005) 407-414.
[59] L. Franchi, V. Fulci and G. Macino, Protein kinase C modulates light responses in Neurospora by regulating the blue light photoreceptor WC-1, *Mol. Microbiol.* 56 (2005) 334-345.
[60] M.S. Freedman, R.J. Lucas, B. Soni, M. von Schantz, M. Munoz, Z. David-Gray and R. Foster, Regulation of mammalian circadian behavior by non-rod, non-cone, ocular photoreceptors, *Science.* 284 (1999) 502-504.
[61] S. Freud, Die Traumdeutung, 1st edn., Franz Deuticke Verlag, Leipzig, Wien, 1900.
[62] T. Fuchs, Melancholia as a desynchronization: towards a psychopathology of interpersonal time, *Psychopathology.* 34 (2001) 179-186.
[63] C. Fukuhara, J.C. Dirden and G. Tosini, Circadian expression of period 1, period 2, and arylalkylamine N-acetyltransferase mRNA in the rat pineal gland under different light conditions, *Neurosci. Lett.* 286 (2000) 167-170.
[64] P.M. Fuller, J.J. Gooley and C.B. Saper, Neurobiology of the sleep-wake cycle: sleep architecture, circadian regulation, and regulatory feedback, *J. Biol. Rhythms.* 21 (2006) 482-493.
[65] P. Gallagher, S. Watson, M.S. Smith, A.H. Young and I.N. Ferrier, Plasma cortisol-dehydroepiandrosterone (DHEA) ratios in schizophrenia and bipolar disorder, *Schizophr. Res.* 90 (2007) 258-265.
[66] M. Gallego and D.M. Virshup, Post-translational modifications regulate the ticking of the circadian clock, *Nat. Rev. Mol. Cell Biol.* 8 (2007) 139-148.

[67] R. Garcia, Stress, hippocampal plasticity, and spatial learning, *Synapse.* 40 (2001) 180-183.

[68] N. Gekakis, D. Staknis, H.B. Nguyen, F.C. Davis, L.D. Wilsbacher, D.P. King, J.S. Takahashi and C.J. Weitz, Role of the CLOCK protein in the mammalian circadian mechanism, *Science.* 280 (1998) 1564-1569.

[69] H. Gerbaldo, L. Demisch and D.P. Cardinali, Light exposure patterns in schizophrenia, *Acta Psychiatr. Scand.* 85 (1992) 94-95.

[70] H. Giedke and F. Schwarzler, Therapeutic use of sleep deprivation in depression, *Sleep Med. Rev.* 6 (2002) 361-377.

[71] M.U. Gillette and J.W. Mitchell, Signaling in the suprachiasmatic nucleus: selectively responsive and integrative, *Cell Tissue Res.* 309 (2002) 99-107.

[72] F. Goetz, J. Bishop, F. Halberg, R.B. Sothern, R. Brunning, B. Senske, B. Greenberg, D. Minors, P. Stoney, I.D. Smith, G.D. Rosen, D. Cressey, E. Haus and M. Apfelbaum, Timing of single daily meal influences relations among human circadian rhythms in urinary cyclic AMP and hemic glucagon, insulin and iron, *Experientia.* 32 (1976) 1081-1084.

[73] P.W. Gold, K.E. Gabry, M.R. Yasuda and G.P. Chrousos, Divergent endocrine abnormalities in melancholic and atypical depression: clinical and pathophysiologic implications, *Endocrinol. Metab. Clin. North Am.* 31 (2002) 37-62, vi.

[74] J.J. Gooley, A. Schomer and C.B. Saper, The dorsomedial hypothalamic nucleus is critical for the expression of food-entrainable circadian rhythms, *Nat. Neurosci.* 9 (2006) 398-407.

[75] M.M. Governale and T.M. Lee, Olfactory cues accelerate reentrainment following phase shifts and entrain free-running rhythms in female Octodon degus (Rodentia), *J. Biol. Rhythms.* 16 (2001) 489-501.

[76] L.D. Grandin, L.B. Alloy and L.Y. Abramson, The social zeitgeber theory, circadian rhythms, and mood disorders: review and evaluation, *Clin. Psychol. Rev.* 26 (2006) 679-694.

[77] C. Hammen, Generation of stress in the course of unipolar depression, *J. Abnorm. Psychol.* 100 (1991) 555-561.

[78] J. Hannibal, Neurotransmitters of the retino-hypothalamic tract, *Cell Tissue Res.* 309 (2002) 73-88.

[79] J. Hannibal, P. Hindersson, J. Ostergaard, B. Georg, S. Heegaard, P.J. Larsen and J. Fahrenkrug, Melanopsin is expressed in PACAP-containing retinal ganglion cells of the human retinohypothalamic tract, *Invest. Ophthalmol. Vis. Sci.* 45 (2004) 4202-4209.

[80] A. Hartmann, J.D. Veldhuis, M. Deuschle, H. Standhardt and I. Heuser, Twenty-four hour cortisol release profiles in patients with Alzheimer's and Parkinson's disease compared to normal controls: ultradian secretory pulsatility and diurnal variation, *Neurobiol. Aging.* 18 (1997) 285-289.

[81] S. Hattar, M. Kumar, A. Park, P. Tong, J. Tung, K.W. Yau and D.M. Berson, Central projections of melanopsin-expressing retinal ganglion cells in the mouse, *J. Comp. Neurol.* 497 (2006) 326-349.

[82] S. Hattar, H.W. Liao, M. Takao, D.M. Berson and K.W. Yau, Melanopsin-containing retinal ganglion cells: architecture, projections, and intrinsic photosensitivity, *Science.* 295 (2002) 1065-1070.

[83] S. Hattar, R.J. Lucas, N. Mrosovsky, S. Thompson, R.H. Douglas, M.W. Hankins, J. Lem, M. Biel, F. Hofmann, R.G. Foster and K.W. Yau, Melanopsin and rod-cone photoreceptive systems account for all major accessory visual functions in mice, *Nature.* 424 (2003) 76-81.

[84] D. Healy and J.M. Waterhouse, The circadian system and the therapeutics of the affective disorders, *Pharmacol. Ther.* 65 (1995) 241-263.

[85] M. Heim, [Bright light therapy in schizophrenic diseases], *Psychiatr. Neurol. Med. Psychol.* (Leipz) 42 (1990) 146-150.

[86] A. Herxheimer and J. Waterhouse, The prevention and treatment of jet lag, *Bmj.* 326 (2003) 296-297.

[87] I.J. Heuser, F. Baronti, C.A. Marin, N. Ma, G.R. Merriam, T.N. Chase and M.M. Mouradian, Growth hormone secretion in Alzheimer's disease: 24-hour profile of basal levels and response to stimulation and suppression studies, *Neurobiol. Aging.* 13 (1992) 255-260.

[88] T. Hirota and Y. Fukada, Resetting mechanism of central and peripheral circadian clocks in mammals, *Zoolog. Sci.* 21 (2004) 359-368.

[89] E. Hogervorst, J. Williams, M. Budge, L. Barnetson, M. Combrinck and A.D. Smith, Serum total testosterone is lower in men with Alzheimer's disease, *Neuro Endocrinol. Lett.* 22 (2001) 163-168.

[90] S. Honma, T. Kawamoto, Y. Takagi, K. Fujimoto, F. Sato, M. Noshiro, Y. Kato and K. Honma, Dec1 and Dec2 are regulators of the mammalian molecular clock, *Nature.* 419 (2002) 841-844.

[91] J. Huang, V. Pashkov, D.M. Kurrasch, K. Yu, S.J. Gold and T.M. Wilkie, Feeding and fasting controls liver expression of a regulator of G protein signaling (Rgs16) in periportal hepatocytes, *Comp. Hepatol.* 5 (2006) 8.

[92] A. Ishida, T. Mutoh, T. Ueyama, H. Bando, S. Masubuchi, D. Nakahara, G. Tsujimoto and H. Okamura, Light activates the adrenal gland: timing of gene expression and glucocorticoid release, *Cell Metab.* 2 (2005) 297-307.

[93] A. Kalsbeek, I.F. Palm, S.E. La Fleur, F.A. Scheer, S. Perreau-Lenz, M. Ruiter, F. Kreier, C. Cailotto and R.M. Buijs, SCN outputs and the hypothalamic balance of life, *J. Biol. Rhythms.* 21 (2006) 458-469.

[94] G. Katz, R. Durst, Y. Zislin, Y. Barel and H.Y. Knobler, Psychiatric aspects of jet lag: review and hypothesis, *Med. Hypotheses.* 56 (2001) 20-23.

[95] L. Kaul and C.D. Berdanier, Effect of meal-feeding on the daily variations of insulin, glucose, and NADP-linked dehydrogenases in rats, *J. Nutr.* 105 (1975) 1132-1140.

[96] C. Kellendonk, P. Gass, O. Kretz, G. Schutz and F. Tronche, Corticosteroid receptors in the brain: gene targeting studies, *Brain Res. Bull.* 57 (2002) 73-83.

[97] D. Kemali, M. Maj, M.G. Ariano, F. Arena and N. Lovero, 24-hour plasma levels of prolactin, cortisol, growth hormone and catecholamines in schizophrenic patients, *Neuropsychobiology.* 14 (1985) 109-114.

[98] J.J. Kim, E.Y. Song and T.A. Kosten, Stress effects in the hippocampus: synaptic plasticity and memory, *Stress.* 9 (2006) 1-11.
[99] Y. Kita, M. Shiozawa, W. Jin, R.R. Majewski, J.C. Besharse, A.S. Greene and H.J. Jacob, Implications of circadian gene expression in kidney, liver and thc effects of fasting on pharmacogenomic studies, *Pharmacogenetics.* 12 (2002) 55-65.
[100] B.F. Kjellman, J. Beck-Friis, J.G. Ljunggren and L. Wetterberg, Twenty-four-hour serum levels of TSH in affective disorders, *Acta Psychiatr. Scand.* 69 (1984) 491-502.
[101] U. Klaus, R. Weinandy and R. Gattermann, Circadian activity rhythms and sensitivity to noise in the Mongolian gerbil (Meriones unguiculatus), *Chronobiol. Int.* 17 (2000) 137-145.
[102] D.C. Klein, R.Y. Moore and S.M. Reppert, Suprachiasmatic Nucleus: T*he Mind's Clock,* Oxford University Press, New York, 1991.
[103] A. Knutsson, Health disorders of shift workers, *Occup. Med.* (Lond) 53 (2003) 103-108.
[104] C.H. Ko and J.S. Takahashi, Molecular components of the mammalian ircadian clock, *Hum. Mol. Genet.* 15 Suppl. 2 (2006) R271-277.
[105] C. Kopp, U. Albrecht, B. Zheng and I. Tobler, Homeostatic sleep regulation is preserved in mPer1 and mPer2 mutant mice, *Eur. J. Neurosci.* 16 (2002) 1099-1106.
[106] B. Kornmann, O. Schaad, H. Bujard, J.S. Takahashi and U. Schibler, System-driven and oscillator-dependent circadian transcription in mice with a conditionally active liver clock, *PLoS Biol.* 5 (2007) e34.
[107] A. Kramer, F.C. Yang, P. Snodgrass, X. Li, T.E. Scammell, F.C. Davis and C.J. Weitz, Regulation of daily locomotor activity and sleep by hypothalamic EGF receptor signaling, *Science.* 294 (2001) 2511-2515.
[108] S. Kraves and C.J. Weitz, A role for cardiotrophin-like cytokine in the circadian control of mammalian locomotor activity, *Nat. Neurosci.* 9 (2006) 212-219.
[109] D.F. Kripke, Light treatment for nonseasonal depression: speed, efficacy, and combined treatment, *J. Affect. Disord.* 49 (1998) 109-117.
[110] P.L. Lakin-Thomas, Transcriptional feedback oscillators: maybe, maybe not, *J. Biol. Rhythms.* 21 (2006) 83-92.
[111] A. Laposky, A. Easton, C. Dugovic, J. Walisser, C. Bradfield and F. Turek, Deletion of the mammalian circadian clock gene BMAL1/Mop3 alters baseline sleep architecture and the response to sleep deprivation, *Sleep.* 28 (2005) 395-409.
[112] N. Le Minh, F. Damiola, F. Tronche, G. Schutz and U. Schibler, Glucocorticoid hormones inhibit food-induced phase-shifting of peripheral circadian oscillators, *Embo J.* 20 (2001) 7128-7136.
[113] C. Lee, V. Parikh, T. Itsukaichi, K. Bae and I. Edery, Resetting the Drosophila clock by photic regulation of PER and a PER-TIM complex, *Science.* 271 (1996) 1740-1744.
[114] H.S. Lee, H.J. Billings and M.N. Lehman, The suprachiasmatic nucleus: a clock of multiple components, *J. Biol. Rhythms.* 18 (2003) 435-449.
[115] E.S. Lein, M.J. Hawrylycz, N. Ao, M. Ayres, A. Bensinger, A. Bernard, A.F. Boe, M.S. Boguski, K.S. Brockway, E.J. Byrnes, L. Chen, L. Chen, T.M. Chen, M.C. Chin, J. Chong, B.E. Crook, A. Czaplinska, C.N. Dang, S. Datta, N.R. Dee, A.L. Desaki, T. Desta, E. Diep, T.A. Dolbeare, M.J. Donelan, H.W. Dong, J.G. Dougherty, B.J.

Duncan, A.J. Ebbert, G. Eichele, L.K. Estin, C. Faber, B.A. Facer, R. Fields, S.R. Fischer, T.P. Fliss, C. Frensley, S.N. Gates, K.J. Glattfelder, K.R. Halverson, M.R. Hart, J.G. Hohmann, M.P. Howell, D.P. Jeung, R.A. Johnson, P.T. Karr, R. Kawal, J.M. Kidney, R.H. Knapik, C.L. Kuan, J.H. Lake, A.R. Laramee, K.D. Larsen, C. Lau, T.A. Lemon, A.J. Liang, Y. Liu, L.T. Luong, J. Michaels, J.J. Morgan, R.J. Morgan, M.T. Mortrud, N.F. Mosqueda, L.L. Ng, R. Ng, G.J. Orta, C.C. Overly, T.H. Pak, S.E. Parry, S.D. Pathak, O.C. Pearson, R.B. Puchalski, Z.L. Riley, H.R. Rockett, S.A. Rowland, J.J. Royall, M.J. Ruiz, N.R. Sarno, K. Schaffnit, N.V. Shapovalova, T. Sivisay, C.R. Slaughterbeck, S.C. Smith, K.A. Smith, B.I. Smith, A.J. Sodt, N.N. Stewart, K.R. Stumpf, S.M. Sunkin, M. Sutram, A. Tam, C.D. Teemer, C. Thaller, C.L. Thompson, L.R. Varnam, A. Visel, R.M. Whitlock, P.E. Wohnoutka, C.K. Wolkey, V.Y. Wong, M. Wood, M.B. Yaylaoglu, R.C. Young, B.L. Youngstrom, X.F. Yuan, B. Zhang, T.A. Zwingman and A.R. Jones, Genome-wide atlas of gene expression in the adult mouse brain, *Nature.* 445 (2007) 168-176.

[116] E. Lenzinger, K. Meszaros, K. Hornik, P. Parzer, E. Hollerer, G. Langer, F. Resch and J.J. Legros, Correlation between vasopressin baseline and TSH-blunting in depressives, *Biol. Psychiatry.* 39 (1996) 341-345.

[117] B.E. Leonard and A. Myint, Changes in the immune system in depression and dementia: causal or coincidental effects?, *Dialogues Clin. Neurosci.* 8 (2006) 163-174.

[118] I. Levin Ia, [Effects of sleep deprivation in patients with parkinsonism], *Sov. Med.* (1991) 9-11.

[119] A.J. Lewy, R.L. Sack, L.S. Miller and T.M. Hoban, Antidepressant and circadian phase-shifting effects of light, *Science.* 235 (1987) 352-354.

[120] P. Linkowski and P.P. Hubain, [Sleep disorders and neuroendocrine regulation in depression], *Encephale* 21 Spec. No 7 (1995) 25-28.

[121] P. Linkowski, E. Van Cauter, R. Leclercq, D. Desmedt, M. Brasseur, J. Golstein, G. Copinschi and J. Mendlewicz, ACTH, cortisol and growth hormone 24-hour profiles in major depressive illness, *Acta Psychiatr. Belg.* 85 (1985) 615-623.

[122] R.Y. Liu, J.N. Zhou, J. van Heerikhuize, M.A. Hofman and D.F. Swaab, Decreased melatonin levels in postmortem cerebrospinal fluid in relation to aging, Alzheimer's disease, and apolipoprotein E-epsilon4/4 genotype, *J. Clin. Endocrinol. Metab.* 84 (1999) 323-327.

[123] A. Lleo, S.M. Greenberg and J.H. Growdon, Current pharmacotherapy for Alzheimer's disease, *Annu. Rev. Med.* 57 (2006) 513-533.

[124] P.L. Lowrey and J.S. Takahashi, Mammalian circadian biology: elucidating genome-wide levels of temporal organization, *Annu. Rev. Genomics Hum. Genet.* 5 (2004) 407-441.

[125] R.J. Lucas, R.H. Douglas and R.G. Foster, Characterization of an ocular photopigment capable of driving pupillary constriction in mice, *Nat. Neurosci.* 4 (2001) 621-626.

[126] R.J. Lucas, M.S. Freedman, M. Munoz, J.M. Garcia-Fernandez and R.G. Foster, Regulation of the mammalian pineal by non-rod, non-cone, ocular photoreceptors, *Science.* 284 (1999) 505-507.

[127] R.J. Lucas, S. Hattar, M. Takao, D.M. Berson, R.G. Foster and K.W. Yau, Diminished pupillary light reflex at high irradiances in melanopsin-knockout mice, *Science.* 299 (2003) 245-247.

[128] S.J. Lurie, B. Gawinski, D. Pierce and S.J. Rousseau, Seasonal affective disorder, *Am. Fam. Physician.* 74 (2006) 1521-1524.

[129] E.G. Marchant and R.E. Mistlberger, Morphine phase-shifts circadian rhythms in mice: role of behavioural activation, *Neuroreport.* 7 (1995) 209-212.

[130] E. Maronde and D. Motzkus, Oscillation of human period 1 (hPER1) reporter gene activity in human neuroblastoma cells in vivo, *Chronobiol. Int.* 20 (2003) 671-681.

[131] E. Maronde and J.H. Stehle, The mammalian pineal gland: known facts, unknown facets, *Trends Endocrinol. Metab.* (2007).

[132] P.S. Masand, L. Culpepper, D. Henderson, S. Lee, K. Littrell, J.W. Newcomer and N. Rasgon, Metabolic and endocrine disturbances in psychiatric disorders: a multidisciplinary approach to appropriate atypical antipsychotic utilization, *CNS Spectr.* 10 (2005) suppl14 11-15.

[133] J.C. Mayo, R.M. Sainz, D.X. Tan, I. Antolin, C. Rodriguez and R.J. Reiter, Melatonin and Parkinson's disease, *Endocrine.* 27 (2005) 169-178.

[134] M. Mazza, M.A. Satta, P. Bria and S. Mazza, Neuroendocrinology of mood disorders, *Clin. Ter.* 155 (2004) 537-541.

[135] P. McNamara, S.P. Seo, R.D. Rudic, A. Sehgal, D. Chakravarti and G.A. FitzGerald, Regulation of CLOCK and MOP4 by nuclear hormone receptors in the vasculature: a humoral mechanism to reset a peripheral clock, *Cell.* 105 (2001) 877-889.

[136] J.H. Meijer, J. Schaap, K. Watanabe and H. Albus, Multiunit activity recordings in the suprachiasmatic nuclei: in vivo versus in vitro models, *Brain Res.* 753 (1997) 322-327.

[137] J.H. Meijer and W.J. Schwartz, In search of the pathways for light-induced pacemaker resetting in the suprachiasmatic nucleus, *J. Biol. Rhythms.* 18 (2003) 235-249.

[138] O.C. Meijer, Coregulator proteins and corticosteroid action in the brain, *J. Neuroendocrinol.* 14 (2002) 499-505.

[139] J. Mendlewicz, P. Linkowski, M. Kerkhofs, D. Desmedt, J. Golstein, G. Copinschi and E. Van Cauter, Diurnal hypersecretion of growth hormone in depression, *J. Clin. Endocrinol. Metab.* 60 (1985) 505-512.

[140] J. Mendoza, Circadian clocks: setting time by food, *J. Neuroendocrinol.* 19 (2007) 127-137.

[141] M. Merrow, C. Boesl, J. Ricken, M. Messerschmitt, M. Goedel and T. Roenneberg, Entrainment of the Neurospora circadian clock, *Chronobiol. Int.* 23 (2006) 71-80.

[142] S. Miche and C.S. Colwell, Cellular communication and coupling within the suprachiasmatic nucleus, *Chronobiol. Int.* 18 (2001) 579-600.

[143] R.E. Mistlberger, Effects of daily schedules of forced activity on free-running rhythms in the rat, *J. Biol. Rhythms.* 6 (1991) 71-80.

[144] R.E. Mistlberger, M.C. Antle, I.C. Webb, M. Jones, J. Weinberg and M.S. Pollock, Circadian clock resetting by arousal in Syrian hamsters: the role of stress and activity, *Am. J. Physiol. Regul. Integr. Comp. Physiol.* 285 (2003) R917-925.

[145] R.E. Mistlberger, J. Belcourt and M.C. Antle, Circadian clock resetting by sleep deprivation without exercise in Syrian hamsters: dark pulses revisited, *J. Biol. Rhythms.* 17 (2002) 227-237.

[146] R.E. Mistlberger and D.J. Skene, Social influences on mammalian circadian rhythms: animal and human studies, *Biol. Rev. Camb. Philos. Soc.* 79 (2004) 533-556.

[147] M.M. Moga and R.Y. Moore, Organization of neural inputs to the suprachiasmatic nucleus in the rat, *J. Comp. Neurol.* 389 (1997) 508-534.

[148] J.M. Monti and D. Monti, Sleep in schizophrenia patients and the effects of antipsychotic drugs, *Sleep Med. Rev.* 8 (2004) 133-148.

[149] R.Y. Moore and V.B. Eichler, Loss of a circadian adrenal corticosterone rhythm following suprachiasmatic lesions in the rat, *Brain Res.* 42 (1972) 201-206.

[150] N. Mrosovsky, Locomotor activity and non-photic influences on circadian clocks, *Biol. Rev. Camb. Philos. Soc.* 71 (1996) 343-372.

[151] N. Mrosovsky, Phase response curves for social entrainment, *J. Comp. Physiol.* [A] 162 (1988) 35-46.

[152] G. Murialdo, F. Nobili, A. Rollero, M.V. Gianelli, F. Copello, G. Rodriguez and A. Polleri, Hippocampal perfusion and pituitary-adrenal axis in Alzheimer's disease, *Neuropsychobiology.* 42 (2000) 51-57.

[153] L. Murri, A. Iudice, A. Muratorio, A. Polleri, T. Barreca and G. Murialdo, Spontaneous nocturnal plasma prolactin and growth hormone secretion in patients with Parkinson's disease and Huntington's chorea, *Eur. Neurol.* 19 (1980) 198-206.

[154] F. Naef, Circadian clocks go in vitro: purely post-translational oscillators in cyanobacteria, *Mol. Syst. Biol.* 1 (2005) 2005 0019.

[155] M. Nakajima, K. Imai, H. Ito, T. Nishiwaki, Y. Murayama, H. Iwasaki, T. Oyama and T. Kondo, Reconstitution of circadian oscillation of cyanobacterial KaiC phosphorylation in vitro, *Science.* 308 (2005) 414-415.

[156] S. Nakamura, K. Koshimura, T. Kato, S. Yamao, S. Iijima, H. Nagata, S. Miyata, K. Fujiyoshi, K. Okamoto, H. Suga and et al., Neurotransmitters in dementia, *Clin. Ther.* 7 Spec No (1984) 18-34.

[157] M. Nakao and E. Yano, Reporting of somatic symptoms as a screening marker for detecting major depression in a population of Japanese white-collar workers, *J. Clin. Epidemiol.* 56 (2003) 1021-1026.

[158] Y. Ohashi, N. Okamoto, K. Uchida, M. Iyo, N. Mori and Y. Morita, Daily rhythm of serum melatonin levels and effect of light exposure in patients with dementia of the Alzheimer's type, *Biol. Psychiatry.* 45 (1999) 1646-1652.

[159] K. Oishi, N. Amagai, H. Shirai, K. Kadota, N. Ohkura and N. Ishida, Genome-wide Expression Analysis Reveals 100 Adrenal Gland-dependent Circadian Genes in the Mouse Liver, *DNA Res.* 12 (2005) 191-202.

[160] M.S. Okun, M.R. DeLong, J. Hanfelt, M. Gearing and A. Levey, Plasma testosterone levels in Alzheimer and Parkinson diseases, *Neurology.* 62 (2004) 411-413.

[161] H. Oster, The genetic basis of circadian behavior, *Genes Brain Behav.* 5 Suppl. 2 (2006) 73-79.

[162] H. Oster, S. Baeriswyl, G.T. Van Der Horst and U. Albrecht, Loss of circadian rhythmicity in aging mPer1-/-mCry2-/- mutant mice, *Genes Dev.* 17 (2003) 1366-1379.

[163] H. Oster, S. Damerow, R.A. Hut and G. Eichele, Transcriptional Profiling in the Adrenal Gland Reveals Circadian Regulation of Hormone Biosynthesis Genes and Nucleosome Assembly Genes, *J. Biol. Rhythms.* 21 (2006) 350-361.

[164] H. Oster, S. Damerow, S. Kiessling, V. Jakubcakova, D. Abraham, J. Tian, M.W. Hoffmann and G. Eichele, The circadian rhythm of glucocorticoids is regulated by a gating mechanism residing in the adrenal cortical clock, *Cell Metab.* 4 (2006) 163-173.

[165] H. Oster, E. Maronde and U. Albrecht, The circadian clock as a molecular calendar, *Chronobiol. Int.* 19 (2002) 507-516.

[166] K. Otake, Y. Oiso, T. Mitsuma, Y. Hirooka and K. Adachi, Hypothalamic dysfunction in Parkinson's disease patients, *Acta Med. Hung.* 50 (1994) 3-13.

[167] R. Overshott, J. Byrne and A. Burns, Nonpharmacological and pharmacological interventions for symptoms in Alzheimer's disease, *Expert Rev. Neurother.* 4 (2004) 809-821.

[168] M.E. Ozcan and R. Banoglu, Gonadal hormones in schizophrenia and mood disorders, Eur Arch Psychiatry *Clin. Neurosci.* 253 (2003) 193-196.

[169] S. Panda, M.P. Antoch, B.H. Miller, A.I. Su, A.B. Schook, M. Straume, P.G. Schultz, S.A. Kay, J.S. Takahashi and J.B. Hogenesch, Coordinated transcription of key pathways in the mouse by the circadian clock, *Cell.* 109 (2002) 307-320.

[170] S. Panda, T.K. Sato, A.M. Castrucci, M.D. Rollag, W.J. DeGrip, J.B. Hogenesch, I. Provencio and S.A. Kay, Melanopsin (Opn4) requirement for normal light-induced circadian phase shifting, *Science.* 298 (2002) 2213-2216.

[171] G. Parker, G. Malhi, D. Hadzi-Pavlovic and K. Parker, Sleeping in? The impact of age and depressive sub-type on hypersomnia, *J. Affect. Disord.* 90 (2006) 73-76.

[172] E.S. Paykel, The evolution of life events research in psychiatry, *J. Affect. Disord.* 62 (2001) 141-149.

[173] E.S. Paykel, R. Ramana, Z. Cooper, H. Hayhurst, J. Kerr and A. Barocka, Residual symptoms after partial remission: an important outcome in depression, *Psychol. Med.* 25 (1995) 1171-1180.

[174] S. Peirson and R.G. Foster, Melanopsin: another way of signaling light, *Neuron.* 49 (2006) 331-339.

[175] P. Pevet, L. Agez, B. Bothorel, M. Saboureau, F. Gauer, V. Laurent and M. Masson-Pevet, Melatonin in the multi-oscillatory mammalian circadian world, *Chronobiol. Int.* 23 (2006) 39-51.

[176] H.D. Piggins, M.C. Antle and B. Rusak, Neuropeptides phase shift the mammalian circadian pacemaker, *J. Neurosci.* 15 (1995) 5612-5622.

[177] F.T. Pinto and D.A. Golombek, Neuroactive steroids alter the circadian system of the Syrian hamster in a phase-dependent manner, *Life Sci.* 65 (1999) 2497-2504.

[178] C.S. Pittendrigh, Temporal organization: reflections of a Darwinian clock-watcher, *Annu. Rev. Physiol.* 55 (1993) 16-54.

[179] N. Pomara, W.M. Greenberg, M.D. Branford and P.M. Doraiswamy, Therapeutic implications of HPA axis abnormalities in Alzheimer's disease: review and update, *Psychopharmacol. Bull.* 37 (2003) 120-134.

[180] J.A. Posener, D. Charles, J.D. Veldhuis, M.A. Province, G.H. Williams and A.F. Schatzberg, Process irregularity of cortisol and adrenocorticotropin secretion in men with major depressive disorder, *Psychoneuroendocrinology.* 29 (2004) 1129-1137.

[181] N. Preitner, F. Damiola, L. Lopez-Molina, J. Zakany, D. Duboule, U. Albrecht and U. Schibler, The orphan nuclear receptor REV-ERBalpha controls circadian transcription within the positive limb of the mammalian circadian oscillator, *Cell.* 110 (2002) 251-260.

[182] M.L. Previti, W. Zhang and W.E. Van Nostrand, Dexamethasone diminishes the pro-inflammatory and cytotoxic effects of amyloid beta-protein in cerebrovascular smooth muscle cells, *J. Neuroinflammation.* 3 (2006) 18.

[183] I. Provencio, G. Jiang, W.J. De Grip, W.P. Hayes and M.D. Rollag, Melanopsin: An opsin in melanophores, brain, and eye, *Proc. Natl. Acad. Sci. U S A.* 95 (1998) 340-345.

[184] S.M. Rajaratnam and J.R. Redman, Social contact synchronizes free-running activity rhythms of diurnal palm squirrels, *Physiol. Behav.* 66 (1999) 21-26.

[185] M.R. Ralph, R.G. Foster, F.C. Davis and M. Menaker, Transplanted suprachiasmatic nucleus determines circadian period, *Science.* 247 (1990) 975-978.

[186] M.L. Rao, G. Gross, B. Strebel, A. Halaris, G. Huber, P. Braunig and M. Marler, Circadian rhythm of tryptophan, serotonin, melatonin, and pituitary hormones in schizophrenia, *Biol. Psychiatry.* 35 (1994) 151-163.

[187] M.L. Rao, B. Strebel, A. Halaris, G. Gross, P. Braunig, G. Huber and M. Marler, Circadian rhythm of vital signs, norepinephrine, epinephrine, thyroid hormones, and cortisol in schizophrenia, *Psychiatry Res.* 57 (1995) 21-39.

[188] M. Reick, J.A. Garcia, C. Dudley and S.L. McKnight, NPAS2: an analog of clock operative in the mammalian forebrain, *Science.* 293 (2001) 506-509.

[189] L. Rensing and P. Ruoff, Temperature effect on entrainment, phase shifting, and amplitude of circadian clocks and its molecular bases, *Chronobiol. Int.* 19 (2002) 807-864.

[190] S.M. Reppert and D.R. Weaver, Coordination of circadian timing in mammals, *Nature.* 418 (2002) 935-941.

[191] C.F. Reynolds, 3rd and D.J. Kupfer, Sleep research in affective illness: state of the art circa 1987, *Sleep.* 10 (1987) 199-215.

[192] G. Richardson and B. Tate, Hormonal and pharmacological manipulation of the circadian clock: recent developments and future strategies, *Sleep.* 23 Suppl 3 (2000) S77-85.

[193] D. Riemann, U. Voderholzer and M. Berger, Sleep and sleep-wake manipulations in bipolar depression, *Neuropsychobiology.* 45 Suppl 1 (2002) 7-12.

[194] F.L. Rocha, C. Hara, C.V. Rodrigues, M.A. Costa, E. Castro e Costa, C. Fuzikawa and V.G. Santos, Is insomnia a marker for psychiatric disorders in general hospitals?, *Sleep Med.* 6 (2005) 549-553.

[195] T. Roenneberg, A. Wirz-Justice and M. Merrow, Life between clocks: daily temporal patterns of human chronotypes, *J. Biol. Rhythms.* 18 (2003) 80-90.

[196] M.N. Rossor, S.P. Hunt, L.L. Iversen, R. Bannister, J. Hawthorn, V.T. Ang and J.S. Jenkins, Extrahypothalamic vasopressin is unchanged in Parkinson's disease and Huntington's disease, *Brain Res.* 253 (1982) 341-343.

[197] F. Rouillon, Efficacy and tolcrance profile of agomelatine and practical use in depressed patients, *Int. Clin. Psychopharmacol.* 21 Suppl 1 (2006) S31-35.

[198] K. Roybal, D. Theobold, A. Graham, J.A. Dinieri, S.J. Russo, V. Krishnan, S. Chakravarty, J. Peevey, N. Oehrlein, S. Birnbaum, M.H. Vitaterna, P. Orsulak, J.S. Takahashi, E.J. Nestler, W.A. Carlezon, Jr. and C.A. McClung, Mania-like behavior induced by disruption of CLOCK, *Proc. Natl. Acad. Sci. U S A.* (2007).

[199] N.F. Ruby, T.J. Brennan, X. Xie, V. Cao, P. Franken, H.C. Heller and B.F. O'Hara, Role of melanopsin in circadian responses to light, *Science.* 298 (2002) 2211-2213.

[200] S.M. Safford, L.B. Alloy, L.Y. Abramson and A.G. Crossfield, Negative cognitive style as a predictor of negative life events in depression-prone individuals: A test of the stress generation hypothesis, *J. Affect Disord.* 99 (2007) 147-154.

[201] P.N. Sakkas, C.R. Soldatos, J.D. Bergiannaki, T.J. Paparrigopoulos and C.N. Stefanis, Growth hormone secretion during sleep in male depressed patients, Prog *Neuropsychopharmacol. Biol. Psychiatry.* 22 (1998) 467-483.

[202] C.B. Saper, G. Cano and T.E. Scammell, Homeostatic, circadian, and emotional regulation of sleep, *J. Comp. Neurol.* 493 (2005) 92-98.

[203] C.B. Saper, T.E. Scammell and J. Lu, Hypothalamic regulation of sleep and circadian rhythms, *Nature.* 437 (2005) 1257-1263.

[204] T.K. Sato, S. Panda, L.J. Miraglia, T.M. Reyes, R.D. Rudic, P. McNamara, K.A. Naik, G.A. FitzGerald, S.A. Kay and J.B. Hogenesch, A functional genomics strategy reveals Rora as a component of the mammalian circadian clock, *Neuron.* 43 (2004) 527-537.

[205] J. Schaap, C.M. Pennartz and J.H. Meijer, Electrophysiology of the circadian pacemaker in mammals, *Chronobiol. Int.* 20 (2003) 171-188.

[206] W. Schulte, Kombinierte Psycho- und Pharmakotherapie bei Melancholikern. In: H. Kranz and N. Petrilowitsch (Eds.), *Probleme der pharmakopsychiatrischen Kombinations- und Langzeitbehandlung,* Karger, Basel, 1966, pp. 150-169.

[207] S. Shibata and R.Y. Moore, Electrical and metabolic activity of suprachiasmatic nucleus neurons in hamster hypothalamic slices, *Brain Res.* 438 (1988) 374-378.

[208] R. Silver, J. LeSauter, P.A. Tresco and M.N. Lehman, A diffusible coupling signal from the transplanted suprachiasmatic nucleus controlling circadian locomotor rhythms, *Nature.* 382 (1996) 810-813.

[209] V. Simonneaux, V.J. Poirel, M.L. Garidou, D. Nguyen, E. Diaz-Rodriguez and P. Pevet, Daily rhythm and regulation of clock gene expression in the rat pineal gland, *Brain Res. Mol. Brain Res.* 120 (2004) 164-172.

[210] S.M. Smith and W.W. Vale, The role of the hypothalamic-pituitary-adrenal axis in neuroendocrine responses to stress, *Dialogues Clin. Neurosci.* 8 (2006) 383-395.

[211] P. Snodgrass-Belt, J.L. Gilbert and F.C. Davis, Central administration of transforming growth factor-alpha and neuregulin-1 suppress active behaviors and cause weight loss in hamsters, *Brain Res.* 1038 (2005) 171-182.

[212] P.S. Sorensen, A. Gjerris and M. Hammer, Cerebrospinal fluid vasopressin in neurological and psychiatric disorders, *J. Neurol. Neurosurg. Psychiatry.* 48 (1985) 50-57.

[213] V. Srinivasan, S.R. Pandi-Perumal, G.J. Maestroni, A.I. Esquifino, R. Hardeland and D.P. Cardinali, Role of melatonin in neurodegenerative diseases, *Neurotox. Res.* 7 (2005) 293-318.

[214] V. Srinivasan, M. Smits, W. Spence, A.D. Lowe, L. Kayumov, S.R. Pandi-Perumal, B. Parry and D.P. Cardinali, Melatonin in mood disorders, *World J. Biol. Psychiatry.* 7 (2006) 138-151.

[215] S.M. Stahl and P.F. Buckley, Negative symptoms of schizophrenia: a problem that will not go away, *Acta Psychiatr. Scand.* 115 (2007) 4-11.

[216] L. Staner, P. Linkowski and J. Mendlewicz, Biological markers as classifiers for depression: a multivariate study, *Prog. Neuropsychopharmacol. Biol. Psychiatry.* 18 (1994) 899-914.

[217] F. Stephan and H. Zucker, Circadian rhythms in drinking behavior and locomotor activity of rats are eliminated by hypothalamic lesions., *Proc. Natl. Acad. Sci. U S A.* 69 (1972) 1583-1586.

[218] F.K. Stephan, The "other" circadian system: food as a Zeitgeber, *J. Biol. Rhythms.* 17 (2002) 284-292.

[219] E.M. Sternberg, Neural regulation of innate immunity: a coordinated nonspecific host response to pathogens, *Nat. Rev. Immunol.* 6 (2006) 318-328.

[220] K.F. Storch, O. Lipan, I. Leykin, N. Viswanathan, F.C. Davis, W.H. Wong and C.J. Weitz, Extensive and divergent circadian gene expression in liver and heart, *Nature.* 417 (2002) 78-83.

[221] M. Stratmann and U. Schibler, Properties, entrainment, and physiological functions of Mammalian peripheral oscillators, *J. Biol. Rhythms.* 21 (2006) 494-506.

[222] S. Suemaru, K. Suemaru, K. Kawai, S. Miyata, K. Nobukuni, Y. Ihara, R. Namba, K. Urakami and K. Hashimoto, Cerebrospinal fluid corticotropin-releasing hormone in neurodegenerative diseases: reduction in spinocerebellar degeneration, *Life Sci.* 57 (1995) 2231-2235.

[223] D.F. Swaab, Reactivation of atrophic neurons in Alzheimer's disease, *Neurol. Res.* 25 (2003) 652-660.

[224] L.W. Swanson, P.E. Sawchenko, R.W. Lind and J.H. Rho, The CRH motoneuron: differential peptide regulation in neurons with possible synaptic, paracrine, and endocrine outputs, *Ann. N Y Acad. Sci.* 512 (1987) 12-23.

[225] M.P. Szuba, J.D. Amsterdam, A.T. Fernando, 3rd, K.A. Gary, P.C. Whybrow and A. Winokur, Rapid antidepressant response after nocturnal TRH administration in patients with bipolar type I and bipolar type II major depression, *J. Clin. Psychopharmacol.* 25 (2005) 325-330.

[226] S. Takekida, L. Yan, E.S. Maywood, M.H. Hastings and H. Okamura, Differential adrenergic regulation of the circadian expression of the clock genes Period1 and Period2 in the rat pineal gland, *Eur. J. Neurosci.* 12 (2000) 4557-4561.

[227] F. Tamanini, K. Yagita, H. Okamura and G.T. van der Horst, Nucleocytoplasmic shuttling of clock proteins, *Methods Enzymol.* 393 (2005) 418-435.

[228] H. Terazono, T. Mutoh, S. Yamaguchi, M. Kobayashi, M. Akiyama, R. Udo, S. Ohdo, H. Okamura and S. Shibata, Adrenergic regulation of clock gene expression in mouse liver, *Proc. Natl. Acad. Sci. U S A.* 100 (2003) 6795-6800.

[229] M. Terman and J.S. Terman, Light therapy for seasonal and nonseasonal depression: efficacy, protocol, safety, and side effects, *CNS Spectr.* 10 (2005) 647-663; quiz 672.

[230] M.A. Tichomirowa, M.E. Keck, H.J. Schneider, M. Paez-Pereda, U. Renner, F. Holsboer and G.K. Stalla, Endocrine disturbances in depression, *J. Endocrinol. Invest.* 28 (2005) 89-99.

[231] C. Torres-Farfan, M. Seron-Ferre, V. Dinet and H.W. Korf, Immunocytochemical demonstration of day/night changes of clock gene protein levels in the murine adrenal gland: differences between melatonin-proficient (C3H) and melatonin-deficient (C57BL) mice, *J. Pineal. Res.* 40 (2006) 64-70.

[232] B.P. Tu and S.L. McKnight, Metabolic cycles as an underlying basis of biological oscillations, *Nat. Rev. Mol. Cell. Biol.* 7 (2006) 696-701.

[233] P.J. Uhlhaas and W. Singer, Neural synchrony in brain disorders: relevance for cognitive dysfunctions and pathophysiology, *Neuron.* 52 (2006) 155-168.

[234] F. Unden, J.G. Ljunggren, B.F. Kjellman, J. Beck-Friis and L. Wetterberg, Twenty-four-hour serum levels of T4 and T3 in relation to decreased TSH serum levels and decreased TSH response to TRH in affective disorders, *Acta Psychiatr. Scand.* 73 (1986) 358-365.

[235] E. Van Cauter, P. Linkowski, M. Kerkhofs, P. Hubain, M. L'Hermite-Baleriaux, R. Leclercq, M. Brasseur, G. Copinschi and J. Mendlewicz, Circadian and sleep-related endocrine rhythms in schizophrenia, *Arch. Gen. Psychiatry.* 48 (1991) 348-356.

[236] O. Van Reeth and F.W. Turek, Stimulated activity mediates phase shifts in the hamster circadian clock induced by dark pulses or benzodiazepines, *Nature.* 339 (1989) 49-51.

[237] D. Vigano, P. Lissoni, F. Rovelli, M.G. Roselli, F. Malugani, C. Gavazzeni, A. Conti and G. Maestroni, A study of light/dark rhythm of melatonin in relation to cortisol and prolactin secretion in schizophrenia, *Neuro Endocrinol. Lett.* 22 (2001) 137-141.

[238] N. Viswanathan, Maternal entrainment in the circadian activity rhythm of laboratory mouse (C57BL/6J), *Physiol. Behav.* 68 (1999) 157-162.

[239] N. Viswanathan and M.K. Chandrashekaran, Cycles of presence and absence of mother mouse entrain the circadian clock of pups, *Nature.* 317 (1985) 530-531.

[240] M.V. Vitiello, D.L. Bliwise and P.N. Prinz, Sleep in Alzheimer's disease and the sundown syndrome, *Neurology.* 42 (1992) 83-93; discussion. 93-84.

[241] M.V. Vitiello and S. Borson, Sleep disturbances in patients with Alzheimer's disease: epidemiology, pathophysiology and treatment, *CNS Drugs.* 15 (2001) 777-796.

[242] E. Walker, L. Kestler, A. Bollini and K.M. Hochman, Schizophrenia: etiology and course, *Annu. Rev. Psychol.* 55 (2004) 401-430.

[243] M. Warnecke, H. Oster, J.P. Revelli, G. Alvarez-Bolado and G. Eichele, Abnormal development of the locus coeruleus in *Ear2*(*Nr2f6*)-deficient mice impairs the functionality of the forebrain clock and affects nociception, *Genes Dev.* 19 (2005) 614-625.

[244] J. Waterhouse, T. Reilly and B. Edwards, The stress of travel, *J. Sports Sci.* 22 (2004) 946-965; discussion 965-946.

[245] T.A. Wehr, A. Wirz-Justice, F.K. Goodwin, W. Duncan and J.C. Gillin, Phase advance of the circadian sleep-wake cycle as an antidepressant, *Science.* 206 (1979) 710-713.

[246] D.K. Welsh, D.E. Logothetis, M. Meister and S.M. Reppert, Individual neurons dissociated from rat suprachiasmatic nucleus express independently phased circadian firing rhythms, *Neuron.* 14 (1995) 697-706.

[247] L.J. Whalley, J.E. Christie, D.H. Blackwood, J. Bennie, H. Dick, I.M. Blackburn and G. Fink, Disturbed endocrine function in the psychoses. I: Disordered homeostasis or disease process?, *Br. J. Psychiatry.* 155 (1989) 455-461.

[248] A. Wirz-Justice, F. Benedetti, M. Berger, R.W. Lam, K. Martiny, M. Terman and J.C. Wu, Chronotherapeutics (light and wake therapy) in affective disorders, *Psychol. Med.* 35 (2005) 939-944.

[249] M. Wittmann, J. Dinich, M. Merrow and T. Roenneberg, Social jetlag: misalignment of biological and social time, *Chronobiol. Int.* 23 (2006) 497-509.

[250] K.P. Wright, Jr. and C.A. Czeisler, Absence of circadian phase resetting in response to bright light behind the knees, *Science.* 297 (2002) 571.

[251] K. Wulff, E. Joyce, B. Middleton, D.J. Dijk and R.G. Foster, The suitability of actigraphy, diary data, and urinary melatonin profiles for quantitative assessment of sleep disturbances in schizophrenia: a case report, *Chronobiol. Int.* 23 (2006) 485-495.

[252] N. Yamaguchi, K. Maeda and S. Kuromaru, The effects of sleep deprivation on the circadian rhythm of plasma cortisol levels in depressive patients, *Folia Psychiatr. Neurol. Jpn.* 32 (1978) 479-487.

[253] S. Yamaguchi, S. Mitsui, L. Yan, K. Yagita, S. Miyake and H. Okamura, Role of DBP in the circadian oscillatory mechanism, *Mol. Cell Biol.* 20 (2000) 4773-4781.

[254] S. Yamazaki, R. Numano, M. Abe, A. Hida, R. Takahashi, M. Ueda, G.D. Block, Y. Sakaki, M. Menaker and H. Tei, Resetting central and peripheral circadian oscillators in transgenic rats, *Science.* 288 (2000) 682-685.

[255] S.H. Yoo, S. Yamazaki, P.L. Lowrey, K. Shimomura, C.H. Ko, E.D. Buhr, S.M. Siepka, H.K. Hong, W.J. Oh, O.J. Yoo, M. Menaker and J.S. Takahashi, PERIOD2::LUCIFERASE real-time reporting of circadian dynamics reveals persistent circadian oscillations in mouse peripheral tissues, *Proc. Natl. Acad. Sci. U S A.* 101 (2004) 5339-5346.

[256] A.H. Young, Antiglucocoticoid treatments for depression, *Aust. N. Z. J. Psychiatry.* 40 (2006) 402-405.

[257] A.H. Young, Cortisol in mood disorders, *Stress.* 7 (2004) 205-208.

[258] J. Zhang, K. Kaasik, M.R. Blackburn and C.C. Lee, Constant darkness is a circadian metabolic signal in mammals, *Nature.* 439 (2006) 340-343.

In: Progress in Circadian Rhythm Research
Editor: Anne-Laure Léglise, pp. 67-82
ISBN: 978-1-60021-876-7

Chapter III

Circadian Sleep-Wake Rhythms in Preterm Infants

Ronny Geva and Ruth Feldman
The Gonda Multidisciplinary Brain Research Center
Bar-Ilan University, Israel

Abstract

Knowledge is recently building regarding the factors that affect the development of an optimal sleep-wake cycle. Fetuses near term experience arousal fluctuations and sleep episodes already in the darkness of the womb. Yet, the factors that allow for a smooth emergence of an optimal sleep wake cycle are the conditions that place newborn at higher risk to develop sleep disorders are not yet fully understood. The current chapter proposes an integrative model for the development of circadian sleep-wake rhythms in preterm infants. This model is based on a comprehensive review of the basic science literature as well as clinical work with at-risk preterm infants. This integrative model proposed a three source risk source: 1) infant dependent neurobiological vulnerability risk, such as brainstem mediated functional disability; 2) a familial source that entails genetic predispositions, parental style and support resources; and 3) an environmental stimulation source, such as excessive NICU related stimulation and handling. These three realms act as main effects and interact with the infants' prenatal and postnatal age and with the infants' self- regulation mechanisms. The manner in which these sources operate and interact lead to directions of interventions that ameliorate the development of circadian sleep-wake rhythm deficit in infants born at-risk for sleep disorders. The chapter details each component of the model and illuminates the complexity of the interactions involved in it to deepen the understanding the mechanisms involved in early organization of sleep-wake rhythms in infants.

Introduction

Sleep-wake rhythms serve as a basis for the modulation of arousal (Feldman et al., 2002; Feldman et al., 2004), which in turn provides the foundation for emotion regulation (Kopp,

1999; Feldman, Weller, Sirota, & Eidelman, 2002), social competencies (Feldman, 2006) attention shifting and maintenance (Aston-Jones, 2005), sustained exploration (Feldman et al., 2002) and cognition (Sadeh, Dark, Vohr, 1996; & Mayes, 1999; Scher, Amir & Tirosh, 2000), by providing a global organizing framework of arousal. The development of circadian sleep – wake rhythms is therefore of central importance in securing adaptive development across infancy.

Noradrenergic locus coeruleus (LC) neurons provide a circadian regulation of the sleep-wake cycle, and the maintenance of LC function depends on light exposure. Studies have shown that noradrenergic LC neurons exhibit a circadian rhythm in impulse activity, which peaks during the active period. This is mediated by an indirect circuit projection from the suprachiasmatic nucleus (SCN) to the LC, via the dorsomedial and paraventricular hypothalamic nuclei, as well as medial and ventrolateral pre-optic areas (Aston-Jones, Chen, Zhu, & Oshinsky, 2001).

Light deprivation induces a loss of noradrenergic fibers, which in turn decreases the amplitude of the sleep-wake rhythm (Gonzalez & Aston-Jones, 2006). The effects of excessive light exposure on sleep wake rhythmicity were recently described in a case study (Doljansky, Kannety & Dagan, 2005). Modifications of this exposure were shown to yield normalization effect of the sleep wake cycle within days. Effects of excessive light exposure, particularly during night hours on the neonatal circadian sleep-wake cycle are presently debated.

Fetal Entrainment

In recent years, experimental data shows that among the significant developmental changes occurring in the 2nd half of pregnancy, sleep related structural and functional developments are among the most central. Recent findings present intriguing findings that fetal endogenous system is capable of generating increasingly differentiated fetal behavioral states and transitions between them in association with fetal development, such clearly defined non-REM and REM conditions (Frank et al., 2006). These differential bio-behavioral states are increasingly evident by short epochs of heart rate acceleration and deceleration that define differential regularity patterns (Kuhnert, Hellmeyer, Stein, Schmidt, 2006); Van Leeuwen, Cysarz, Lange & Gronemeyer, 2006)

Halpern and colleagues in their review (Halpern, Maclean and Baumeister ,1995) noted that the chronological age at which binary sleep – wake states first appear are reported to occur at different phases, depending upon methodological differences. Studies conducted in the 70s noted that active sleep emerges at about 32-35 weeks gestation (Dreyfus-Brisac, 1970;Parmelee, Wenner, Akiyama, Schultz & Stern, 1967) and quiet sleep emerges a bit later at about 32-35 weeks of gestation (Dreyfus-Brisac, 1970; Prechtl, 1974). However, studies conducted in the 90s showed that both active and quite sleep can be reliably detected in both healthy and medically vulnerable preterm infants that are 28-31 weeks old (Curzi-Dascalova et al., 1988; Holditch-Davis, 1990; Stefanski, Schulze, Bateman, Kairam, Pedley, Masterson, & James, 1984).

Yet more recent studies from the last 5 years indicate that cyclicity of neonatal sleep behaviors emerges for most at 25 to 30 weeks' post-conceptional age, reflecting an ultradian biologic rhythm during the early perinatal stage of brain development (Scher, Johnson, Holditch-Davis, 2005).

It has been recently shown that the fetal sleep-wake system is also capable of producing in vivo intrauterine circadian rhythms during the third trimester (Frank et al., 2006). This development is to some degree endogenously driven (Mirmiran, Baldwin, Ariagno, 2003). It is also operating via responding to maternal entrainment signals of day-night rhythms. This circadian rhythmicy is measurable via recordings of fetal heart rate fluctuations in human fetuses for various periods of times. This capacity is not an expression of a free running internal clock, despite the relative darkness, typical of uterine environment. It represents effects related to fetus' capability of sensing the maternal diurnal rhythm, through reception of rhythmic auditory changes, bowel movement changes, HR changes, and more. Analysis of the results is indicative of some preliminary rhythmicity.

In primates, maternal melatonin has been suggested as a candidate to entrain fetal circadian rhythms, including the SCN rhythms of metabolic activity. Recently maternal melatonin has been indeed shown to be a Zeitgeber for the fetal supra charismatic nucleus of the hypothalamus, but probably not for the adrenal gland (Torres-Farfan, Rocco, Monso, Valenzuela, Campino, Germain, et al., 2006).

Neonatal Entrainment

McGrow and colleagues (1999), following the emergence of entrainment of a healthy male infant in Texas from birth to 6 months of age in conditions of a household with a fixed schedule and natural light environment showed that the circadian rhythm of temperature appeared first, during the first week of life. The wake circadian rhythm appeared second, attaining significance at day 45; approximately at the same time that increased melatonin concentration began to occur at sunset. The sleep circadian rhythm appeared last, attaining significance after day 56. Ninety to 120 minute zones of sustained wakefulness first appeared in the second month of life subsequent to awakening and prior to sleep onset. The infant's nocturnal sleep-onset was coupled to sunset before day 60 and subsequently to family bedtime, giving evidence of initial photic entrainment followed by social entrainment. McGraw and colleagues concluded that Circadian rhythms appeared much more rapidly in this infant than previously reported; their rapid appearance was probably facilitated by maximal exposure to sunlight, and presence of social routine cues, such as schedules of family meals, bath time and such, that were closely tied to daylight cyclicity. Thus McGraw's work underscores the importance of exposure to artificial light and irregular handling schedules as risk factors that delay normal circadian rhythm attainment (McGraw, Hoffmann, Harker & Herman, 1999).

An Integrative Model for the Development of Circadian Sleep-Wake Rhythms in Preterm Infants

Based on reviewed literature that will be discussed in detail below, we hypothesize an integrative model that integrates neurobiological vulnerability, maturation and stimulation factors that eventually pattern circadian sleep-wake rhythms in neonates. This model is based on an extensive literature review of both basic science models as well as on clinical studies with normally developing and at-risk human neonates.

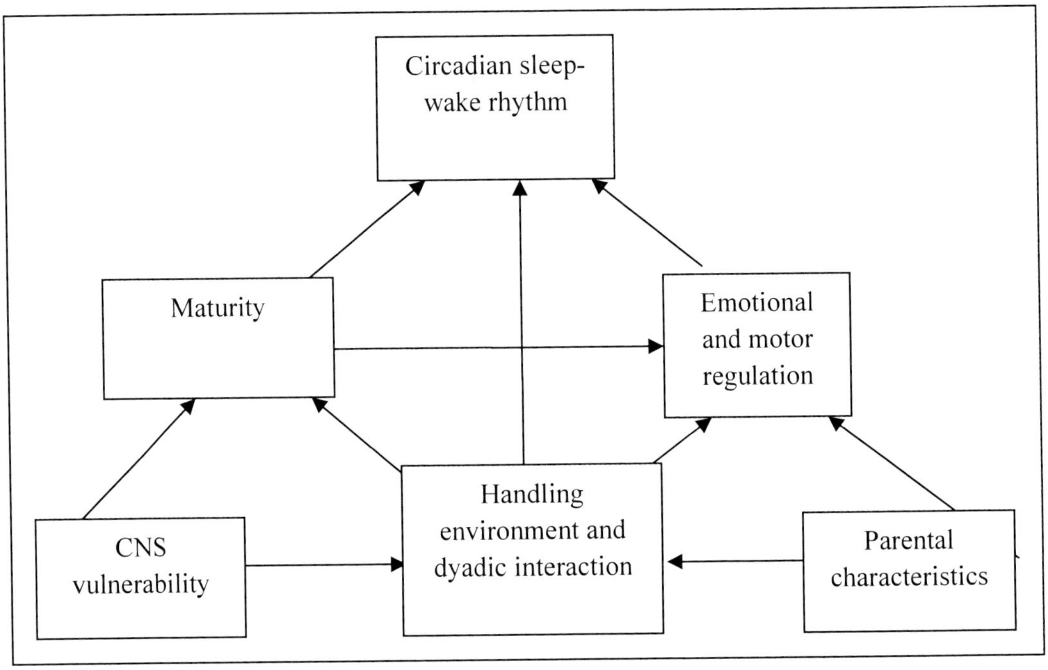

Figure 1: An integrative model for the development of circadian sleep-wake rhythms in preterm infants

This integrative model proposed a three source risk source: 1) infant dependent neurobiological vulnerability risk, such as brainstem mediated functional disability; 2) a familial source that entails genetic predispositions, parental style and support resources; and 3) an environmental stimulation source, such as excessive NICU related stimulation and handling. These three realms act as main effects and interact with the infants' prenatal and postnatal age and with the infants' self- regulation mechanisms. The later then mediate the development of infants' circadian sleep-wake rhythm. Further understanding of the manner in which these sources operate and interact would lead to development of proper interventions to eliminate the development of circadian sleep-wake rhythm deficit in infants born at-risk for sleep disorders. In the following we detail each component of the model

Prematurity - A Unique Window to Study Preliminary Circadian Rhythm Development

Maternal illness and premature birth adversely affect the development of circadian rhythms (Mirmiran & Ariagno, 2000). These effects seem to be strongest around the developmental mile stone at which the adverse process, .i.e., preterm birth, occurred (Mirmiran & Ariagno, 2000). The authors note this process disrupts the initial mother-infant process which is also vital for proper entrainment.

Prematurely born infants offer a unique window to study the circadian rhythm related developmental progressions. However, studies on sleep patterns of preterm infants at the age of expected birth have shown a fairly variable pattern of results (Ingersoll & Thoman, 1999). For example, on some patterns, preterm neonatal measures appeared to be precaucial relatively to neonates born at term, while on other parameters, an opposite direction was apparent (Booth Thoman & Leonard 1980; Holditch-Davis & Thoman, 1987). So that around birth, preterm neonates showed a higher frequency of REM in active sleep, and their sleep was characterized by longer quite sleep episodes.

It is worth noting that some similarities in circadian behaviors occurred in the development of sleep-wake cycle in preterm infants in the extra-uterine environment and full-term infants in the intrauterine environment (de Vries, Visser, & Prechtl, 1982; Diambra & Menna-Barreto, 2004; Olischar, Klebermass, Kuhle, Hulek, Kohlhauser, Rucklinger, Pollak & Weninger, 2004). However, recent data point to a vital transitional milestone that occurs around 31 weeks of gestation. This milestone, points to a possible additional risk related to preterm birth prior to 31 weeks of gestation. Following the development of sleep-wake cyclicity in premature infants born at 25 to 29 weeks gestation, we found that a developmental leap in the organization of the sleep-wake cycle occurred at 31 weeks gestational age and that those infants who did not show a timely maturation of the biological clock showed deficiencies in social interactions and arousal modulation capacities in later infancy (Feldman, 2006).

Overall, as noted in Halpern, Maclean & Baumeister's review (Halpern et al., 2006) the variability in sleep parameters in preterm infants is much greater than that reported in infants born at term. This tendency holds also in preterm infants born extremely early and or with extremely small weights (Leitner, Bloch, Sadeh, Neuderfer,Tikotzky, Fattal-Valevski et al., 2002). Researchers thus continue to search for additional risk factors that would explain significant portions of this large variability. Among the variables recently explored both infants factors and environmental measures are studied.

CNS Perinatal Susceptibility as a Risk Factor

Brainstem structures regulate and modulate sleep and arousal (Losier & Semba, 1993). Structures in the brainstem influence higher brain structures in the midbrain and cortex. The ascending reticular activating system that transverses through the brainstem stimulation to the thalamus, hypothalamus, and basal forebrain, results in excitation of the cortex. Its cholinergic, glutamic and catacholinergic neurons are stimulated by afferent sensory input, such as audition and vision (Harris, 2005). Wakefulness is manifested by an active cortex that exhibits a characteristic pattern of desynchronized EEG

Mechanisms related to the nucleus tractus solitarius, which project on to midbrain and forebrain structures, inhibit activity in the reticular system, resulting in activation of inhibitory thalamocortical projections to the cortex. During a state of decreased activation, the cortex exhibits a pattern of synchronized EEG that is manifested as various forms of Non-REM and REM sleep (Harris, 2005).

Other brainstem structures regulate activities known to affect sleep and arousal indirectly, such as arousal and attention (Lai, Shalita, Hajnik, Wu, Kuo, Chia, et al., 1999; Wong & Wong, 1991; McNamara, Wulbrand & Thach, 1999; Karmel, Gardner & Freedland, 1996) and the processing of sensory information (Anthony & Graham, 1997; Nattie, 2000), such as visual and auditory perception (Larson & Yajima, 1994; Jiang, & Wilkinson, 2006; Aitkin, 1986; Davidson & Bender, 1991). Still, direct effects of brainstem dysfunction on sleep and arousal regulation early on in development are not yet fully understood.

Neonatal brainstem dysfunction has been suggested as a risk factor for various neural-developmental deficits (Karmel, Gardner, Zapulla, Magnano & Brown, 1988; Darnall, Ariagno, Kinney, 2006): neurobehavioral deficits (Karmel, Gardner, Zappulla, Magnano, & Brown, 1988), orodigestive and cardiorespiratory difficulties (Abadie, Morisseau-Durand, Beyler, Manach & Couly, 2002; Darnall, Ariagno, Kinney, 2006), and neurological dysfunction (Yilmaz, Degirmenci, Akdas, Kulekci, Ciprut, Yuksel, et al., 2001). We propose that brainstem dysfunction may be directly related to an aberrant development of sleep- wake cycles, particularly in preterm neonates for whom the maturation of higher-order structures is still under way. For such infants, the sensitive circadian system may have been compromised by the pre-term birth.

It has been shown that Cholinergic and aminergic brainstem neurons can concurrently modulate the activity of neurons in the thalamus and basal forebrain during cortical arousal, through dual projections (Losier & Semba, 1993- in the rat model). Structural damage in brainstem structures, specifically in the ventral mesencephalon and rostroventral Pons have been shown to results in deficits in sleep- wake organization (Lai, Shalita, Hajnik, Wu, Kuo, Chia & Siegel, 1999- in the cat model). It is thus hypothesized that brainstem-related neonatal dysfunctions would result in abnormalities in the development of the sleep-wake circadian rhythms.

NICU Environment as a Risk Factor

The environment of the NICU presents ample demands on the preterm infants (Field, 1990). The NICU's great life-saving potential comes often at some cost. The NICU environment is different from maternal uterus in numerous fashions, some of particular importance to circadian rhythm development.

The NICU presents various levels and types of lights that are significantly different from natural lights. Solar lights are presented un-baffled by uterine wall and artificial light, at various wave lengths, are often intense and are presented at prolonged phases, way beyond characteristics of solar light phase (Niessen, 2006).

The NICU environment also presents the infant with excessive noise (Kent, Tan, Clarke & Bardell, 2002; Byers, Waugh & Lowman, 2006; Niessen, 2006). It is fraught with multiple noises, monitor beeping, ventilator alarms, low spectrum noises of wheeling incubators and various carts during the day and the night and stuff activity related noises. Some are fairly constant through the day and the night, some are impulsive and unexpected. Overall activity levels are higher than other environments to which infants are brought to. Still, recent measurements of sound pressure levels at NICU indicated the highest levels were recorder at 6-7 AM and around noon (Kruege, Wall, Parker, & Nealis, 2005). Activity at NICU is maintained at relatively high levels during the day and also is significantly high during the night, though typically lower than that of the day.

The effects of loud noise on the preterm infants are evident on a physiological level by increased heart and respiratory rates and decreased oxygen saturation (Bremmer, Byers, & Kiehl, 2003). Its effects on sleep –wake rhythms are not yet fully understood. However, animal models have demonstrated via simulating continuous lights and sounds of a typical nursery and exposing newborn animals to these conditions, showed that this constellation of stimuli schedule results in permanent damage to the biological clock, in differences in sensory processing of stimuli, and in immature habituation to stimuli in the young animal (Hao & Rivkees, 1999- in a baboon's model; Philbin, Ballweg, & Gray, 1994 and Sleigh & Lickliter, 1998- in chicks models). The studies point to the potentially damaging effects of nursery conditions on the developing infant in general and have a potential harmful effect on its developing systems that regulate circadian activity.

Experimental light manipulation has shown, that the circadian sleep-wake rhythm in the preterm infants entrained after a similar time of exposure to an environment with daily time cues but at an earlier post-conceptional age when compared with the term group (McMillen, Kok, Adamson, Deayton & Nowak, 1991). While McMillen's group in Australia has concluded that maturity of the premature infant is less of a determinant in attaining circadian sleep, but rather that the length of time spent in a cycles day-night environment is vital for this purpose (McMillen, Kok, Adamson, Deayton & Nowak, 1991).

Environmental variability at the NICU is often less directly related to the status of the infants, but rather to necessary management of neighboring infants and the fixed routine procedures of the NICU, irrespective of the infants' current status. This factor is also very different from uterine environment where by the uterine environment is almost solely responsive to the mother-infant dyad capacity, changes and needs. It may very well be that a chaotic, non-circadian environment of the NICU adversely affects initiation and stabilization of circadian rhythms of the preterm infants (Mirmiran & Ariagno, 2000).

Attempts to support this hypothesis were so far only partly successful, since circadian and sleep development in preterm infants occurs independently from the influences of environmental lighting (Mirmiran, et al., 2003). Entrainment of preterm infants using a different lighting environment during the night (a day-night cycled room), seem to be not effective or at least not sufficient to differentially effect entrainment during the first 3 months of the preterm infants post natal life (Mirmiran, et al., 2003). Similarly, phototherapy at the NICU has been shown to have no short-term effects, on neonatal sleep-wake cycles when compared with preterm infants who have not received photic intervention. Nor does it effect sleep-wake rhythms or saliva cortisol levels at two years of age (Shimada, Segawa, Higurashi, Kimura, Oku, Yamanami et al., 2003).

At the same time, analysis of activity levels recorded by Actiwatches, cycled light (7PM-7AM use of crib covers) has been shown to affect activity levels during the first month post preterm birth (Rivkees, 2003; Rivkees, Mayes, Jacobs & Gross, 2004). Exposure of premature infants to low-intensity cycled lighting, rather than to a fixed dim light level, in the hospital nursery induces distinct patterns of rest-activity that are apparent within 1 week after discharge. The data thus shows that day-night rhythms in activity patterns can be detected shortly after discharge to home in premature infants and that the circadian clock of developing infants is entrained by cycled lighting.

Harris (2005) in his review summarizes that the amount of time spent sleep changes throughout the life span. Such, that at the neonatal-period the infant sleep most of time. Sleep periods of 1-4 hours are distributed evenly during the circadian cycle. By 6 months of age well defined sleep-activity states appear. Sleep episodes lengthen and the proportion of REM sleep declines (Harris, 2005)

Birth Mode as a Risk Factor

Effects of specific risk factors, which typically accompany preterm birth on circadian sleep wake rhythms is yet significantly understudied. Of the various significant risk factors for which some preliminary data has been collected, one can find: Delivery mode, that seem to be of no consequence on the amount of sleep but rather has an effect on the development of circadian rhythms in neonates (Korte, Hoehn & Siegmund, 2004). It was found that the majority of vaginally born neonates showed a distinct circadian frequency in their spectra, which is characterized by more sleep during the night. In contrast, both groups of neonates born by C-section showed significantly less distinct circadian frequencies in their spectra. The authors did not find these differences in infants born preterm.

Others find that c-section deliveries are related to differences in sleep- wake organization during the first few days post birth.

In Finland, postnatal adaptation after c-section was found to be related to changes in sleep organization, and increases in oxyhaemoglobin saturation and frequency of body movements in the first few days after birth (Nikkol, Kirjavainen, Ekblad, Kero & Salonen, 2002). Similar findings were reported in the US, where by full term neonates born via vaginal delivery had more wakefulness, shorter mean sleep periods and shorter longest-sleep periods during the daytime on the first two days post birth (Freudigman & Thoman, 1998) . Thus, perinatal interventions such as stress and pain (Nikkol, Kirjavainen, Ekblad, Kero & Salonen, 2002) and surgical birth (Freudigman & Thoman, 1998) seem to disrupt neonatal circadian rhythm of full term neonates. The mechanism/s of these effects is not yet understood.

Prematurity as a Risk Factor

Effects of prematurity of sleep-wake rhythms seem to be complex. On the one hand one may hypothesize greater vulnerability of a younger system in handling excessive interventions. At the same time greater plasticity to aberrations is to be expected under certain conditions. Generally, preterm infants exhibit more active sleep and indeterminate sleep states than full-term infants (Curzi-Dascalova, Peirano, & Morel-Kahn, 1988; High & Gorski, 1985; Holditch-Davis, 1990; Holditch-Davis & Edwards, 1998; Parmelee, Wenner, Akiyama, Schultz, & Stern, 1967). At the same time, there is a significant debate regarding indications of sleep-wake rhythmicity in preterm neonates.

Following these robust findings, some presume a greater weight for prematurity than to its related complications, such as illness severity and medical treatments, handling and hospitalization (Holditch-Davis, Scher, Schwartz, Hudson-Barr, 2004). Indeed, large scale studies, such as that performed by Sadeh, show that sleep of infants born at later gestational ages is characterized of greater organization levels, indicated by increased percent of quiet and motionless sleep (Sadeh, et al., 1996). Yet, it is not know whether the effect of prematurity on sleep-wake rhythms is a main effect one, or rather that risk factors interact with gestational age to affect sleep-wake cycles during the neonatal period. An example for such an interaction effect is maternal glucose values during pregnancy (Sadeh, et al., 1996) - which is related to sleep-wake patterns in preterm infants, but not in the controls. Another factor recently discovered that interacts with prematurity to affect sleep characteristics in preterm infants is administration of antenatal magnesium sulfate and corticosteroid therapy (Black, Holditch-Davis, Schwartz & Scher, 2006). It was found infants exposed to magnesium sulfate had more active sleep without rapid eye movement, indicating poorly organized active sleep. Investigators seem to concur that the mechanism involved in the majority of these processes is some subtle affect on CNS in these vulnerable infants.

The Development of Circadian Sleep-Wake Rhythm in Preterm Infants as a Factor of Parenting and Environmental Stimulation

Postnatal entrainment goes through marked changes during the first few weeks of life. Entrainment to maternal schedule occurs irrespective of prematurity around the post conceptual age of 45 weeks in 75% of infants (Shimada, Takahashi, Segawa, Higurashi, Samejim & Horiuchi, 1999). During the transition periods infants go through either ultradian of irregular sleep-wake cycles (Shimada, Takahashi, Segawa, Higurashi, Samejim & Horiuchi, 1999).

Neonatal cyclicity measures are reported to be negatively related to indices of advanced perinatal status as well as 6-month mental scores; however, at 6 months, the cyclicity measures from infant born preterm were positively related to perinatal and neonatal measures as well as mental scores (Borghese, Minard & Thoman, 1995; Feldman, 2006). These seemingly contradicting results may indicate the necessity for different interpretations of periodicity at the preterm and later age (Borghese, Minard & Thoman, 1995).

Sleep disorders have been reported in children born preterm as they grow up. A study conducted at 20 months corrected age has shown that the quality of night-time sleep of preterm toddlers is more disrupted on average than that of toddlers born at term (Gossel-Symank, Grimmer, Korte, Siegmund, 2004). The daytime rest duration of preterm infants at 20 months is significantly shorter than that of full-term infants. The sleep quality of preterm infants is significantly lower than that of full-term infants, since the preterm infants have a larger percentage of less restful nighttime sleep (Gossel-Symank, Grimmer, Korte, & Siegmund, 2004).

A theoretical framework that may account for the complex circadian sleep-wake rhythm of neonates proposes an interaction between components that are more directly related to CNS vulnerability and other components that are more prone to handling and stimulation effects (Anders, Keener, Kraemer, 1985)

Literature cited by Halpern (Halpern et al., 1995) shows that during the 90s it was thought that the first 6 months of life are not yet accompanied by changes in the awaked state during nighttime (Anders & Keener, 1985). Halpern summarizes reports of observational studies of infants during the day time during the first 6 months of postnatal life, in which a non-linear increase in the percent of day time wakefulness, and longer periods of wakefulness during the day (Halpern, 1995; Wolff, 1984). Comparisons of wakefulness at daycare settings and in the home revealed that wakefulness in the home setting was maintained at earlier ages and for longer periods of times (Wolff, 1984) This difference was attributed to the level of stimulation provided in the home relatively to relatively lack of it at the day care setting (Thoman & Whitney, 1989). Stimulation effect on wakefulness was also found as a factor of maternal company, parental co-sleeping (Thoman, 2006); as opposed to a lonely setting.

The age of early marks of circadian influences on sleep-wake patterns is still in debate (Halpern, 1995). Some reported on first appearance between 6-8 weeks of age (Thoman & Whitney, 1989), even though gradual increase in daylight wakefulness emerges earlier, at around 4-5 weeks of age in the home setting (Thoman & Whitney, 1989). Significant sleep episodes during the night are typically fairly common at 3 months of age (Coons & Giullenault, 1984).

Effects of Touch Interventions for Preterm Infants on Sleep-Wake Circadian Rhythmicity

The development of sleeping and waking behaviors in preterm infants is reactive to handing and to social encounters right from the neonatal period (Ingersoll & Thoman, 1999). Nursing care affects sleep states already at the nursery and during the first three months of post natal life (Brandon, Holditch-Davis & Beylea, 1999).

Of special interest are the effects of touch-and-contact interventions on sleep-wake rhythmicity in neonates.

Among the central benefits of touch-and-contact intervention to the fragile premature infant is the organization of the biological clock. Animal studies point to the important role of the mother's physical proximity and specific body cues for the maturation of the biological clock in the pup during the first post-birth days (Hofer, 1995). In addition to their immature

CNS, premature infants are deprived of the soothing, regulating, and organizing presence of the mother's body and closeness (Feldman, 2004). Two types of touch-and-contact interventions were found to have an impact on sleep-state organization; Kangaroo Care and massage.

Kangaroo Care (KC), a method developed in Bogota, Columbia during a period of incubator shortage, involves skin-to-skin contact between mother and her premature infant. Several studies showed both concurrent and long-lasting effects to the KC intervention on state organization and sleep-wake cyclicity. During contact, infants spent more time in quiet sleep, heart rate was lower and more stable, apnea and bradycardia decreased, body temperature was maintained, and oxygenation improved. Moreover, the effects of skin-to-skin contact seemed to persist even after contact was ended. Following kangaroo contact premature infants slept longer and sleep was more restful and organized. Interestingly, KC improved not only sleep states but also infant alertness, as shown by the findings that infants spent longer periods in alert states after KC. It is thus possible that KC organizes the biological clock by increasing the time infants spend in the two ends of the arousal continuum—quiet sleep and alert wakefulness—while reducing transitory states and active sleep. An addition contribution of kangaroo contact to state organization was the decrease in transitory, indeterminate sleep states among infants born preterm who underwent KC as compared to matched controls (Feldman & Eidelman, 2003).

Kangaroo care similarly contributed to the organization of sleep-wake rhythmicity. Following two weeks of daily kangaroo contact, we found that the amplitudes of sleep-wake cyclicity as measured in 10-second epochs over four consecutive hours were higher for treated infants compared to controls (Feldman et al., 2002). Better organization of the biological clock, in turn, was related to improved arousal and emotion regulation capacities across infancy and up to age five. Infants with better sleep-wake organization had higher thresholds to negative stimuli, were better able to modulate arousal, and showed more exploratory behavior across the first year. During the toddler and preschool age, these children scored higher on neuropsychological tests that required attention shifting and maintenance and displayed higher executive functioning at age five. In the social-emotional domains, children with better organization of the biological clock were more able to show restrained and self-regulation during social and nonsocial tasks that requires inhibition and regulation of anger and frustration (Feldman, 2007).

Massage therapy, active stroking of the premature infants following a specific regime similarly showed that premature infants who received massage therapy spent more time in alert wakefulness and less time in active sleep states, were discharged earlier from the hospital, and showed more mature neurobehavioral profiles (Field, 1995).

Overall, recent developments demonstrate intervention effects of stimulation, on sleep-wake state organization and the development of sleep-wake circadian rhythms of infants born prematurely, via careful modifications of infant's exposure to light, sound, and touch in early phases of development. Developing multi-modality intervention protocols that are more closely tied to circadian mechanisms of the infant and its CNS capacities may re-direct development of infants born at risk into more adaptive sleep-wake circadian rhythms.

Conclusion

We reviewed and critically tested the primary issues affecting the development of sleep-wake circadian rhythms in infants: 1) CNS vulnerability, as indexed by specific brain structures known to be involved in regulation of sleep-wake functions, such as *perinatal brainstem dysfunction*; 2) *Aberrant maturation caused by premature birth*; and, 3) *Suboptimal environmental conditions.* This topic relates to the unnatural neonatal environment and excessive exposure to light sound and touch in early development.

We propose that an integrative view that takes into account the vulnerability of autonomic system, neonatal neural maturation, and environmental conditions provide the foundation for the emergence of circadian sleep patterns in early development. Finally, this source has been proposed as a valuable intervention tool, in the forms of specific message and KC in ameliorating the adverse effects of CNS related vulnerability, and excessive stimulation early on in development and redirecting a more optimal development of neonatal circadian sleep-wake cycles.

References

Abadie, V., Morisseau-Durand, M.P., Beyler, C., Manach, Y., Couly, G. (2002). Brainstem dysfunction: a possible neuroembryological pathogenesis of isolated Pierre Robin sequence. *European Journal of Pediatrics, 161*, 275-280.

Aitkin, L.M., Byers M., & Nelson J.E. (1986). Brain stem auditory nuclei and their connections in a carnivorous marsupial, the northern native cat (Dasyurus hallucatus). *Brain Behavior and Evolusion, 29*, 1-16.

Anders, T.F., Keener, M.A.,& Kraemer, H. (1985). Sleep-wake state organization, neonatal assessment and development in premature infants during the first year of life. II. *Sleep, 8*, 193-206.

Aston-Jones, G., Chen, S., Zhu, Y., & Oshinsky, M.L.(2001). A neural circuit for circadian regulation of arousal. *Nature Neuroscience, 4*, 732-738.

Aston-Jones, G. (2005). Brain structures and receptors involved in alertness. *Sleep Medicine, 6*, S3-7.

Atkinson, E., Vetere, A., Grayson, K. (1995). Sleep disruption in young children. The influence of temperament on the sleep patterns of pre-school children. *Child: Care, Health and Development, 21*, 233-246.

Black, B., Holditch-Davis, D., Schwartz, T., & Scher, M.S. (2006). Effects of antenatal magnesium sulfate and corticosteroid therapy on sleep states of preterm infants. *Research in Nursing & Health, 29*, 269-280.

Borghese, I.F., Minard, K.L., Thoman, E.B. (1995). Sleep rhythmicity in premature infants: implications for development status. *Sleep, 18*,523-530.

Brandon, D.H., Holditch-Davis, D., Beylea, M. (1999). Nursing care and the development of sleeping and waking behaviors in preterm infants. *Research in Nursing and Health, 22*, 217-229.

Bremmer, P., Byers, J.F., Kiehl, E. (2003) Noise and the premature infant: physiological effects and practice implications. *Journal of Obstetric, Gynecologic and Neonatal Nursing, 32*, 447-454.

Byers, J.F., Waugh, W.R., Lowman, L.B. (2006). Sound level exposure of high-risk infants in different environmental conditions. *Neonatal Network, 25,* 25-32.

Curzi-Dascalova, L., Peirano, P., & Morel-Kahn, F. (1988). Development of sleep states in normal premature and fullterm newborns. *Developmental Psychobiology, 21*, 431–444.

Darnall, R.A., Ariagno, R.L., Kinney, H.C. (2006). The late preterm infant and the control of breathing, sleep, and brainstem development: a review. *Clinics in Perinatology, 33(4)*, 883-914.

Davidson, R.M., & Bender, D. B. (1991). Selectivity for relative motion in the monkey superior colliculus. *Journal of Neurophysiology, 65*, 1115-1133.

de Vries, J.I., Visser, G.H., Prechtl, H.F. (1982). The emergence of fetal behaviour. I. Qualitative aspects. *Early Human Development, 7*, 301-322

Diambra, L., & Menna-Barreto, L. (2004). Infradian rhythmicity in sleep/wake ratio in developing infants. *Chronobiology International, 2*, 217-227.

Doljansky, J.T., Kannety, H., Dagan, Y. (2005). Working under daylight intensity lamp: an occupational risk for developing circadian rhythm sleep disorder? *Chronobiology International, 22*, 597-605.

Feldman, R. (2004). Mother-infant skin-to-skin contact and the development of emotion regulation. S. P. Shohov (Ed.). *Advances in psychology research*. Vol. 27 (pp. 113-131). Hauppauge, NY; Nova Science.

Feldman, R. (2006). From biological rhythms to social rhythms: Physiological precursors of mother-infant synchrony. *Developmental Psychology, 42*, 175-88.

Feldman, R. (2007). Mother-infant contact and child development; Insights from the Kangaroo intervention. In L. L'Abate, D. D. Embry, & S. M. Baggett (Eds.) *Handbook of low-cost interventions to promote physical and mental health.* (pp. 323-351). New York; Springer

Feldman, R. & Eidelman, A. I. (2003). Mother-infant skin-to-skin contact (Kangaroo Care) accelerates autonomic and neurobehavioral maturation in premature infants. *Developmental Medicine and Child Neurology, 45*, 274-280.

Feldman, R., Weller, A., Sirota, L., & Eidelman, A. I. (2002). Skin-to-Skin contact (Kangaroo care) promotes self-regulation in premature infants: sleep-wake cyclicity, arousal modulation, and sustained exploration. *Developmental Psychology, 38,* 194-207.

Field, T. (1990). Alleviating stress in newborn infants in the intensive care unit. *Clinics in Perinatology, 17*, 1-9.

Field, T. (1995). Massage therapy for infants and children. *Journal of Behavior and Developmental Pediatrics, 16*, 105-111.

Frank, B., Frasch, M.G., Schneider, U., Roedel, M., Schwab, M., & Hoyer, D. (2006). Complexity of heart rate fluctuations in near-term sheep and human fetuses during sleep. *Biomedizinische Technik Biomedical engineering_ ,51*, 233-236.

Freudigman, K.A., Thoman, E.B. (1998). Infants' earliest sleep/wake organization differs as a function of delivery mode. *Developmental Psychobiology, 32*, 293-303.

Gonzalez, M. M., Aston-Jones, G. (2006). Circadian regulation of arousal: role of the noradrenergic locus coeruleus system and light exposure. *Sleep. 1, 29*, 1327-1336.

Gossel-Symank, R., Grimmer, I., Korte, J., & Siegmund, R.(2004). Actigraphic monitoring of the activity-rest behavior of preterm and full-term infants at 20 months of age. *Chronobiology Internation, 21*, 661-671.

Halpern, L. F., Maclean, W. E., & Baumeister, A. A. (1995). Infant Sleep-Wake Characteristics: Relation to Neurological Status and the Prediction of Developmental Outcome, *Developmental Review, 15*, 255-291.

Harris, C. D. (2005). Neurophysiology of sleep and wakefulness. *Respiratory Care Clinics of North America, 11*, 567-586.

Hao H, Rivkees SA. (1999). The biological clock of very premature primate infants is responsive to light. *Proceedings of the National Academy of Sciences of the United States of America, 2, 96*, 2426-2429.

High, P.C., & Gorski, P.A. (1985). Recording environmental influences on infant development in the intensive care nursery: Womb for improvement. In A.W. Gottfried & J.L. Gaiter (Eds.), Infant stress under intensive care: Environmental neonatology
(pp. 131–155). Baltimore: University Park.

Hofer, M. A. (1995). Hidden regulators: Implications for a new understanding of attachment, separation, and loss. In S. Golberg, R. Muir, & J. Kerr (Eds.), *Attachment theory: Social, developmental, and clinical perspectives* (pp. 203-230). Hillsdale, NJ: Analytic Press.

Holditch-Davis, D. (1990). The development of sleeping and waking states in high-risk preterm infants. *Infant Behavior and Development, 13*, 513–531.

Holditch-Davis, D., Scher, M, Schwartz T, Hudson-Barr D. (2004). Sleeping and waking state development in preterm infants. *Early Human Development, 80*, 43-64.

Ingersoll, E. W. & Thoman, E. B. (1999). Sleep/wake states of preterm infants: Stability, developmental change diurnal variation and relation with care-giving activity. *Child Development, 70*, 1-10.

Jiang, Z.D., & Wilkinson, A.R. (2006). Does peripheral auditory threshold correlate with brainstem auditory function at term in preterm infants? *Acta Oto-Laryngologica, 126*, 824-827.

Karmel, B.Z., Gardner, J.M., & Freedland, R.L. (1996). Arousal-modulated attention at four months as a function of intrauterine cocaine exposure and central nervous system injury. *Journal of Pediatric Psychology, 21*, 821-832.

Karmel, B. Z., Gardner, J. M., Zapulla, R. A., Magnano, C. L. & Brown, E. G. (1988). Brain-stem auditory evoked responses as indicators of early brain insult. *Electroencephalography and Clinical Neurophysiology, 71*, 429-442.

Keener, M.A., Zeanah, C.H., Anders, T.F. (1988). Infant temperament, sleep organization, and nighttime parental interventions. *Pediatrics, 81*, 762-771.

Kent, W.D., Tan, A.K., Clarke, M.C., & Bardell, T. (2002). Excessive noise levels in the neonatal ICU: potential effects on auditory system development. *Journal of Otolaryngology, 31*, 355-60.

Kleitman, N. (1957). Sleep, wakefulness and consciousness. *Psychological Bulletin, 54*, 354-359.

Kopp, C., Vogel, E., Rettori, M.C., Delagrange, P., Renard, P., Lesieur, D., & Misslin, R. (1999) Regulation of emotional behaviour by day length in mice: implication of melatonin. *Behavioral Pharmacology,10*, 747-752.

Korte, J., Hoehn, T., Siegmund, R. (2004). Actigraphic recordings of activity-rest rhythms of neonates born by different delivery modes. *Chronobiology International, 21*, 95-106.

Krueger, C., Wall, S., Parker, L., Nealis, R. (2005). Elevated sound levels within a busy NICU. Neonatal Network, 24, 33-37.

Lai, Y. Y., Shalita, T., Hajnik, T., Wu, J.P., Kuo, J.S., Chia, L.G., & Siegel J.M. (1999) Neurotoxic N-methyl-D-aspartate lesion of the ventral midbrain and mesopontine junction alters sleep-wake organization. *Neuroscience, 90*, 469-483.

Larson, C.R., Yajima, Y., & Ko, P. (1994). Modification in activity of medullary respiratory-related neurons for vocalization and swallowing. *Journal of Neurophysiology,71*, 2294-2304.

Leitner, Y., Bloch, A.M., Sadeh, A., Neuderfer, O., Tikotzky, L., Fattal-Valevski, A., & Harel, S.(2002). Sleep-wake patterns in children with intrauterine growth retardation. *Journal of Child Neurology, 17*, 872-876.

Losier, B.J., & Semba K. (1993). Dual projections of single cholinergic and aminergic brainstem neurons to the thalamus and basal forebrain in the rat. *Brain Research, 26, 604*, 41-52.

McGraw, K., Hoffmann, R., Harker, C., Herman, J.H. (1999). The development of circadian rhythms in a human infant. *Sleep, 1 , 22*, 303-310.

McMillen, I.C., Kok, J.S., Adamson, T.M., Deayton, J.M., & Nowak R.(1991). Development of circadian sleep-wake rhythms in preterm and full-term infants. *Pediatric Research, 29*, 381-384.

McNamara, F., Wulbrand, H., Thach, B.T. (1999). Habituation of the infant arousal response. *Sleep, 1, 22*, 320-326.

Mirmiran, M., Ariagno, R. L. (2000). Influence of light in the NICU on the development of circadian rhythms in preterm infants. *Seminars in Perinatology, 24*, 247-257.

Mirmiran, M., Baldwin, R. B., Ariagno, R.L. (2003). Circadian and sleep development in preterm infants occurs independently from the influences of environmental lighting. *Pediatric Research, 53*, 933-938.

Nattie, E. (2000). Multiple sites for central chemoreception: their roles in response sensitivity and in sleep and wakefulness. *Respiration Physiology, 122*, 223-235.

Niessen, F. (2006). [Neurosensory visual development of the foetus and newborn and neonatal intensive care units]. *Archives de Pediatrie, 13*, 1178-1184. [Article in French]

Nikkol, E. M., Kirjavainen, T. T., Ekblad, U. U., Kero, P. O., Salonen, M.A.(2002). Postnatal adaptation after Caesarean section or vaginal delivery, studied with the static-charge-sensitive bed. *Acta Paediatrica, 91*, 927-933.

Olischar, M., Klebermass, K., Kuhle, S., Hulek, M., Kohlhauser, C., Rucklinger, E., Pollak, A., & Weninger, M. (2004). Reference values for amplitude-integrated electroencephalographic activity in preterm infants younger than 30 weeks' gestational age. *Pediatrics, 113*, e61-66.

Parmelee, A. H., Jr., Wenner, W. H., Akiyama, Y., Schultz, M., & Stern, E. (1967). Sleep states in premature infants. *Developmental Medicine and Child Neurology, 9*, 70–77.

Philbin, M. K., Ballweg, D. D., Gray, L. (1994). The effect of an intensive care unit sound environment on the development of habituation in healthy avian neonates. *Developmental Psychobiology, 27*, 11-21.

Rivkees, S.A. (2003). The management of hyperthyroidism in children with emphasis on the use of radioactive iodine. *Pediatric Endocrinology Reviews, 1*, 212-21.

Rivkees, S. A., Mayes, L., Jacobs, H., & Gross, I. (2004). Rest-activity patterns of premature infants are regulated by cycled lighting. *Pediatrics, 113*, 833-839.

Sadeh, A., Dark, I., Vohr, B. R. (1996). Newborns' sleep-wake patterns: the role of maternal, delivery and infant factors. *Early Human Development, 23*, 44, 113-126.

Scher, M.S., Johnson, M.W., Holditch-Davis, D. (2005). Cyclicity of neonatal sleep behaviors at 25 to 30 weeks' postconceptional age. *Pediatric Research, 57*, 879-882.

Shimada, M., Takahashi, K., Segawa, M., Higurashi, M., Samejim, M., & Horiuchi, K.(1999). Emerging and entraining patterns of the sleep-wake rhythm in preterm and term infants. *Brain & Development, 21*, 468-473.

Shimada, M., Segawa, M., Higurashi, M., Kimura, R., Oku, K., Yamanam,i S., & Akamatsu, H. (2003). Effects of phototherapy in neonates on circadian sleep-wake and saliva cortisol level rhythms. *The Journal of Perinatal and Neonatal Nursing, 17*, 222-231.

Sleigh, M.J., & Lickliter, R. (1998). Timing of presentation of prenatal auditory stimulation alters auditory and visual responsiveness in bobwhite quail chicks (Colinus virginianus). *Journal of Comparative Psychology, 112*, 153-60.

Torres-Farfan, C., Rocco, V., Monso, C., Valenzuela, F.J., Campino, C., Germain, A., Torrealba, F., Valenzuela, G.J., & Seron-Ferre, M. (2006). Maternal melatonin effects on clock gene expression in a nonhuman primate fetus. *Endocrinology, 147*, 4618-4626.

Van Leeuwen, P., Cysarz, D., Lange, S., & Gronemeyer, D. (2006). Increase in regularity of fetal heart rate variability with age. *Biomedizinische Technik Biomedical engineering, 51*, 244-247.

Wong, V., & Wong, S.N. (1991). Brainstem auditory evoked potential study in children with autistic disorder. *Journal of Autism and Developmental Disorders, 21*, 329-340

Yilmaz, Y., Degirmenci, S., Akdas, F., Kulekci, S., Ciprut, A., Yuksel, S., Yildiz, F., Karadeniz, L., & Say, A. (2001). Prognostic value of auditory brainstem response for neurologic outcome in patients with neonatal indirect hyperbilirubinemia. *Journal of Child Neurology, 16*, 772-775.

In: Progress in Circadian Rhythm Research
Editor: Anne-Laure Léglise, pp. 83-122

ISBN: 978-1-60021-876-7

Chapter IV

Pineal Gland and Circadian Rhythm Regulation

V.I. Arav, E.V. Slesareva and S.M. Slesarev
Department of Common Biology, Ul'yanovsk State University, Russia

Abstracts

Research of influence pinealectomy and epithalamin injection to pinealectomized animals on circadian rhythm proliferation epithelium in jejunum crypts and the spermatogenesis rhythm in albino rats were studied. Epithalamin is melatoninless extract of pineal oligopeptides. Animals were kept under 12/12h day/night regimen. Researches were carried out on day 40-41 after pinealectomy.

Epithalamin (in dose 2.5 mg/kg) was injected before beginning night period of day since 26 till 40–th day after pinealectomy.

Circadian rhythm of mitotic index epithelium in jejunum crypts, which disappeared after pinealectomy and restored after epithalamin injection to pinealectomized animals, was observed in intact animals. Daily dynamic of the spermatogonia mitotic index and stage frequency of spermatogenic cycle, which was cyclic change of cell combination spermatogenic coat in process of spermatogenesis, were carried out for studied of spermatogenesis rhythmicity. Circadian rhythm of mitotic index type B spermatogonia, the duration of reproduction period and differentiation intermediate type spermatogonia and also meiosis was observed, that it is shown on spermatogenesis circadian rhythm existence in general. Pinealectomy caused disappearance the circadian rhythms of spermatogenesis and epithalamin injection restored circadian rhythms.

According our results the pineal gland is the necessary part in formation of circadian rhythm of various tissue proliferations. The involvement of pineal gland peptides in regulation of proliferation circadian rhythm shows the necessity of extraction their in clean form, testing of biological activity and usage ability in desynchronizes correction, accompanying various diseases.

Keywords: circadian rhythm, pineal gland, pinealectomy, pineal peptides, proliferation epithelium in jejunum crypts, spermatogenesis.

Introduction

In the basis of alive system organization is the rhythm its function. Biorhythms are the fundamental alive substance property and are marked on all levels its organization:

- on the molecular-genetic level it is periodicity of replication and transcription processes, oscillation of molecule-biopolymers concentration;
- the processes biosynthesis albumen changes rhythmically on the cellular and subcellular levels;
- on the tissue level it are biorhythms of output tissue regulators proliferation and the process of cells proliferation;
- on the organ organism levels it are the rhythmicity physiological functions: rhythms peptides synthesis, alteration systole and diastole heart, the rhythm sleep and cheerfulness;
- biorhythms are marked in dynamic there quantity, migration, seasons reproduction on the levels of population and species;
- the most important condition of stability the biospheres and its ecosystems are substance circulations.

The organism biorhythm classification, was found on duration of oscillation period, got a wide application. According this classification there are high –frequency rhythms (period of oscillation is less 0.5 h.), rhythms with middle oscillation frequency: ultradian (period is 0.5-20 h.), circadian (period is 20-28 h.), infradian (period is 28-60 h.), and also low-frequency rhythms with oscillation period more 2.5 days.

Period, oscillation amplitude and acrophase are known the important properties of biorhythms. Period is the shortest time interval, from which oscillation system returns to initial state. Amplitude is the largest divergence at zero, oscillating to definite rule. Acrophase is the time of day, in which at most size is marked.

Circadian rhythms are evidence of organism adaptation to condition of existence. The environment of organism existence is exposed to rhythmically changes, which set rhythm to dynamic activity, nourishment, functions nervous and endocrine systems, which form the rhythm other function. Only duration of light day constantly changes in time out of all the rest organism function. Therefore circadian rhythms of organism are synchronized by photoperiod duration. The pineal gland is the main synchronizer of organism function, it transforms nervous signal, which is occurred by eye photoreceptor irritation, to endocrine.

Lighting changes during year and day. The first changes lead to appearance the season rhythms, the second- to circadian.

The role of pineal gland was the most in detail studied in mammals in regulation season rhythms by example the season reproduction. The pinealectomy in Syrian hamster turns it out of season reproduction animal into constant reproduction animal [78]. The pineal gland denervation by extraction upper neck hanglion or destruction suprachiasmatic nuclei lead to so effect. D.A. Lincoln researches evidence that the pineal gland denervation of sheep influences on the reproduction cycle its species turning sheep into polyestrous animals. The involvement of pineal gland in formation of circadian rhythm various functions has more interest and carrying out by influence on peripheral endocrine glands. The pineal gland regulates both the cortex [23] and medullary adrenal coats and also thyroid [6]. Endocrine gland activity is characterized by circadian rhythmicity, which could be conditioned by synchronous pineal gland effect. The pinealectomy changes the daily rhythm of glucocorticoids activity rats' adrenal. Injection of melatonin causes chronomodulate action on production corticosteroids [84]. It is found out receptors to melatonin in various peripheral endocrine glands [78], the pineal gland control at function of peripheral glands could carry out by activation of their. It carries out also by influence on hypothalamo-hypophysis system. Hence, the injection pineal extract leads to increasing the level of somatoliberin in hypothalamus female albino rats and inhibits the synthesis of phollitropin and lutropin in hypophysis [92].

The pineal gland produces bioactive substances two biochemical types: biogenic amine (melatonin and serotonin) and complex of oligopeptides.

Circadian rhythm for production by pineal gland biological active substances is conditioned by suprahiasmatic nucleus of hypothalamus, which regulates the activity of noradrenergetic neurons upper neck ganglia. Their fibroses reach adrenergetic receptors of pinealocytes, the activity of which is characterized by circadian rhythmicity [68].

The circadian rhythm is conditioned for production biological active substances apart from species accessories and mode of animal live (night or daily). The circadian rhythm of melatonin secretion is revealed in people. The activity of pineal fermentative systems, which are answerable for hormone synthesis, is characterized by cycle. In detail, it is known circadian rhythm of N-acetiltranspherase activity in the pineal of rats [46], and also decreasing hydrocsy-indol-O-metiltranspherase [8] with night peak of activity. The last leads to increasing the melatonin level in bloodstream at the middle of night and decreasing its level in blood by day. Circadian rhythm is characterized for synthesis and secretion some pineal peptide hormone (for example, arginine-vasotocine). However, unlike of melatonin, the circadian rhythm of arginine-vasotocine maintenance is characterized increasing its concentration in iron at the light hours [15].

Pineal Gland and Proliferation Regulation

Biorhythms of proliferation are one kind of rhythmic organization of organism systems. The proliferation of cell – is unique and the so time universal natural phenomenon, being on the basis of alive properties as reproduction, growth, homeostasis, the heredity and eth. Cell proliferation also determines the key role for providing individual organism development, physiological and reparative regeneration, supporting structural homeostasis of tissues, development malignant neoplastics. Alive organisms exist in changing external environment constantly. The state of internal environment so is not constant, that determines the necessity detailed adaptation of processes cell reproduction to these changes. Therefore, one of fundamental question in investigation of proliferation is research of regularities its regulation in general, so and research the regulation its biorhythms.

For the first time circadian periodicity of proliferation was described on plants. Later on circadian cyclic changes of mitotic activity were discovered and in animal tissues. Results of the most researches evidence, that one of evidence cell proliferation is circadian biorhythm reproduction cells. It were shown periodical daily oscillations of mitotic activity epithelium cornea [63], tongue [95], esophagus [81], stomach [29], epithelium of all intestine parts [80], exocrine cells of pancreas [28], cells of adenohypophysis [89], sebaceous glands [53], ostein medulla [67] and alveolar macrophages-monocytes [66].

Engel P. (1934) at first discovered that pinealectomy intensifies proliferation of tumour cell in experimental animals. These results were confirmed by following works Kitay J., Alschule M. (1954), Kallenbach H., Malz W. (1957).

It was shown in experiments on animals that pinealectomy could lead to increasing proliferative activity of jejunal cryptic epithelial [16], cortex cells [85] and chromaphine cells medullary adrenal coats [49]. In albino rats Yoker-256 carcinoma metastasizes more intensively in pinealectomized animals than in control, the duration of pinealectomized albino rat life with carcinoma transplantate was on 30% smaller, the middle diameter of tumour was on 33% more than in control group [79]. Analogous mode pinealectomy influences on the speed of prolifertation and growth inoculated melanoma in Syrian hamster [24].

Heterotransplantation rat pineal to mice with induced cancer of mammary leaded to delay of tumour growth almost in all animals [47]. In so conditions heterotransplantation in liver, spleen and hypophysis did not influence on tumour growth in mice. V.Lapin (1974) showed pinealectomy stimulates tumour growth in animals were induced chemical cancerogenes.

Results some researches evidences about increasing of time illumination or darkening on the pineal functional activity. Constant animal maintenance in darkness intensified synthetic and secretorial pineal activity, while day and night illumination leaded to weakening physiological status the gland up to functional pinealectomy [19]. At the same time, maintenance Vistar`s albino rats by day and night illumination during 15 weeks with following inoculation it the Yoker-256 carcinoma caused visible increasing tumour inoculation. Yoker-256 tumour inoculation decreased and duration of life sick animals increased if animals, long time were keeping in darkness.

Thus, according references returns, pineal suppression and also the gland extirpation intensify proliferation and speed up the growth of neoplastic diseases were inoculated, induced by cancerogenes, spontaneous. The keeping of animals in darkness stimulates pineal activity and caused some antineoplastic effect.

At the present time in references have a lot of works about influence pineal biological active substances on proliferation. It was connected, at the first time, with possible perspective using these substances for correction damages proliferous processes by on oncopathologies. A number of articles were devoted this problem [10; 11; 14; 60; 93]. Researches some authors devoted to influence of melatonin on neoplastic growth, were noting inhibitory melatonin effect on blastomogenesis. In particular, it is noted melatonin protects the effect pinealectomy, stimulating cancerogenesis in mammary. Injection of melatonin to animals, were keeping in conditions of day and night illumination, suppressed the growth of mammary adenocarcinomata: its frequency decreased since 100 to 25% [7; 10].

Bartsch H. and Bartsh C. (1981) shown the effect of melatonin influence depended on photoperiod and time of day, when preparation was injected. The growth ascites fibrosarcoma and Erlich tumour in mice was intensified by injection of melatonin in the morning and delayed by evening injection significantly.

It is necessary to note sometimes melatonin caused antineoplastic effect also in vitro, this fact attested on its cytostatic properties. Inhibitory influence physiological and pharmacological melatonin doses on proliferation in vitro were shown on example malignant trophoblast cells [88], on culture of cancer ovarium cells [71]. In culture adenocarcinoma cells such was noted only by processing high melatonin doses [34].

Antineoplastic melatonin effect, was finding as regards to some types of tumours, was more expressed in pinealectomy or suppression pineal function by constant illumination [69]. In Karasek M. et al. (1987) experiments absence melatonin effect on regeneration adrenal cortex in rat with intact pineal gland and its suppressing action after pinealectomy stimulating proliferation was shown. It is in accord with references returns about significant suppressing melatonin effect on proliferation neoplastic cells by damages pineal function [85].

Thus, melatonin possesses inhibitory action as regards in tumour and normal tissues both in vivo and, it is possible, in culture. The inhibitory melatonin effect is determined by action some factors: doses, time of injection, type of tissue, duration melatonin exposition and etc.

At the same time with indol junctions the pineal gland syntheses regulatory peptides. However the role of biological active pineal peptides is beyond doubt. It is conditioned by present time doesn't identify the most polypeptide gland products and in experiments complex extract, cleaned indol junctions, continue is used. Single researches, pointing out on important role of biological active pineal peptides in realization its function, devote to study properties of some pineal peptide components.

The influence on proliferation polypeptide pineal extract, was called by epithalamin, was investigated the most good [1; 2; 3]. Epithalamin is the preparation, containing complex polypeptides were extracted from cattle pineal glands by way the acetone extraction with following homogenization, dissolve and liophilization. The epithalon preparation has worked out on the basis of amino aced composition and has researching. It is synthetic tetrapeptide (Ala-Glu-Asp-Glu) [52], which possesses more biological activity than its natural analogy.

The functions of pineal peptides were beginning to research only. It is known a few quantities of researches about influence pineal peptides on proliferation. So, epithalamin injection to mice in doze 2mg/kg delays the growth of cervix cancer on 55-87% [26]. Considerable decreasing mitosis quantities and nucleus sizes in tumour cells of hepathoma 22-A were observed by epithalamin influence. Besides, epithalamin significantly inhibited blastomogenesis in rats, induced by 7,12 dimethilbenz(a)anthracene, causing the increasing of percent growth tumours and a great number tumour rise [26].

In conclusion it is necessary to note:

- activity of various organism systems is characterized by functional rhythmicity;
- proliferation tissue activity characterizes by circadian biorhythmic organization;
- photoperiodic external environment changes are the main reading of time for circadian rhythm of functions;
- important role belongs to the pineal gland in photoperiodic circadian biorhythms control, the meaning of which remains unknown in formation circadian rhythm of proliferation;

- pineal gland produces complex bioactive peptides and melatonin hormone , which could synthesize by extrapineally;
- pineal production of biological active substances is characterized by circadian rhythmicity;
- pineal biological active substances influence on proliferation both in vivo, and in vitro;
- meaning of biological active pineal peptides in organism is beyond doubt, that could depend on nonidentification the most components of polypeptide complex.

Pineal Gland and Spermatogenesis

Early it was noted participation of pineal gland in formation of circadian rhythms many physiological organism functions. Spermatogenesis is the function biorhythmic organization of which remains unknown absolutely. The duration of spermatogenesis is different in people and various animal species, in people this process lasts about 74 days, in albino rats – 47 days. Such duration of the process suggests the possibility of its circadian time dynamics. This supposition is confirmed by circadian rhythm of maintenance in blood gonadotropins and testosterone, regulating the process of spermatogenesis [33; 35].

Spermatogenesis is characterized by regulated in time complex processes differentiation developing reproductive cells. The duration of spermatogenesis is constant for each animal species [83] that is evidences about its conservatism. The spermatogenesis is under the hard control endocrine, nervous and immune organism systems. Many facts attest to coordination reproductive function with the level of external environment illumination. The season rhythmicity of animal reproduction is the most investigated and is aimed to choice the favourable season for reproductive activity. The pineal gland is probable synchronizer stages of spermatogenesis with photoperiod phases. Reproductive function of monoestrous animals is characterized by seasonal rhythm of spermatogenesis formed with participation of the pineal gland. The dependence of polyestrous animal's reproduction on season is less pronounced.

Facts about pineal influence on reproductive system reflect the role of pineal gland in formation reproduction in ontogenesis and its seasonal changes [22; 36]. The influence of pineal gland on reproduction polyestrous animals is not investigated.

In process of spermatogenesis on each part of seminal tubule are row following one after another cell combination. P.Leblond and Y.Clermont (1952) divided 14 stages in albino rats, which formed spermatogenic cycle, duration of its 12 days. Its role in organization of spermatogenesis remains still unknown.

Thus, the existence of circadian rhythm of spermatogenesis and pineal gland participation in its regulation are the actual for understanding the reproductive function and need in according to research.

Problems of Research

The aim this research was determine the role of pineal biological active oligopeptides in formation of circadian rhythm proliferation somatic and reproductive cells and revealing the circadian structure of spermatogenesis. Achievement this aim was found on decision following problems:

1. daily dynamic mitotic index of jejunal cryptic epithelial in intact and pinealectomized albino rats to study;
2. influence of pineal peptides injection to pinealectomized albino rats on circadian rhythm mitotic index of jejunal cryptic epithelial formation to reserch;
3. circadian dynamics of mitotic index type B spermatogonia and frequency stages spermatogenic cycle in intact and pinealectomized animals to study;
4. influence of pineal peptides injection to pinealectomized albino rats on circadian dynamics of mitotic index type B spermatogonia and frequency stages spermatogenic cycle to research.

Materials and Methods

Experiment was carried out on 270 male albino rats by weight 160-200 g. the animals were adapted to illumination regimen for 20 days: darkness 12:12 (illumination from 6 to 18 h). The access to food and water was ad libitum during all experiment. After adapted period animals were divided into 3 experimental groups: intact control (n=90), pinealectomized (n=90) and pinealectomized +epithalamin treatment (n=90).

We did not use false undergone surgery animals for control of effect pineal peptides injection on dynamics proliferative processes in jejunal cryptic epithelial and spermatogonia, and also processes of reproductive cells differentiation because research was carried out on day 40 after pinealectomy. To this time total healing of wound take place and undergone surgery animals don't differ from intact. At the same time, addition this animals group increases the volume of work, doesn't improving pinealectomy and injection biological active pineal peptides results. To our opinion injection biological active substances to intact animals has not sense, these animals could have circadian biorhythm of proliferation. The absence of effect injection pineal biological active peptides doesn't permit to do conclusions. At the same time, in works El-Domeiri A.A., Das Gupta T.K. (1967) daily melatonin injections during 3 weeks didn't influence on proliferative activity of melanoma in Syrian hamster with intact pineal, but injection hormone to pinealectomized animals delayed tumour growth. Besides, melatonin injection didn't cause influence on formation circadian locomotor rhythms of intact rat. Its ability to synchronization daily periodicity locomotor activity was noted in constant darkening conditions, then in conditions are leading to breach of normal physiological pineal activity [5].

Pinealectomy was carried out on original method. All painful manipulations with animals were carried out by Helsinki agreement "About humane treatment with experimental animals" Narcosis was accomplished by intraperitoneum thiopental natrium injection in dose

50 mg/kg. Portable surgical table for small laboratory animals was able to fix hardly head behind auricles and maxilla. Auricle keepers with square basis and conic tip were carried out. Auricle keepers were taken courage by counters, having general foot, which conditioned the hardness of construction. Compatibility longitudinal axes of auricle keepers were provided by counters construction, having prismatic slot for fixation auricle keepers. Strengthening of maxilla was carried out by plate putting under upper incisors. Animal head was fixed hardly behind auricles and maxilla. Hair cover was removed in region cut of skin, being surface was worked up by alcohol. Cut of skin by 1,5 centimeters length carried out in the region of head middle line. Slot by 3 mm diameter was drilled in occipital on 4 mm caudal of intersection sagittal and parietal- occipital sutures after naked cover skull. In this slot needle was injected and cut hard cerebral coat.

Pinealectomy was carried out by resection a part of bone with being subject the pineal gland by special made biter. Its lower branch has conic form with curve edge and flat surfaces. Its thickness of working part is 1mm, width is 2,6 mm. On upper surface is hollow like ditch with sharp edges for separated from bone part. The upper branch in anterior part has window like lower branch form with small split that is following make not breakage but carving bone.

Bone resection was carried out by lower upper branch biter on lower branch. In the slot towards intersection bone sutures between lower surface of bone and brain surface injected lower branch biter and cut part of bone with fragment of sinus and with being subject the pineal gland. The control of pinealectomy was carried by visually. Removed pineal gland had cycle form with 1-1.5×1-1.5 mm in size. A part of pineal glands were fixed in formalin (6 hours), after were washed by water, removed water in alcohols and placed in paraffin. Cut of pineal gland 5-6 mkm thickness were painted by hematoksilin and eosin. The fact pinealectomy was confirmed on histological preparations.

Hemorrhage, arising for venous sinus damage, was stopped by cotton tampon. Waxen paper was laid on slot in the bone for protection following junction connective tissue with brain tissue, and then wound was sewed up. As disinfection means for wound powder on basis stperptoced and mumie were used. Sutures were removed on 7 day after pinealectomy. Survival animals after surgery were 60%. Pinealectomized animal were kept under 12/12h day/night regimen (illumination at 6.00-18.00 h.) Animals with displeasure state, expressing in damage of coordination activity, weight increasing, damage of intestine functions, inflamed processes were not used in following experiment.

As model for research role pineal biological active substances in formation circadian rhythm of proliferation somatic cells jejunum cryptic epithelial was chose, which classic model for physiological regeneration research were. The advantage of this model is determined by high proliferative activity of tissue and expressed spatial-time organization cells proliferation (cells being in the proliferative phase are forced out general row interphase cells, placed as a rule in end criptic part and are differed on preparation easy), that proved with easy determination and calculation mitosis figures. The calculation of quantity dividing cells was examined in the end section 20 longitudinally cut jejunum crypts on part in 25 cells since each side crypt to its bottom. Mitotic index was determined as attitude of quantity cells dividing to 1000 examined.

The role of pineal gland and its biological active peptides in formation spermatogenesis circadian structure was estimated at determining absolute weight of testes, circadian dynamic of mitotic index of type B spermatogonia at the spermatogenic cycle stage VI and frequency of spermatogenic cycle stages finding. Circadian dynamics of the mitotic index of type B spermatogonia was studied at stage VI of spermatogenesis cycle, during which their division is the most intensive. For that a total of 1000 spermatogonia and preleptotene spermatocytes on this stage from each animal were examined.

Spermatogenic cycle is main form spatial organization spermatogenic tissue. C.P. Leblond and Y. Clermont introduced the notion of spermatogenic cycle, that is description total series cell combination and follow each other. There are 14 stages in rat spermatogenic cycle. The incidence of spermatogenic cycle stages was evaluated as their relative duration, which estimated by the formula:

$$t=(P\backslash k)*N,$$

where t is duration of stage in hours;

P - duration of spermatogenic cycle in hours;
k – number of studied objects;
N – incidence of stages.

The determination of absolute stages duration in hours was based on facts about duration spermatogenic cycle [20].

Epithalamin (2.5mg/kg) dissolved in physiological solution was injected to pinealectomized animals since 26 till 41 days after pinealectomy at 18 h (it is the time of change photoregimen from day phase to night). For injections new prepared solution of oligopeptides was used only.

The experiment was carried out since February till April. On day 40-41 after pinealectomy the animals were sacrificed by ether narcosis every 3 h ($18^{\underline{00}}$, $21^{\underline{00}}$, $24^{\underline{00}}$, $3^{\underline{00}}$, $6^{\underline{00}}$, $9^{\underline{00}}$, $12^{\underline{00}}$, $15^{\underline{00}}$, $18^{\underline{00}}$, $21^{\underline{00}}$, $24^{\underline{00}}$, $3^{\underline{00}}$, $6^{\underline{00}}$, $9^{\underline{00}}$, $12^{\underline{00}}$) for 2 days, that was provided with all studying properties during two periods circadian rhythm. Organs were fixed in Carnua liquid.

Circadian and ultradian rhythms revealing and their period's definition were determined by two methods:

1) of spectral analysis. The period duration was estimated by the formula:

$$T=\frac{2*\Delta t*100}{f},$$ where T is period of rhythm proliferative system; Δt is time interval in hours; f – is spectral frequency according to the maximal mean of amplitude oscillation.

2) of the least squares method, which is allow to get smoothed curve of dynamic mean, to define existence biorhythm and its period and to draw up a biorhythm diagram also.

Analyzing daily biorhythms determined following parameters:

1. mesor (middle-day mean) –is quotient of sum means the parameter on quantity measuring during day;
2. acrophase – is the time of day on which maximal registered mean of parameter was noted;
3. active phase (AP) – is the time (in hour) during which means of parameter more mesor.

Role of Pineal Gland in Regulation Circadian Rhythms of Proliferation Epithelium in Jejunum's Crypts

One of displays of regulation of cellular division is circadian biorhythm of cells reproduction. Results of the overwhelming majority of works testify the presence of periodic daily fluctuations of mitotic activity in tissues of various localization and origin.

Results of our researches also confirm the conventional law. Dynamics of mitotic index (MI) of epithelium in jejunum's crypts in intact albino rats has a daily rhythm of monophase character (figure 1).

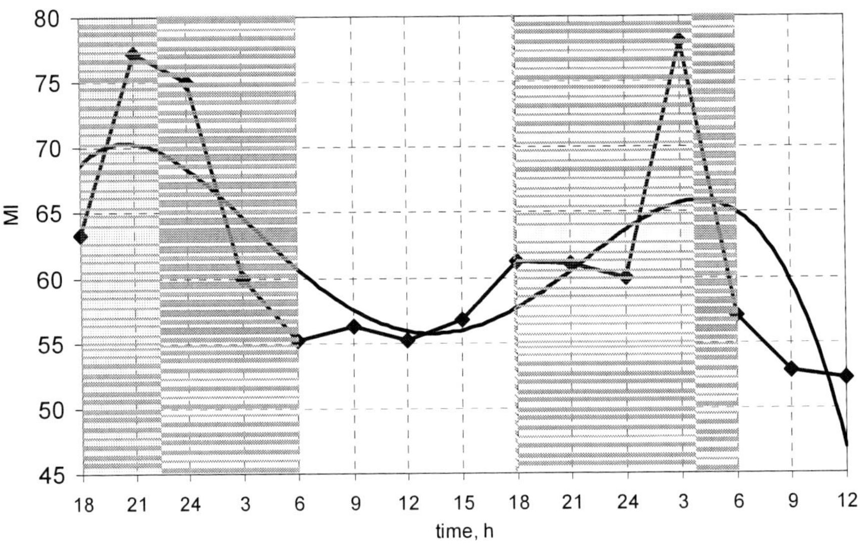

Figure 1. Dynamics of MI epithelium in jejunum's crypts in intact control animals (here and later dark time marked by gray color).

Data of the spectral analysis testify the rhythmic organization of mitotic activity of epithelium in jejunum's crypts in intact animals with the period 21h. Besides the results of the spectral analysis of mitotic index of dynamics have shown the presence of ill-defined ultradian component of a rhythm of proliferation with the period of about 8 hours (figure 2).

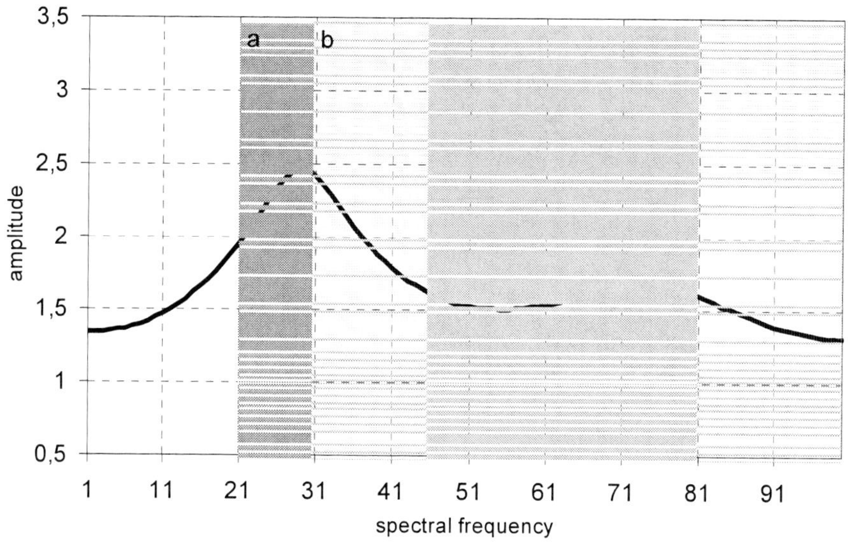

Figure 2. Results of the spectral analysis of MI dynamics of epithelium in jejunum's crypts in intact animals of control group; a – low frequency area (circadian rhythms area), b – high frequency area (ultradian rhythms area).

The active phase of circadian biorhythm MI fell to night hours and had various durations in the first and second day of experience (table 1). In day 1 it began right after the changes of the light period into the dark period (6 pm) and proceeded until 3 am. In day 2 the active phase proceeded from 12 am till 6 am. Values MI were maximal at 9 pm (the first day) and at 3 am (the second day), authentically exceeding minimal values MI in 12h ($P < 0{,}001$). Mitotic activity of epithelium in jejunum's crypts in intact animals in dark period to a phase (66,8±4,7 ‰) the photoperiod exceeded those in a light phase (55,2±4,4 ‰; $P < 0{,}05$).

Table 1. Parametrs of circadian rhythms MI of epithelium in jejunum's crypts in intact control and in intact control and pinealectomized animals

Intact control			**Pinealectomy**	
Parameter	Day 1	Day 2	Day 1	Day 2
Mesor,‰	61.38±2.21		59.36±1.99	
Acrophase,h	21.00 *(18,3)	3.00 (24,6)	Rhythm is absent	
AP, h	18.00-3.00	24.00-6.00		
Duration of AP, h	9	6		
Middle of AP, h	22.30	3.00		
AA,‰	22	25.8		

Note: *-reliable difference (P‹0,05) from value in hours citing in brackets.

These data received in conditions of strict alternation of illumination and darkness, testify that acrophases rhythm position is not strictly dated to any point within the 24-hour scale. Noted specifies the certain lability of circadian rhythm of proliferation. The passage of the 24- hour scale, within the limits of which the position of a daily maximum parameter varies can be characterized as "a zone of wandering acrophase". As a rule, absence of rigid synchronization of rhythm of proliferation when it is not equal to 24 hours is marked in growing organisms, and also during period of season changes in the spring and in the fall. Alongside with it we track interrelation of mitotic activity rise with the certain time interval within 24 hours. For a maximum of mitotic epithelium in jejunum's crypts the arrangement of "a zone of wandering" is a necessary activity on the dark period of a photomode, and the active phase of a rhythm can be referred to the beginning of the light period. Thus, circadian biorhythm of dynamics MI is characterized by lability in a certain "a zone of wandering" which precisely corresponds to the conditions of a photomode.

In intact animals dynamics MI of epithelium in jejunum's crypts has character of a monophase daily rhythm during both the first, and the second day of experience. The following fact is worthy of attention, that works of some authors establish two-topmost character of MI rhythm [82] the first rise of which was carried out at night, and the second was registered during a light phase of the photoperiod. In formation of circadian rhythm of various functions, the major signal of time is photoperiodicity, which defines a secretory activity rhythm of pineal gland. Synthesis of biologically active substances is carried out by pineal gland in dark period. The retina represents the first part of regulatory to circuit in circadian rhythms of an organism that is proved by the absence rhythm of melatonin synthesis in a daily and its predecessors among animals with a hereditary degeneration of a retina [73]. The synthesis of biologically active substances carried out by the pineal gland during the daytime sleep most likely causes occurrence in some cases of two-topmost curves of epithelium mitotic activity in jejunum's crypts. It corresponds to the results of research by Kosichenko L.P. (1961) who has shown formation of two-topmost curve of cornea epithelium mitotic activity during the change of illumination and darkening within every six hours.

The adaptation of rhythmic systems of an organism to the environment is a constant process of approaching the system rhythm structure to the environmental cyclic factors of rhythm structure. However, as well as in other cases of adaptation, this approach is never a 100 % identity. Such discrepancy is defined by environmental cyclic factors. Actually, there is no gauge of time which characteristics would be absolutely constant. Within a year the length of light day, pressure, temperature, etc. continuously change besides circadian rhythms do not remain strictly constant from day by day. An organism being an open system is constantly influenced the effects of factors, capable, this way or that way, of influencing the biorhythm character. Absence of absolute identity in a diet, sleep, mental loadings and so forth excludes an opportunity of a constancy phase architectonics of circadian systems of an organism. Nevertheless, steady displacement of a phase to borders of a zone of wandering testifies to redistribution of elements of adaptation and disadaptation in favour of last. Differently any infringement of regulation circadian biorhythm, both internal, and external will begin with such steady displacement of a phase to borders of a zone of its wandering and will end with disappearance circadian biorhythm as those.

As well as regulation circadian biorhythm of any function of an organism, regulation of a biorhythm of proliferation is incorporated in hierarchically organized system [55]:

- the part connecting the time organization of biological system with other systems and with an environment;

- the part adjusting the time organization of biological system;
- the part perceiving signals of regulation;
- the effector part which expresses an end result of activity of biological system.

At formation circadian biorhythm mitotic activity periodically changing conditions of environment (first of all change of day and night) define a biorhythm of functioning organisms regulators of proliferation (for example, gland internal secretion) which, apparently, influence development or activity of fabric regulators of cell fission (chalones, factors of growth), that finally causes occurrence circadian biorhythm of proliferation in the given concrete kind of a tissues.

The constancy and change of cellular structure of tissue in ontogenesis in many respects are provided owing to functionally multicomponent system of regulation which separate parts belong to various levels of the organization of an organism. Well-known, that in the control over cellular duplication takes part nervous, immune and endocrine systems. Within last decades researches of influence of factors of growth and histotype-specific inhibitor of proliferation especially roughly develop.

In maintenance of the photoperiodic control of daily and seasonal biorhythms of various functions of an organism, the important role belongs pineal gland [22]. However, the role of pineal gland in formation circadian rhythm of proliferation is investigated not full.

In this connection by us, it has been lead pinealectomy in male albino rats and dynamics mitotic index of epithelium in jejunum's crypts in pinealectomyzed animals. Pinealectomy has led to disappearance circadian rhythm of proliferation of epithelium in jejunum's crypts. Fluctuations MI have not found out any communication with a photomode. The smoothed curve has got a linear kind (figure 3). On the general monotonous background of changes MI it was observed statistically authentic three rises ($P < 0{,}05$): in 9h (the first day), in 18h (the beginning of the second day) and in 9h (the second day) (figure 3). Maximum MI is noted in 18h the second day, a minimum - in 3h the first day ($P < 0{,}01$) (figure 3). The level of proliferation of epithelium in jejunum's crypts in pinealectomyzed animals was same as in intact, that was expressed in absence of distinctions ($P > 0{,}05$) between daily average values MI (61,38±3,21 ‰ and 59,36±1,99 ‰). Proliferation activity in light and dark phases of the photoperiod had no statistically authentic ($P > 0{,}05$) differences, making 61,4±5,4 ‰ and 57,3±4,1 ‰ during day time and night time accordingly.

Results of the spectral analysis have shown absence circadian rhythm MI (figure 4). At the same time, it was found out precisely expressed ultradian a rhythm with the period nearby 9h. Similar results are received at research of proliferation of epithelium a mucous mouth of the person [65] and meiotic activity spermatocytes mice [76].

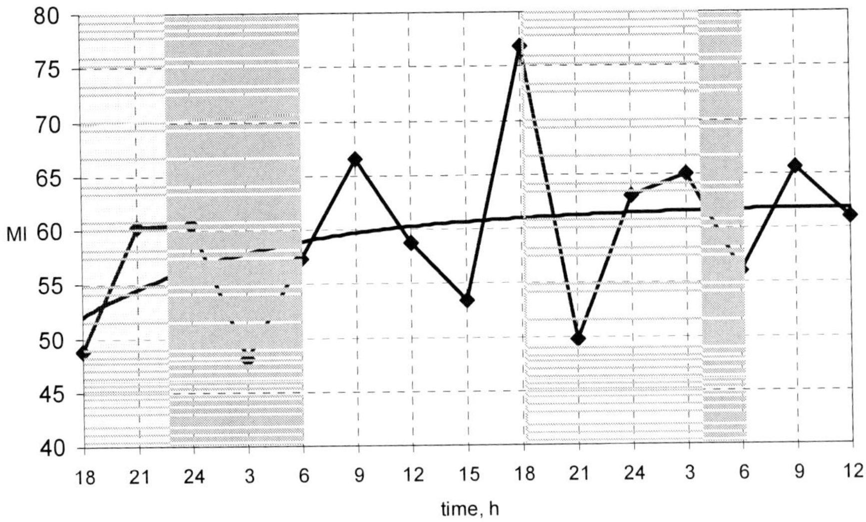

Figure 3. Dynamics of MI epithelium in jejunum's crypts in pinealectomyzed animals.

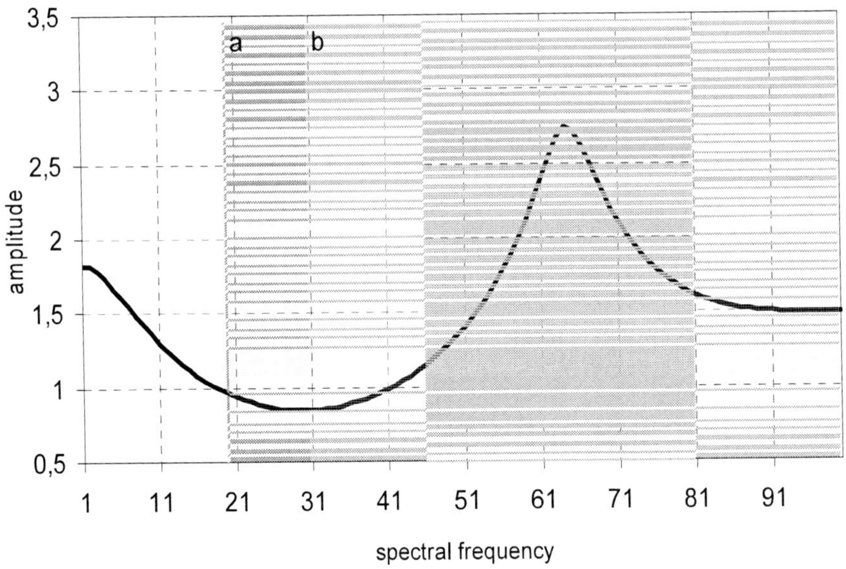

Figure 4. Results of the spectral analysis of MI dynamics of epithelium in jejunum's crypts in pinealectomyzed animals a – low frequency area (circadian rhythms area), b – high frequency area (ultradian rhythms area).

From resulted above data follows, that the organization of proliferation systems is characterized by hierarchical structure. Features of mutual relations of biorhythms of various levels remain insufficiently studied. Possibly, primary it is necessary to consider endogenous ultradian rhythms based on which at participation of external synchronizers are formed circadian rhythms.

Occurrence ultradian biorhythms, apparently, is caused by cyclicity of metabolic processes in a tissue. It proves to be true following results: the culture of a fabric hepatocytes of the rats, created of two populations of cells with initially antiphase rhythms of synthesis of fiber, found out in the further uniform per houses a rhythm of synthetic processes. Thus, the group of cells can possess independent oscillator ability of formation of a rhythm of the activity [12].

Strictly speaking, pineal gland the role neuroendocrine the converter of nervous impulses in a hormonal signal [56] is allocated. In a control system circadian the complex "suprachiasmatic nuclei – pineal gland" provides with rhythms harmonious adaptation to varying factors of environment, in particular, to light.

Results of our experiment confirm the mentioned function chart, testifying about the important role of pineal gland in the photoperiodic control of daily rhythms of proliferation activity of fabrics of an organism. It specifies disappearance circadian rhythm mitotic index of epithelium in jejunum's crypts in pinealectomyzed animals. Besides experiments, Ding C.H. et al. (1995) reveal infringement circadian rhythm of proliferation lymphocytes spleens in pinealectomyzed animals. According to Haldar C. et al. (1992), pinealectomy leads to infringement circadian rhythm of proliferation glandulocytes of testicle and macrophages of a bone brain. Alongside with it, pinealectomy leads to infringement circadian rhythm of activity ornithinedecarboxylase (the enzyme, being a parameter of intensity of synthesis of fiber in a cell) in a liver and in lymph nodes rats [17; 18].

Synchronization of functions of an organism with conditions of illumination is carried out pineal gland, which produces two groups of biologically active substances: indoleamines and oligopeptides complex. As it has been shown above, a role biologically active peptides in regulation of biorhythms of an organism including biorhythms proliferation continues to remain obscure. In this connection, we investigated influence of introduction epithalamin on formation circadian rhythm of proliferation of epitheliums in jejunum's crypts in pinealectomyzed animals.

Introduction pineal gland peptides of animals after pinealectomy have led to formation circadian rhythm mitotic index of epithelium in jejunum's crypts. By means of the spectral analysis, two components of rhythm MI pinealectomyzed animals by which it was entered epithalamin have been revealed. The curve describing rhythm of functioning proliferation system differs substantial growth of amplitude of fluctuations in the field of low frequencies and less expressed amplitude in the field of high frequencies (figure 6). Thus, are revealed circadian rhythm MI of epithelium in jejunum's crypts with the period, which is coming nearer to 24h, and ultradian with the period nearby 8h.

Circadian rhythm MI pinealectomyzed animals after introduction epithalamin had character of a monophase rhythm with a maximum of cell fission in 24h and a minimum in 12h ($P < 0{,}001$) as during the first, and the second day (table 2). Increase mitotic activity occurred at night and in the early morning during, both the first, and the second day of experience (figure 5). Duration of an active phase of a rhythm was restored up to a level of parameters of animals of control group. Acrophase a rhythm observed in the middle dark period of a photomode (figure 5).

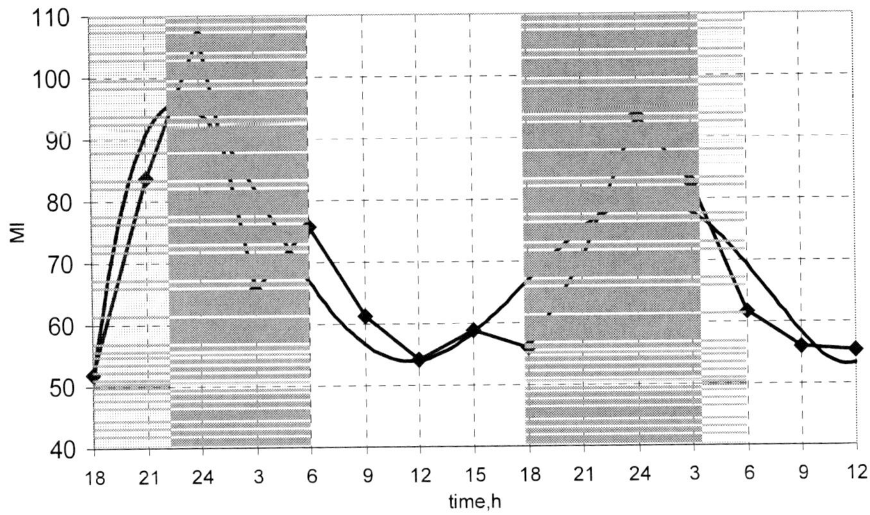

Figure 5. Dynamics of MI epithelium in jejunum's crypts in pinealectomyzed animals after injecting epithalamin; (— MI, ---- smoothed curve).

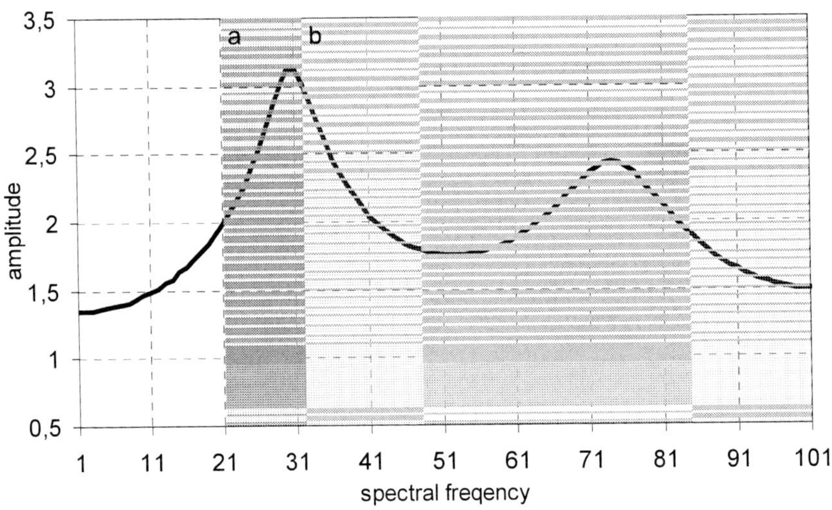

Figure 6. Results of the spectral analysis of MI dynamics of epithelium in jejunum's crypts in pinealectomyzed animals after injecting epithalamin a – low frequency area (circadian rhythms area), b – high frequency area (ultradian rhythms area).

Displacement of an active phase of a rhythm in relation to those intact has made animals 0,5h in the first day and 2,5h in the second day of experience, and displacement acrophase – 3h. It is necessary to note substantial growth of peak parameters of a rhythm and factor of synchronization in pinealectomyzed animals after introduction epithalamin (table 2) in comparison with intact animals (table 1).

Table 2. Parameters of circadian rhythms MI of epithelium in jejunum's crypts in pinealectomyzed animals after injecting epithalamin

	Epithalamin	
Parameter	Day 1	Day 2
Mesor,‰	68.82±4.22	
Acrophase,h	24.00 (18,21)	24.00 (21,6)
AP, h	20.00-2.00	20.00-5.00
Duration of AP, h	6	9
Middle of AP, h	23.00	0.30
AA,‰	54.75	38.17

The parameter of absolute amplitude has increased in relation to corresponding parameter intact animals for 49 % that specifies a greater degree of synchronization of the introduction of cells in mitosis. Features biorhythmologycal process found out similarity to those at intact animals that proves to be true also character of the smoothed curve (figure 1,5). At animals after pinealectomy, which entered epithalamin daily average value, MI had no statistically authentic differences (P > 0,05) from values of corresponding parameters for intact and animals after pinealectomy. At the same time dynamics MI, unlike animals after pinealectomy was characterized by presence circadian rhythm. A share of the cells sharing in dark phase to a phase of the photoperiod, it is authentic (P<0,05) exceeded those in a light phase, making accordingly 82,2±5,6‰ and 58,3±3 ‰.

Thus, introduction pineal gland peptides animals after pinealectomy to animals lead to restoration circadian rhythm of proliferation of epithelium in jejunum's crypts.

Noted testifies to direct participation pineal gland peptides in regulation of circadian biorhythm proliferation processes of an organism.

Stated allows to conclude, that formation circadian rhythm of proliferation is carried out at participation of pineal gland. It proves to be true as disappearance in pinealectomyzed animals circadian rhythm MI, and its restoration after introduction epithalamin. Pineal gland produces greater group biologically active oligopeptides that allows assuming existence, unlike melatonin, them cutting-off and tissue specificity. Besides the opportunity of processing peptides and clearings peptid fragments providing specific reactions, not connected with action of initial forms is not excluded. Features of interaction melatonin and pineal gland oligopeptides in processes of regulation of daily periodicity of various functions of an organism demand statement of special researches.

One of key questions experimental chronobyology is finding-out of mechanisms of regulation of a daily rhythm proliferation processes in tissue. It is shown, that rhythmic of proliferation activity systems is supervised at various levels, but action of any factor finally is realized by means of interstitial regulators, on what specifies presence cutting-off and tissue specificity of proliferation rhythms.

To one of participants of interstitial regulation of proliferation processes are chalones - endogenous histotype-specific physiological inhibitors of cellular division [13]. It is confirmed by a strict coordination of organization G2-chalones of system in Ehrlich's ascites tumors with a rhythm of division of its cells, and also participation G2-chalones of system in formation of a daily rhythm of proliferation epitheliocytes a gullet. Alongside with it the researches testifying to daily rhythms of production G2-chalones [4] are executed. It is shown, that phases of a daily rhythm of mitosis, sensitivity of cells action containing chalones a preparation and biological activity chalon are interconnected among themselves in time. The highest sensitivity of cells action chalones and its most expressed biological activity coincide with the period of decrease in number of sharing cells in a daily rhythm [31].

Thus, the system of regulation proliferation processes represents a multilevel complex. Based on results of own research and considering the given literatures, it is possible to conclude, that formation circadian rhythm of proliferation is provided with participation of following structures:

1) suprachiasmatic nuclei perceiving the information from a retina of an eye about a mode of illumination and setting a rhythm secret of pineal gland activity;
2) pineal gland, the transforming nervous pulse answer on light irrotant in endocrine signal;
3) peripheral endocrine glands, modulating a biorhythm of production of tissue regulators of proliferation;
4) endogenous tissue inhibitors of proliferation are chalones.

Role of Pineal Gland in Formation of Circadian Structures in Spermatogenesis

Data about influence of pineal gland on reproductive system have unilateral character and reflect a role pineal gland in formation of a reproduction in ontogenesis and its seasonal changes. Effects of pineal gland on sexual gland are described basically as inhibitors and show participation pineal gland in regulation of puberty [78].

Influence pineal gland on weight of testicles and additional sexual gland is most expressed at animals with seasonal duplication. So, the increase in weight sexual gland at golden hamsters after pinealectomy during a season of duplication is described. Introduction of melatonin blocks effect pinealectomy [74]. Besides the works executed on hamsters, weight reduction of testicles and additional sexual gland after introduction melatonin during a season of duplication is revealed at polecats, weasels, hares, sanderlings, goats and the pony [22].

At polyestrous animals, data about influence of pineal gland on weight of testicle are ambiguous. According to H. Simoonet et al. (1951) it is direct after pinealectomy in albino rats the insignificant increase in weight of testicle is observed. However later works F. Turek et S. H. Losee (1978) have shown, that introduction of melatonin causes an atrophy seed plant in two kinds which reproductive cycles are adjusted by a light mode, - at golden and scorpion- hamsters, however does not render influence on weight gonads at two kinds at which reproductive cycles closely are not connected with length of a light interval of day, - in albino rat and the house mouse.

In our experiment, pinealectomy has not rendered influence on testicles weight. On the average the weight of testicles in intact to control group of animals has made 1,33±0,02g. In

40 days after pinealectomy weight of gonads did not differ from values intact animals and made 1,32±0,02g. At animals after pinealectomy the animals received epithalamin weight of testicle on 9-11 % was above, than at intact animals (P <0,05) and has made accordingly 1,44±0,02

There is an opinion, that pineal gland secrete biologically active substances according to a mode of illumination at all kinds, however sensitivity them hypothalamus at kinds with seasonal duplication and polyestrous animals is various [22]. At animals to seasonal duplication outside of a season of reproductive activity occurs destruction seed canalicus, carried out at participation of pineal gland [75]. With it, sexual gland at animals change of weight is connected with seasonal duplication and absence of visible changes at polyestrous animals in experiences with pinealectomy and introduction biologically active pineal gland peptides.

For spermatogenesis mammals, greater duration of current is characteristic. In albino rats, this process proceeds 48 days. Such extent gives the basis to believe an opportunity of existence of daily dynamics of current of spermatogenesis. So synthesis of DNA in spermatogonia mice is characterized circadian by rhythm [39]. The greatest number of DNA-synthesizing of cells, namely spermatogonia and preleptotenic spermatocytes, according to the author, is necessary on a time interval 8 - 10 h, and the least on 2 h.

Our results of studying of dynamics mitotic index type B spermatogonia in intact albino rats would show presence circadian rhythm of mitotic activity. The average level of mitotic activity has made 120,28+3,26 ‰ (table 3). The smoothed curve of daily dynamics MI of spermatogonia looked like a sinusoid (figure 7). The active phase fell to early morning hours of each day of experiment. In day 1, it began shortly before switching a mode with dark phases on light (5) and proceeded up to the middle of the light period (12). In 2 day the active phase proceeded with 7 up to 12 h, that is borrowed the beginning of the light period of time. Acrophase dynamics MI of spermatogonia fell 9 h in every day. The amplitude of fluctuations MI for a two-daily interval of time has made 45‰.

Table 3. Mitotic index of spermatogonia (‰)

	Animal groups		
	Intact animals	Pinealectomyzed animals	Pinealectomyzed animals after injecting epithalamin
Mesor	120,28±3,26	139,45±3,68	134,55±6,98
Day time	127,98±2,48*	138,41±3,26	162,08±1,64*
Night time	112,17±4,04*	135,34±4,52	112,43±4,38*

Daytime value MI authentically exceeded (P <0,05) value MI at night (table 3). Results of the spectral analysis of proliferative dynamics spermatogonia specify to activity presence circadian rhythm MI with the period nearby 24h and ultradian a rhythm with the period nearby 8h (figure 8). Presence ultradian rhythms is caused by cyclicity of metabolic reactions in a tissue.

After pinealectomy authentic increase of daily average values of mitotic activity type B spermatogonia up to 139,45 + 3,68 ‰ (P <0,01, table 3) Would be observed, that corresponds to the developed representations about inhibitors influence pineal gland on proliferative processes [77; 87].

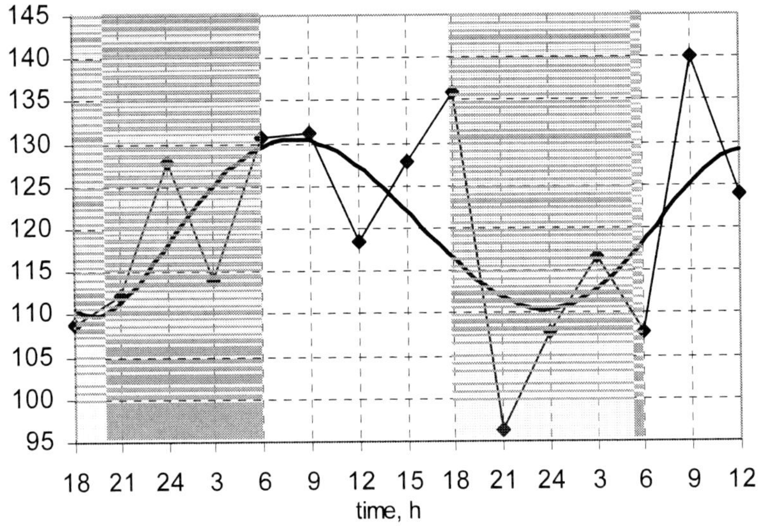

Figure 7. Dynamic of mitotic index of type B spermatogonia in intact animals.

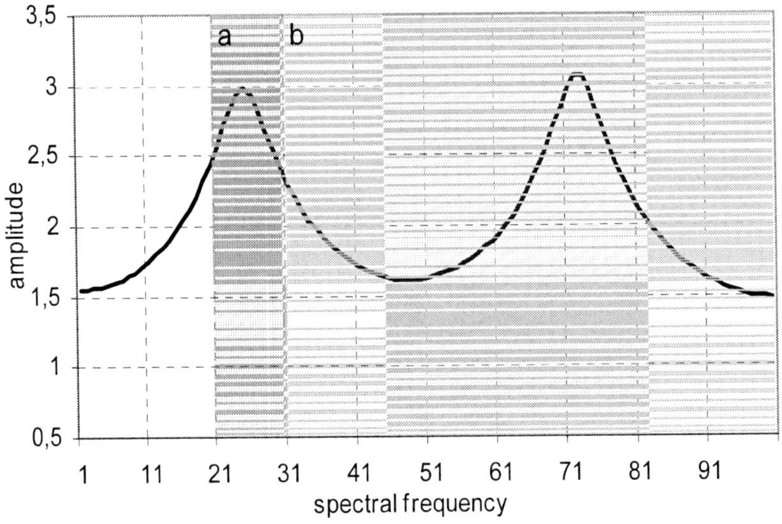

Figure 8. Results of the spectral analysis of dynamic of mitotic index of type B spermatogonia in intact animals (a – circadian rhythms area), (b –ultradian rhythms area).

Despite raised mitotic activity, in our experience the weight of testicles after pinealectomy as it was specified above, has not changed. Under E. Roosen-Range's (1980) certificate, the quantity formed spermatozoons from one rotting spermatogonia for the given kind of animals always constantly also does not depend from any regulations influences. The number formed spermatozoons is defined by quantity of divisions rotting spermatogonia during a differentiation. However, it always below, than it is possible to assume by simple multiplication. The given decrease is provided with process of programmed cellular destruction to which the part of differentiated sexual cells owing to what in an organism biological balance between processes proliferation of spermatogonia and an output mature spermatozoons is reached is exposed. Possibly, after pinealectomy, there is both increase proliferation activity of spermatogonia, and an increase of percent of the cells entering in apoptosis therefore the weight of body does not change. However, this assumption demands the further experimental acknowledgement.

At animals after pinealectomy animals, disappearance circadian rhythm of proliferation type B spermatogonia would be observed. The smoothed curve of daily dynamics MI spermatogonia had the form of a direct line (figure 9). The active phase in the first day was observed during the light period of time - with 6 up to 14 h, and in the second it was marked as in dark period, and borrowed the beginning of light time - with 24 up to 8 h. Displacement acrophase, led loss of its communication with light-dark a cycle came to light. So, in the first day of experiment acrophase it was observed in 12 h (at intact animals in 9), in the second - in 6 h (at intact animals in 9). The amplitude of fluctuations MI spermatogonia in pinealectomyzed animals has increased up to 60 ‰, however average values MI during daytime and night time authentically did not differ (table 3). The spectral analysis has revealed only ultradian rhythmic fluctuations with the period of 17 hours (figure 10).

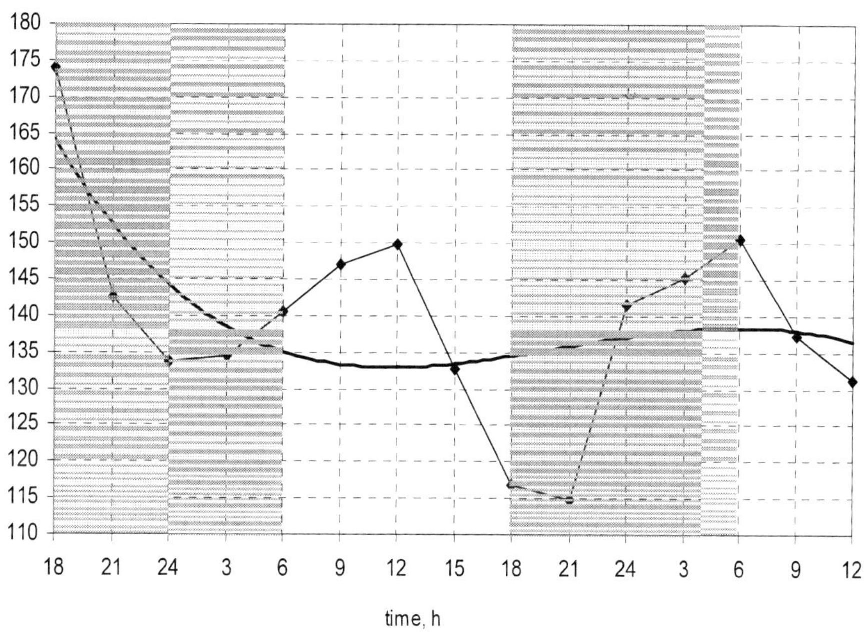

Figure 9. Dynamic of mitotic index of type B spermatogonia in pinealectomyzed animals.

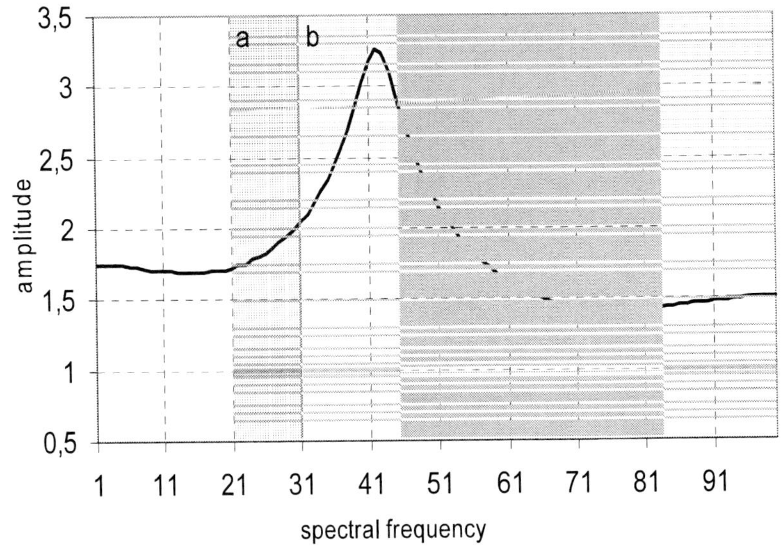

Figure 10. Results of the spectral analysis of dynamic of mitotic index of type B spermatogonia in pinealectomyzed animals (a – circadian rhythms area), (b –ultradian rhythms area).

Thus, pinealectomy has caused growth proliferation activity spermatogonia and their mismatch proliferation with conditions of light exposure that has found reflection in disappearance circadian rhythm MI spermatogonia. At the same time, the amplitude of fluctuations MI in comparison with intact animals has increased.

Pineal gland produces melatonin and group biologically active peptides. Plurality pineal gland peptides, and also their distant actions and functional specificity testify to display peptides properties of hormones. Introduction epithalamin to pinealectomized animals has led to restoration circadian rhythm of mitotic activity of B spermatogonia. Inektsii's type of a peptides complex were spent at change of a light phase of the photoperiod on dark period the Smoothed curve received by a method of the least squares, was characterized by sine wave fluctuations (figure 11), similar to dynamics MI intact animals (figure 7).

In the first day of experience active phase MI of spermatogonia has been fixed with 9h up to 18 h, in the second with 6 up to 12 h, that is the active phase borrowed a light time interval in every day experiment. It acrophase in the first and second day fell to 9 and 12 hours accordingly. During too time the amplitude of fluctuations MI of spermatogonia has remained, as well as at pinealectomy, at high enough level, making 71 ‰. Excess MI of spermatogonia in the afternoon concerning night (table 3) was restored. Distinctions became more expressed, than at intact animals. By means of the spectral analysis are revealed circadian fluctuations of dynamics MI of spermatogonia with the period 24 h (figure 12).

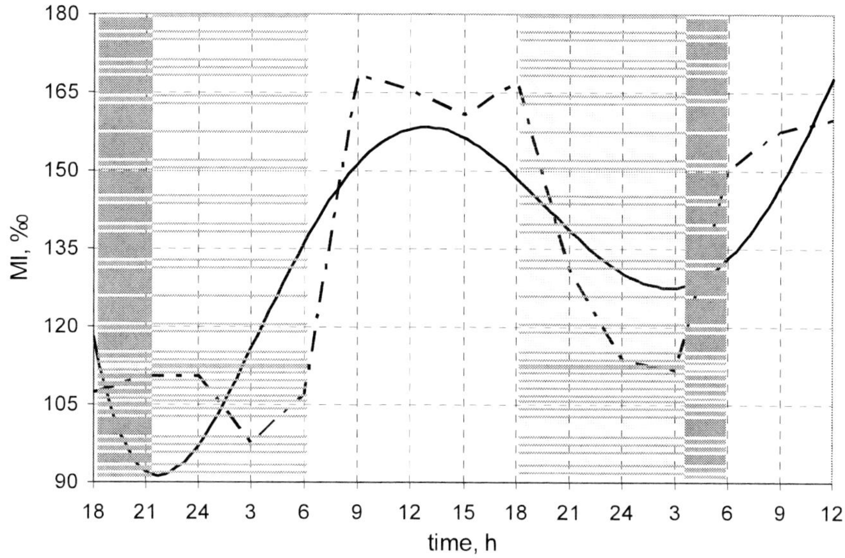

Figure 11. Dynamic of mitotic index of type B spermatogonia in pinealectomyzed animals after injecting epithalamin.

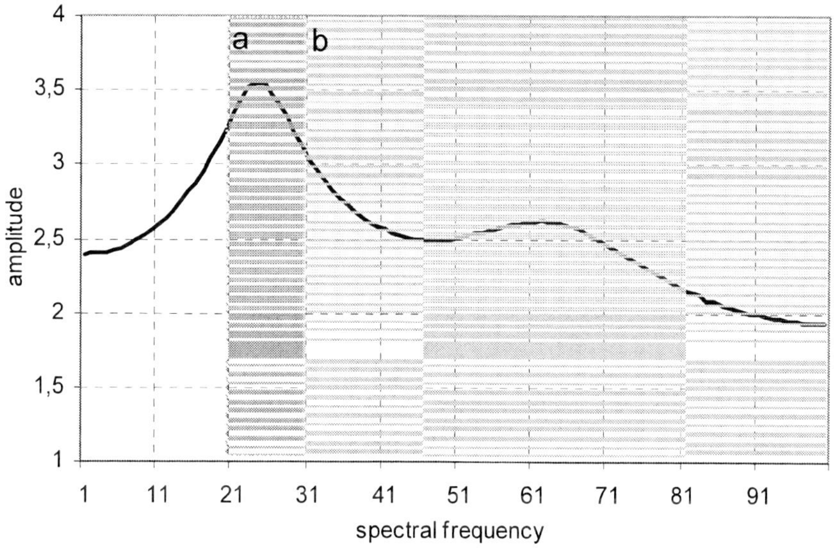

Figure 12. Results of the spectral analysis of dynamic of mitotic index of type B spermatogonia in pinealectomyzed animals after injecting epithalamin. (a – circadian rhythms area), (b –ultradian rhythms area).

Higher daily average value mitotic index of spermatogonia in pinealectomized animals received epithalamin in comparison with intact animals (134,55 ± 6,98‰ and 120,28 ± 3,26‰ accordingly, table 3), apparently, is caused by insufficient concentration of peptides in a blood-groove, owing to unitary introduction within day epithalamin, and also insufficiency of regulation proliferation only pineal gland oligopeptides. Antigonad effect nonmelaton extract of pineal gland is shown in works R.J. Reiter et al. (1981), W.B. Quay et al. (1982), Panzer A., Viljoen M., (1997.) data obtained by us will be coordinated with results of the specified researches and define participation of pineal gland peptides in formation of circadian rhythm proliferation activity of spermatogonia.

Thus, proliferation of type B spermatogonia it would be characterized circadian by rhythm in which formation pineal gland takes part. It proves to be true disappearance circadian rhythm mitotic activity spermatogonia after pinealectomy and its restoration in pinealectomyzed animals after introduction oligopeptides of pineal gland.

Pineal gland produces greater group biologically active peptides [59]. It allows to assume, unlike melatonin, them organic and tissues specificity. It is established, that proliferated spermatogonia type A3 - A4 have receptors to FSH and are its targets [64]. Besides insufficiency, luteotrophin and testosterone would cause a delay of transformation type B spermatogonia in spermatocytes [48]. Thus, pineal gland peptides, possibly, influence on proliferation of spermatogonia by means of inhibition of secretion gonadotropins and testosterone.

Duplication of spermatogonia - the initial stage of process spermatogenesis. Periodicity of its current shows, that spermatogenesis except for seasonal changes is characterized and circadian by rhythm, which, apparently, is shown in change of speed of current of separate stages spermatogenesis during day. Influence of pineal gland on daily dynamics of the maintenance in blood gonadotroph hormones and testosterone [33], separate stages adjusting current spermatogenesis, gives the basis to believe participation of pineal gland in regulation of daily dynamics spermatogenesis as a whole.

Spermatogenesis process includes 4 periods, each of which is characterized by the certain cellular types. Seed of canaliculus the periods spermatogenesis which current defines natural change of cellular combinations are peculiar to any site all. Depending on type of the cells forming combinations, at white rats 14 stages have been allocated. In 1952 year, C.P. Leblond et Y. Clermont has entered into a science representation about spermatogenesis cycle consisting of stages, consistently replacing each other. On a course seed of canaliculus, stages of a cycle settle down consistently one after another, forming spermatogenetic wave. During convoluted seminiferous tubule is up to 12 spermatogenetic waves that has led to formation of representations about wavy character of the spatial organization spermatogonia to a tissue. In everyone convoluted seminiferous tubule simultaneously various stages spermatogenic cycle and the different periods spermatogenesis precede, that provides a continuity of production of gametes in time and a constant opportunity of realization of reproductive function.

Representation about spermatogenic cycle and spermatogenic to wave has made spermatogenesis combination of various rhythms. Thus, on the one hand, spermatogenesis represents change of the periods of duplication, growth, maturing and formation, and on the other hand, all these processes should be included in cycle of spermatogenesis and wave of spermatogenesis. If the cycle and a wave are a part spermatogenesis their current should be subject regulatory to influences. In all animals, spermatogenesis process depends on the photoperiod: in monoestrous animals, this dependence is more expressed, than in polyestrous animals. Thus, current spermatogenesis, and cycle of spermatogenesis and waves in polyestrous animals should have the daily dynamics depending on duration of the photoperiod.

Frequency of occurrence of stages cycle of spermatogenesis is directly proportional to its duration. Fourteen stages of spermatogenesis cycle intact can be divided rats on frequency of occurrence into three groups. To the first group concern I, VII and VIII stages with frequency from 100 up to 190 ‰ and duration from 35 up to 55 h. The second group includes II, III, IV, V, XII, XIII and XIV stages with frequency of 48-75 ‰ and duration 14-20 h. It is possible to carry to the third group VI, IX, X and XI stages with frequency of 30-40 ‰ and duration 9-14 h. The stages reflecting the period of formation have the greatest duration. Stages of the second group characterize a differentiation spermatogones and the period of maturing. The final group of stages would reflect division type B spermatogonia and leptotene. Pinealectomy and introduction pinealectomyzed animals pineal gland's oligopeptides have not affected daily average values of frequency of occurrence of stages spermatogenesis cycle.

In the work lead by us as a method of the spectral analysis are revealed circadian rhythms of daily dynamics IV, VI, XIII and XIV stages spermatogenesis cycle (figures 13-16). Schedules of the spectral analysis IV, VI and XIII stages find out similarity to those for XIV stage. Dynamics of frequency occurrences II, III, V, IX, X, XI, XII stages had no circadian periodicity and was characterized only ultradian by a rhythm with the period nearby 12 h for V, IX, XI stages and nearby 18 h for X, XII stages. The period approximately 14 h was registered by duration at II stage and 8 h at III stage. Daily dynamics I, VII and VIII stages had no periodic changes.

Average values of frequency of occurrence of stages spermatogenesis cycle during dark and light time of day had authentic distinctions on IV, VI, VII, VIII, XII and XIV stages of a cycle (table 4).

Table 4. Frequency rate of stages spermatogenesis cycle in intact animals (M±m) ‰

Stage cycle	Mesor	Light time	Dark time	Duration in hours
I	105,14±2,14	107,83±6,09	104±2,46	30,28±0,62
II	48,9±1,06	49,78±0,76	48,82±2,03	14,08±0,31
III	74,52±2,13	80,38±3,36	73,03±2,83	21,46±0,61
IV	59,28±1,92	66,03±2,26*	56,31±3,3*	17,07±0,55
V	57,46±1,71	59,27±2,7	59,21±3,5	16,55±0,49
VI	32,77±1,45	34,56±2,7*	31,9±1,46*	9,44±0,42
VII	187,96±3,58	198,04±5,19*	181,15±4,13*	54,13±1,03
VIII	132,38±4,42	118,81±4,33*	136,5±6,09*	38,13±1,27
IX	40,74±1,89	37,4±2,56	40,73±3,36	11,73±0,54
X	30,27±1,13	31,37±1,67	29,4±2,36	8,72±0,33
XI	36,74±1,56	36,13±1,96	39,26±2,43	10,58±0,45
XII	71,26±2,32	65,2±2,33*	75,62±3,36*	20,52±0,67
XIII	65,02±1,92	67,9±4,4	64,6±2,4	18,73±0,55
XIV	55,29±2,43	47,45±3,53*	62,04±2,93	15,92±0,7

Note: * - difference is reliable under comparison value in light and dark time, P<0,05.

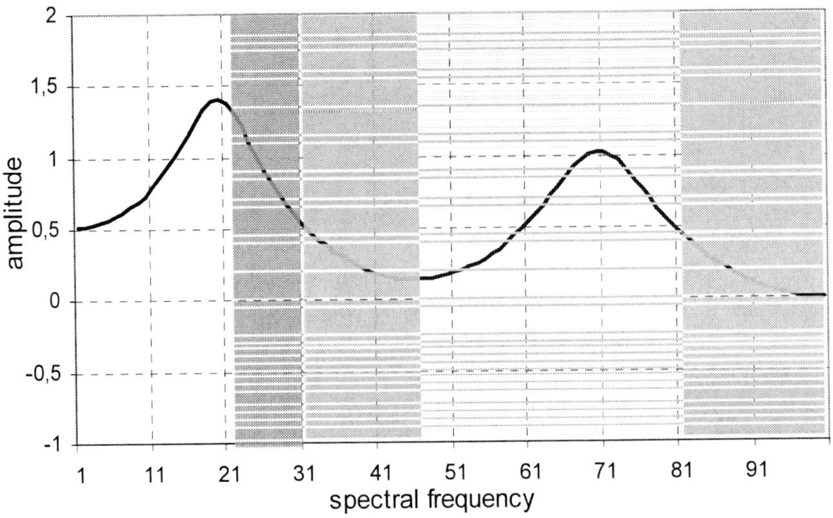

Figure 13. Results of the spectral analysis of popularity rate IV stages of spermatogenesis cycle in intact animals (a – circadian rhythms area), (b –ultradian rhythms area).

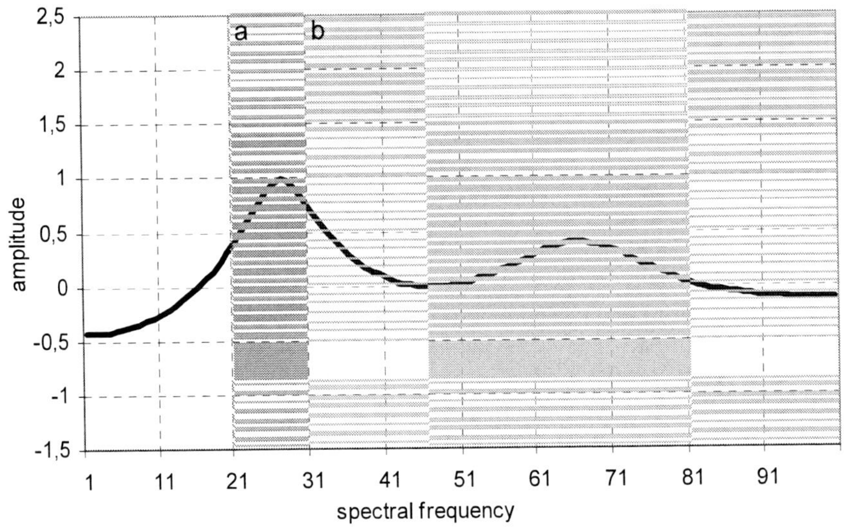

Figure 14. Results of the spectral analysis of popularity rate VI stages of spermatogenesis cycle in intact animals (a – circadian rhythms area), (b –ultradian rhythms area).

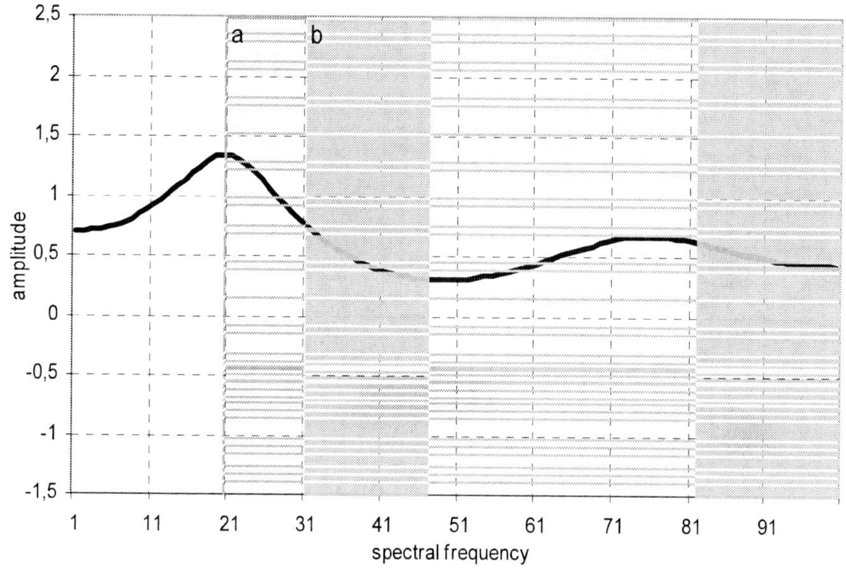

Figure 15. Results of the spectral analysis of popularity rate XIII stages of spermatogenesis cycle in intact animals (a – circadian rhythms area), (b –ultradian rhythms area).

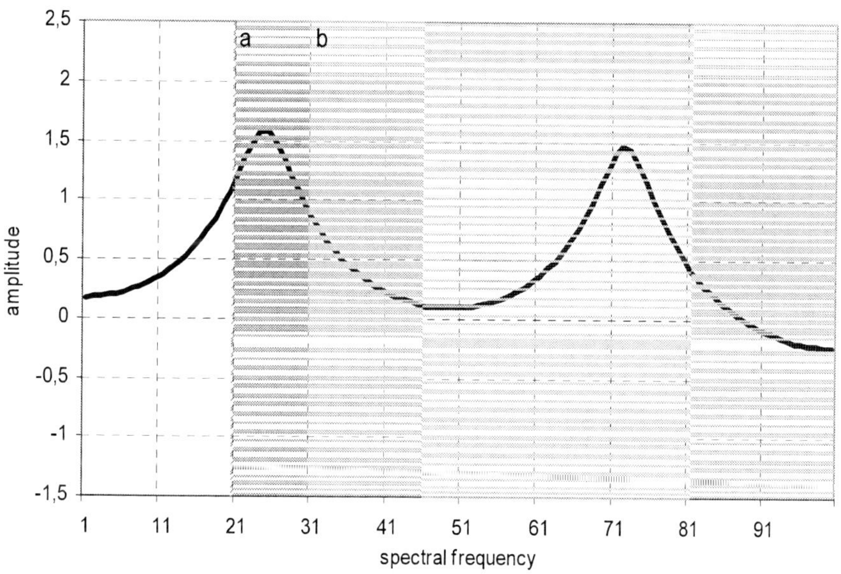

Figure 16. Results of the spectral analysis of popularity rate XIV stages of spermatogenesis cycle in intact animals (a – circadian rhythms area), (b –ultradian rhythms area).

However, the smoothed curves only on IV, VI and XIV stages spermatogenesis cycle were characterized by a correct sinusoid with the expressed amplitude (figures 17-19). In all other cases, the smoothed curves under the form came nearer to a straight line, or were characterized by a sinusoid with insignificant amplitude. It is necessary to notice, that mathematical methods of revealing of periodic fluctuations have greater accuracy if duration of supervision makes about 10 periods. At reduction of the periods of supervision reliability of the analysis considerably decreases.

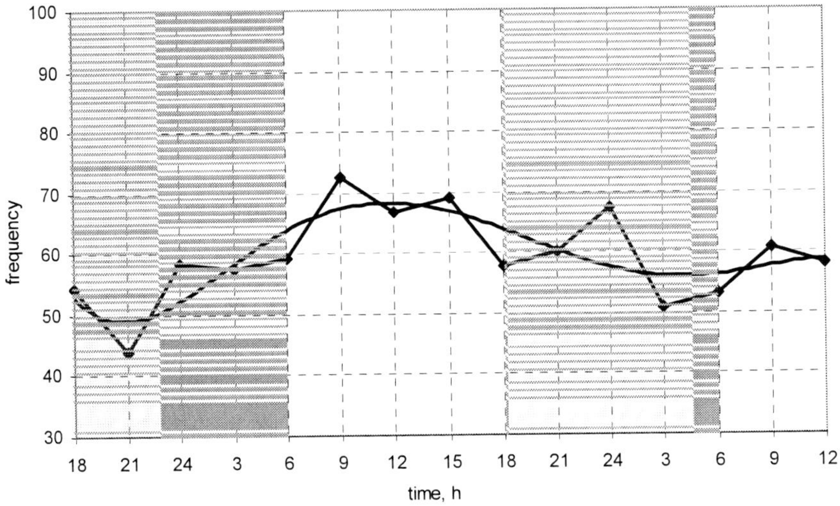

Figure 17. Dynamic of popularity rate IV stages of spermatogenesis cycle in intact animals (a – circadian rhythms area), (b –ultradian rhythms area).

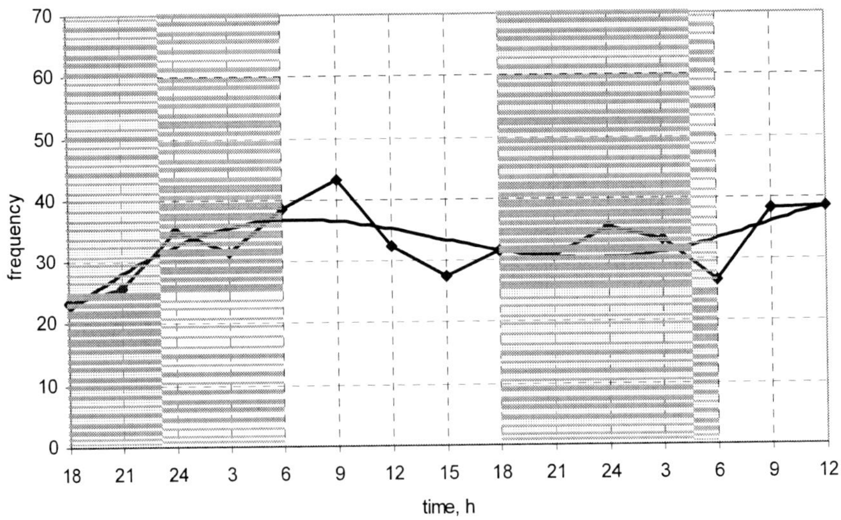

Figure 18. Dynamic of popularity rate VI stages of spermatogenesis cycle in intact animals (a – circadian rhythms area), (b –ultradian rhythms area).

At the same time, in experimental chronobiological researches of supervision during the long period - rare exception as they demand a plenty of experimental animals and are labor consuming. The combination of various methods of the analysis (the spectral analysis, the smoothed curve, comparison of processes during light and dark time of day) allows concretize a conclusion. Set of these criteria specifies authentic presence circadian rhythm only at three stages from fourteen - IV, VI and XIV (figures 20 - 22).

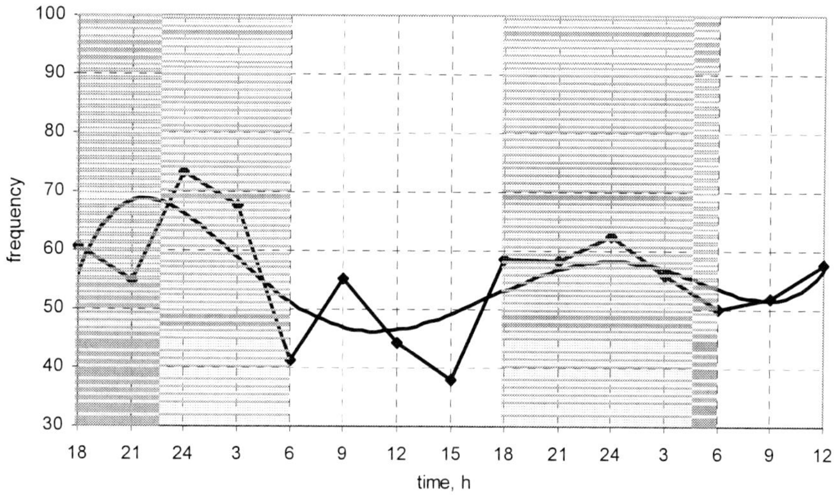

Figure 19. Dynamic of popularity rate XIV stages of spermatogenesis cycle in intact animals (a – circadian rhythms area), (b –ultradian rhythms area).

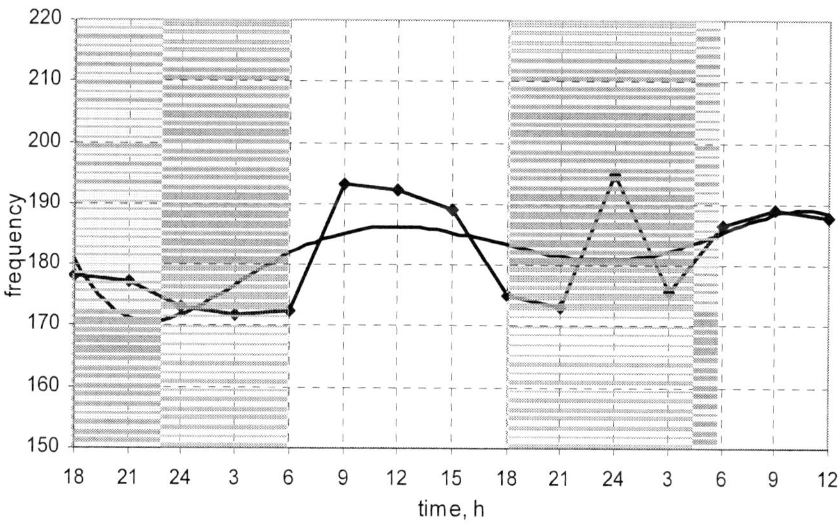

Figure 20. Dynamic of popularity rate differentiation spermatogonia of intermediate type in intact animals (a – circadian rhythms area), (b –ultradian rhythms area).

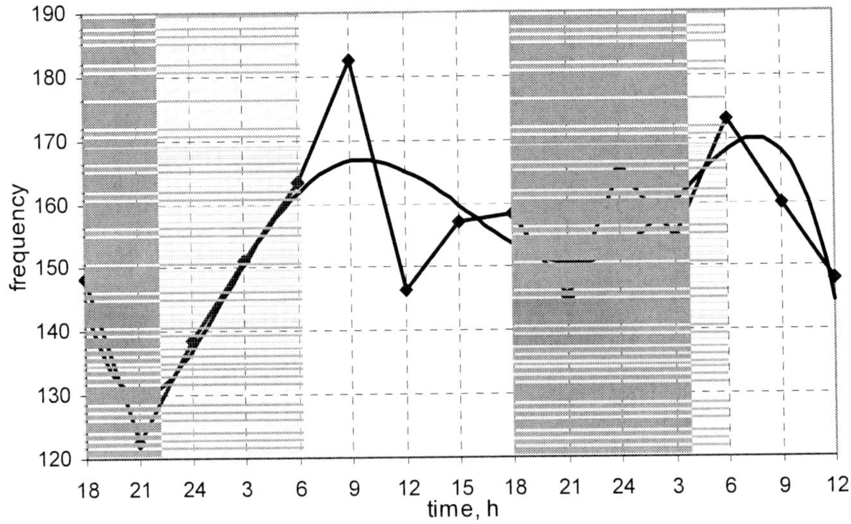

Figure 21. Dynamic of popularity rate differentiation type B spermatogonia in intact animals (a – circadian rhythms area), (b –ultradian rhythms area).

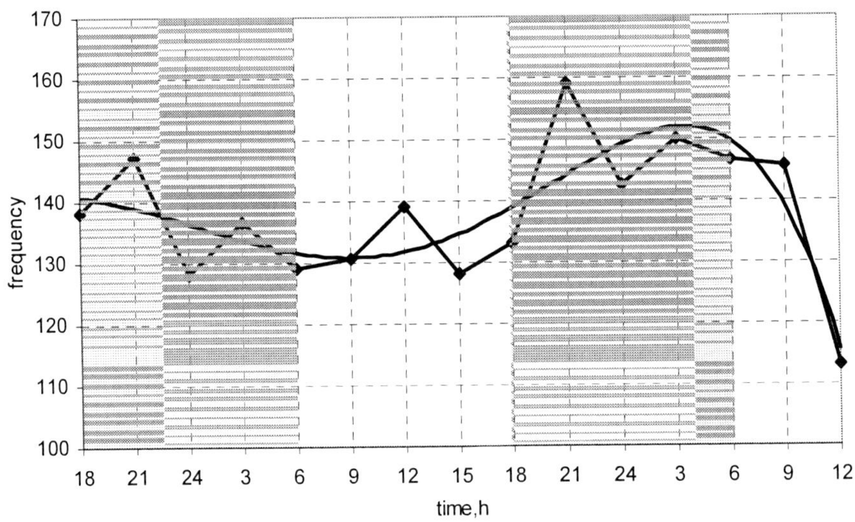

Figure 21. Dynamic of popularity rate the stages including zygotene reducing divisions of merosis in intact animals (a – circadian rhythms area), (b –ultradian rhythms area).

Cellular types, characteristic for the certain period of spermatogenesis, meet at once at several next stages spermatogenesis cycle. If to unite frequency of occurrence of the stages, reflecting separate stages spermatogenesis such group finds out circadian a rhythm. This rhythm came to light at association of the stages reflecting a differentiation spermatogonia of intermediate type (II, III, IV stages), a differentiation type B spermatogonia (IV, V, VI stages), and the stages including zygotene reducing divisions of merosis (XII, XIII stages). Besides, if any spermatogenesis stage begins and ends within the limits of one stage of a cycle current of this stage finds out circadian rhythm. It is IV stage - duplication of spermatogonia intermediate type, VI stage - duplication of type B spermatogonia, XIV stage - reduction and equational division merosis (table 5, figures 20-22).

Current spermatogenesis is adjusted by gonadotropic hormone of hypophysis and testosterone, which level in blood changes not only seasonally, but characterized and circadian rhythm [33; 35]. Presence of daily periodicity regulatory mechanisms should be accompanied by corresponding periodicity of adjustable spermatogenesis process. However, in hypophysectomyzed rats frequency of occurrence on cuts of various combinations of differentiated sexual cells, which is a parameter of relative duration of stages of a cycle, does not differ from those at control intact rats [21]. Hypophysectomy does not render influence on duration spermatogenesis cycle, certain by means of 3H- thymidine [42]. Introduction testosterone, chorionic gonadotropic hormone, FSH and luteotrophin to hypophysectomyzed rats also does not affect speed of passage spermatogenic cells of stages of a [25; 42].

The results received by us testifying to absence of rhythmic fluctuations of duration of current of stages, not described the finished process, and also the data stated above showing absence regulatory of influences on spermatogenesis cycle and a wave, allow to consider, that cellular combinations and consequently, spermatogenesis cycle and a wave, arise because on any site seed of canalicus all the periods of spermatogenesis which current has natural periodic dynamics simultaneously proceed. Besides spermatogenesis cycle and a wave are found out not in all kinds of animals. Supervision of some authors [75] testify to absence of the strict order in an arrangement of sexual cells in testicle the person. Researches of a plenty of serial cuts have allowed Y. Clermont (1963) to find out the tendency various generation of spermatogenic cells to form groups, which have been described as six cellular combinations, to corresponding six stages spermatogenesis cycle. But, besides typical cellular combinations, in seed канальцах the person often there are heterogeneous combinations when cellular elements of two stages of a cycle mix up, or incomplete combinations in which those or other categories of cells appear passed [75]. Spermatogenesis wave also is not found out in the person [45]. In view of these data, absence circadian rhythms of daily dynamics of stages spermatogenesis cycle allows to draw a conclusion, those cellular combinations of stages are derivative result of processes spermatogenesis, and spermatogenesis cycle is not the form of the spermatogenesis organization.

At various animals, the unequal quantity of stages spermatogenesis cycle is allocated. Therefore, 12 stages in spermatogenesis cycle of the mouse, 11 stages in spermatogenesis cycle of a bat, 10 stages in a cycle of a reindeer, 8 stages in a cycle of the rabbit, 7 stages in a cycle of a camel, 6 stages in spermatogenesis cycle of the person have been certain. At different kinds of animals the beginning and the termination of spermatogenesis is necessary on various stages of a cycle. As a result, same number of a stage at animal different kinds reflects various stages of spermatogenesis, that does impossible comparison of the data received at various experimental animals, without knowledge of the detailed scheme of division of spermatogenesis at a stage. It complicates an estimation of the received results and demands from the researcher of decoding of data of the literature.

Table 5. The characteristic of dynamics frequency occurrence of some stages spermatogenesis cycle in intact and experimental animals

Parametr		Control			Pinealectomy			Epithalamin		
		IV	VI	XIV	IV	VI	XIV	IV	VI	XIV
Period		31; 9	23	24; 8	-	-	-	-	24	21; 8
Amplitude		28,6	20	35,3	37,5	20	31,6	19	20	32
Active phase	Day I	6-18	3-12	21-4	Rhythm is absent	Rhythm is absent	Rhythm is absent	Rhythm is absent	5-14	23-4
	Day II	-	7-12	18-3					9-12	1-7
Acrophase	Day I	9	9	24					6	3
	Day II	-	9	24					12	6

Apparently, spermatogenesis cycle and spermatogenesis wave it only one of displays of the spatial organization spermatogenesis tissue which arises owing to simultaneous course in one plane seed canalicus the periods of duplication, growth, maturing and the formation, differing from each other on duration and having specific distinctions. During too time, process of spermatogenesis it is characterized circadian rhythm.

In formation circadian, the important role belongs to rhythm of various functions of an organism of pineal gland. Disappearance specifies participation of pineal gland in formation circadian spermatogenesis after pinealectomy circadian rhythms of frequency of occurrence of stages and groups of the stages reflecting processes spermatogenesis to rhythm of current, and also restoration of daily rhythm of the specified groups after introduction biologically active pineal gland peptides (table 5). Noted testifies to direct participation pineal gland peptides in formation of circadian structures of spermatogenesis. It proves to be true data of the literature about biorhythmogenic effect pineal gland oligopeptides, in V.D.Slepushkina's opinion and co-author. (1983) at functionally active condition of a hypophysis, epithalamin operates as inhibitor. In old male rats, epithalamin reestablish regular estrale cycles and sensitivity of hypothalamus to estrogen [44]. R.I. Kovalenko (1993) has shown, that introduction of low-molecular fraction pineal gland peptides leads to decrease in a hypophysis and plasma of blood prolactin, FSH and luteotrophin. That fact, that restoration circadian rhythms of frequency of occurrence has occurred not at all stages and groups of the stages reflecting processes spermatogenesis on which were found out circadian rhythms in intact animals, apparently, speaks unitary introduction within day epithalamin and absence in a blood-groove pinealectomyzed animals during 40 days melatonin. A number of experimental works specifies on reciprocals mutual relations of melatonin and pineal gland peptides. Studying of character of secretion melatonin at macaques of Rhesus factors has shown, that introduction in the evening epithalon - is artificial synthesized oligopeptide, consisting of four amino acids, stimulating an effect has on synthesis melatonin [38]. On the other hand, melatonin can act in a role of a regulator of synthesis and secretion of some pineal gland peptides [43]. Melatonin and others indoles can change permeability of cellular membranes pinealocyte for synthesized in them oligopeptides [72]. Apparently, reciprocal mutual relations melatonin and pineal peptides peptides are carried out and at a level of tissue-targets, acknowledgement of that is plurality pineal gland peptides and presence at them distant actions. In intact animals action melatonin and oligopeptides is carried out in common, that leads to formation complete circadian structures of processes and functions of an organism.

Conclusions

Proliferation epithelium in jejunum's crypts has circadian rhythm in which formation the great value has pineal gland. It specifies disappearance of a rhythm after pinealectomy and its restoration after introduction pinealectomyzed animals epithalamin (a complex pineal gland oligopeptides).

Current spermatogenesis in white rats is characterized circadian rhythm in which formation pineal gland takes part. It would prove to be true presence circadian rhythm of proliferation spermatogonia type B, duration of the period of duplication and a differentiation spermatogonia intermediate type and current merosis. Pinealectomy introduction epithalamin to their restoration leads to disappearance circadian rhythms of spermatogenesis, and.

Cellular combinations of stages spermatogenesis cycle grow out current of stages spermatogenesis, instead of the form of its organization.

References

[1] Anisimov V.N., Bondarenko L.A., Khavinson V.Kh. Effect of pineal peptide preparation (epithalamin) on life span and pineal and serum melatonin levels in old rats *//Ann.N.Y.Acad.Sci.* – 1992. – Vol.673 – P.53-57.

[2] Anisimov V.N., Khavinson V.K., Morozov V.G. Carcinogenesis and aging. IV. Effect of low-molecular-weight factors of thymus, pineal gland and anterior hypothalamus on immunity, tumor incidence and life span of C3H/Sn mice *//Mech. Ageing Dev.* – 1982. – Vol.19. – P.245-258.

[3] Anisimov V.N., Khavinson V.K., Morozov V.G. Twenty years of study on effect of pineal peptide preparation: epithalamin in experimental gerontology and oncology *//Ann. N. Y. Acad. Sci.* – 1994. – Vol.719. – P.483-493.

[4] Antohin A.I., Romanov J.A., Bogoeva M.V., Torshina I.A., Matsak N.J., Semenova M.V., Barishev V.P. *Chalones mechanism of formation of a daily rhythm of mitosises / Chronobiology and chronomedicine: thesis report* . On II symposium of the USSR. - Tyumen. - 1982. - P.78.

[5] Armstrong S.M. Melatonin and circadian control in mammals // *Experientia.* – 1989. – Vol.45, №10. – P.932-938.

[6] Arushanjan E.B., Elbekjan K.S. Value of pineal gland attitudes with sympathico-adrenal system for depressogene actions of reserpine // *The Bulletin of experimental biology.* - 1996. - 6 - P.651-653.

[7] Aubert C., Janiaud P., Lecalvez J. Effect of pinealectomy and melatonin on mammary tumor growth in Sprague-Dawley rats under different conditions of lighting *//J. Neural Transmiss.* - 1980. - Vol.47. - P.121-130.

[8] Axelrod J., Wurtman R.J., Snyder S.H. Control of Hydroxyindole-O-Methyltransferase activity in the rat pineal gland by environmental hgthing *//J.Biol.Chem.* - 1965. - V.240. - P.949-955.

[9] Bartsch H., Bartsch C. Effekt of melatonin on experimental tumors under different photoperiods and time of administration *//J. Neural Transmiss.* - 1981. - Vol.52, № 4. – P.269-279.

[10] Blask D.E. The pineal gland: an oncostatic gland? */Comprehensive Endocriology: The Pineal Gland.* - Reiter R.J., ed. New York, Ravon Press. – 1984. - 284p.

[11] Blask D.E., Hill S.M. Melatonin and cancer: basic and clinical aspects. *//Melatonin. Clinical Perspectives.* Miles A., Philbrick D.R.S. (Eds.) Oxford: Oxsford Univ. Press. – 1988. – P.128-173.

[12] Brodsky V.J. Perihourses rhythms in a cellular population. A problem of synchronization // *The Bulletin of experimental biology.* - 1997. - T.124, 1. - P.604-609.

[13] Bullough W.S., Laurence E.B. The diurnal cycle in epidermal mitotic duration and its relation to chalone and adrenalin //*Expti.Cell Res* - 1966. – Vol.43. - P.343-350.
[14] Buswell R.S. The pineal and neoplasia //*Lancet.* - 1975 - Vol.1, № 7897. – P.34-35.
[15] Calb M., Goldstein R., Pavel S. Diurnal rhythm of vasotocin in the pineal of the male rat //*Acta endocr., Copenh.* – 1977. - Vol.84. - P.523-526.
[16] Callaghan B.D. The effect of pinealectomy on the crypt of defunctioned rat colon //*J. Pineal Res.* – 1997. - Vol.23, №3. – P.117-122.
[17] Cardinalli D.P., Cutrera R., Castrillon P., Esquifino A.I. Diurnal rhythms in ornithine decarboxylase activity and norepinephrine and acetilcholine synthesis and acetylcholine synthesis ofrat submaxillary lymph nodes: effect of pinealectomy, superior cervical ganglionectomy and melatonin replacement //*Neuroimmunomodulation.* - 1996. - Vol.3, №2-3. - P.102-111.
[18] Cardinalli D.P., Cutrera R., Castrillon P., Esquifino A.I. Effect of pinealectomy, superior cervical ganglionectomy, or melatonin treatment on 24-hour rhythms in ornithine decarboxylase and tyrosine hydroxylase activities of rat spleen //*J. Pineal Res.* – 1997. - Vol.22, №4. – P.210-220.
[19] Chazov E.I., Isachenkov V.A. Pineal gland: a place and a role in system neuroendocrin regulation - M.: *the Science,* - 1974. - 238 p.
[20] Clermont Y. The cycle of seminiferous epithelium in man // *Amer. J. Anat.* – 1963. – Vol.112. – P.35-52.
[21] Clermont Y., Morgentaler H. Quantitative study of spermatogenesis in the hypophysectomized rat // *Endocrinilogy.* – 1955. – Vol.57. – P.369-382.
[22] D.A. Lincoln. Pineal gland. *Hormonal regulation of duplication at mammals.* M.: the World. - 1987. - P.71-99.
[23] Damian E., Janas O., Badescu J. Decrease of plasma and adrenal corticosterone after pineal polypeptide administration to rats // *Rev. Roum. Med.-Endocrinol.* – 1976. – Vol.14, №1. – P.27-30.
[24] Das Gupta T.K., Terz J. Influence of pineal gland on the growth and spread of melanoma in the hamster //*Cancer Res.* - 1967. - Vol.27, №7. - P.1306-1311.
[25] Desclin J., Ortavant R. Influence des hormones gonadotropes sur la duree desprocessus spermatogenetiques chez le rat // *Ann. biol. anim., biochim., biophys.* – 1963. – Vol.3. – P.329-342.
[26] Dilman V.M., Anisimov V.N., Ostroumova M.N. Study on the antitumor effect ofpolypeptide pineal extract //*Oncology.* - 1979.- Vol.30. - P.274-280.
[27] Ding C.H., Wei W., Xu S.Y. Effects of pineal body and melatonm on lymphocyte proliferation and dinoprostone production in rat spleen //*Zhongguo Yao Li Xue Bao.* - 1995. – Vol.16, № 1. - P.54-57.
[28] Dobrohotov B.H., Bystrenina N.G., Podderyugina G.I. Influence colchamine on rates of cellular duplication in fabrics of animals // *Bull. of exp. biology.* - 1983. - 4. - P.84-85.
[29] Dolinskij G.A., Fedchenko S.N., Mammoth S.G. Daily's rhythm and intensity of proliferation cells epithelium a mucous membrane of a stomach at introduction pantothenate calcium // *Bull. of exp. biology.*-1993. - 5. - P.521-524.

[30] El-Domeiri A.A., Gas Gupta T.K. The influence of pineal ablation and administration of melatonin on growth and spread of hamster melanoma // *J.Surh.Oncol*.- 1976. – Vol.8, №3. – P.197-205.

[31] Elgjo K., Clausen O.P. Proliferation - dependent effect of scm extracts (chalone) on mouse epidermal cell flux at the G_1-S, S-G_2 and G_2-M transitions // *Virchows Arch*. - 1983. - Vol.42, №2. – P.143-151.

[32] Engel P. Űber den Einfluss von Hypophysenvorderlappen-Hormonen und Epi-Hormonen auf das Washstum von Impftumoren. Z. Krebsforsch., - 1934. - №41. – P.281-291.

[33] Esquifino A.I., Arce A., Villanua M.A., Cardinali D.P. Development of 24-hour rhythms in serum prolactin and luteinizing hormone levels in rats neonatally administered melatonin // *Chronobiol. Int*. – 1998. – Vol.15, №1. – P.21-28.

[34] Farriol M., Venereo Y., Orta X., Castellanos J.M., Sigovia-Silvestre T. In vitro effects of melatonin on cell proliferation in a colon adenocarcinoma line //*J. Appl. Toxicol*. – 2000. – Vol.20, №1. – P.21-24.

[35] Garcia-Bonacho M., Esquifino A.I., Castrillon P.O., Toso C.R., Cardinali D.P. Age-dependent effect of Freund's adjuvant on 24-hour rhythms in plasma prolactin, growth hormone, thyrotropin, insulin, follicle-stimulating hormone, luteinizing hormone and testosterone in rats // *Life Sci*. – 2000. – Vol.66, №20. – P.1969-1977.

[36] Goldman B.D. Parameters of the circadian rhythm of pineal melatonin secretion affecting reproductive responses in Siberian hamsters /B.D. Goldman //*Steroids*. - 1991. - Vol.56, № 5. - P.218-225.

[37] Golikov P.P., Bobkova A.S. Influence of pinealectomy on a daily rhythm glucocorticoid, androgenic, estrogenic and gestagenic functions of endocrine glands // *Problems of endocrinology*. - 1976. - №5. - P.77-81.

[38] Goncharova N.D., Havinson V.H., Lapin B.A. Adjust influence epitalon on production of melatonin and cortisol at old monkeys // *the Bulletin of experimental biology*. - 2001. - №4. - P.466-468.

[39] Hacker-Klom U.B. Diurnal changes in murine spermatogenesis // *Z. Naturforsch*. C. – 1994. – V.49, №7-8. – P.522 – 525.

[40] Halberg G.F., Reinberg A. Rythmes circadiens et rythmes de basses frequences en physiologie humaine // *J. Physiol*. – 1967. - №1. – P.117-200.

[41] Haldar C., Haussler D., Gupta D. Effect of the pineal gland on circadian rhythmicity of colony forming units for granulocytes and macrophages (CFU-GM) from rat bone marrow cell cultures //*J. Pineal Res*. – 1992. – Vol.12, №2. – P.79-83.

[42] Harvey S.C., Clermont Y. The duration of the cycle of the seminiferous epithelium of normal and hypophysectomized hormone treated albino rat // *Anat. Rec*. – 1962. – Vol.142. – P.239-248.

[43] Hastings M.H., Herbert J. Neurotoxic lesions of the paraventriculo-spinal projection block the nocturnal rise in pineal melatonin synthesis in the Syrian hamster // *Neurosci. Lett*. – 1986. – Vol.69. – P.1-6.

[44] Havinson V.H., Morozov V.G. *Preparations of pineal gland and thymus in gerentology*-St.-Peterburg. - 1992 – 50p.

[45] Heller C.G. Y. Clermont Kinetics of the qerminal epithelium in man // *Recent Proqr. Hormone res.*-1964. - Vol.20.- P.545-571.

[46] Ilnerova H. Resetting of the mammalian circadian clock through lervering of the amplitude: rat pineal N-acetyltransferase rhythm as a model (editorial) //*Physiol. Res.* - 1992. - Vol.41, №5. - P.33-44.

[47] Jacobelli G. Gentilini L. A new surgical treatment (pineal transplantation) for disseminated breast cancer //*Proc. 18 - th World Congr. Int.Coll. Surg.* Rome, 1972. Amsterdam, New York, - 1973. - P.758-759.

[48] Jeyakumar M., Suresh R., Krishnamurthy H.N., Moudgal N.R. Changes in testicular function following specific deprivation of LH in the adult male rabbit // *J. Endocrinol.* – 1995. – Vol.147, №1. – P.111-120.

[49] Kachi T., QuayW.B., Banerji T.K., Imagawa T. Effect of pinealectomy on the mitotic fctivity of adrenomedullary chromaffin cells in relation to time of day //*J. Pineal Res.* – 1990. – Vol.8, №1. – P.21-34.

[50] Kallenbach H., Malz W. Űber den Einfluss der Epiphyse auf das Wachstum transplantabler Impftumoren //*Endokrinologie* – 1957. - №5/6. – P.293-295.

[51] Karasek M., Zielenowski J., Nowakowska-Jankiewicz B. Failure of melatonin to inhibit the mitotic activity of regenerating adrenal cortex in rats //*J. Pineal Res.* – 1987. – Vol.4, №2. – P.197-200.

[52] Khavinson V.Kh., Golubev A.G. Aging of the pineal gland //*Adv. Gerontol.* – 2002. – Vol.9. – P.67-72.

[53] Kinzhibulatova R.ZH. Daily rhythm of proliferation processes in sebaceous glands // *Archives of anatomy, histology, embryology.* - 1975. - 3. - P.86-91.

[54] Kitay J.I., Altschule M.D. The Pineal Gland. A Review of the Physiologie Literature //*Harvard Unib.* Press, Cambridge, - 1954. - P.280.

[55] Komarov F.I., Rapoport S.I., Malinovskaja N.K., Voznesenskaja L.A., Sharov A.A., Vettenberg L. Production melatonin in patients stomach ulcer of a duodenal gut in various stages of current of disease // *Clinical medicine.* - 1998.3. - P.15-18.

[56] Korf H.W. The pineal organ as a component of the biological clock: Phylogenetics and ontogenetic considerations //*Ann.N.Y.Acad.Sci.* - 1994. - Vol.719. - P.13-42.

[57] Kosichenko L.P. Daily periodicity mitosises and a role of a light mode in its definition: *Instrument of experimental patholgic and therapies AMS the USSR.* - 1961. – 12p.

[58] Kovalenko R.I. Neuroendocrine "bodies" of a brain vertebrates. 1. Pineal gland // *«Neuroendocrinology»* / Under the direction. A.L.Polenov (the Basis of modern physiology). A part 1. The book the second. The Russian Academy of Science: SPb. 1993. - P.300-324.

[59] Kuznik B.I. Cytomedines and their role in regulation of physiological functions/B.I. Kuznik, V.G. Morozov, V.H. Havinson // *Successes we shall tell lies.*-1995. 3.-P.360-367.

[60] Lapin V. Einfluss der Simultanen Thymectomie und epiphise ctomie auf das Tumorwachstum bei Ratten //*Obst. Arzterztg.* - 1974. - Vol.29, №23. - P.1380-1382.

[61] Lapin V. The pineal and neoplasia //*Lancet.* –1975. – Vol.1, №1902. – P.341.

[62] Leblond C.P., Clermont Y. Spermatogenesis in rat, mouse, hamster and guinea pig as revealed by the periodic acid-Schiff-fuchsin sulfureus acid technique // *Amer. J. Anat.* – 1952. – V.90. – P.167-215.

[63] Mamontov S.G., Arion V.J., Kremli S.M., Breusov JU.N. Cellular proliferation in tissues of mice after thymectomy and introductions T - activin // *The bulletin experimental biology*. - 1986. - 1. - P.102-104.

[64] Meachem S., Von Schonfeldt V., Schlatt S. Spermatogonia: stem cells with a great perspective // *Reprod.* – 2001. – Vol.121, №6. – P.825-834.

[65] Metzger G., Li X.M., Mormont C., Levi F. Rhythmic component of twelve hours //*Pathol. Biol.* - 1996. - №3. - P.179-182.

[66] Mladovskaja T.B., Romanova L.K., Pokrovskaja M.S., Kulikova G.V. Daily of fluctuation mitotic activity of alveolar macrophages- monocytes from bronchoalveolars washouts of mice //*Bulletin of experimental biology*. - 1993. - №10. - P.430-431.

[67] Moskalik K.G. Daily periodicity mitotic activity, synthesis of DNA and duration mitosis in cells of a bone brain of mice // *The Bulletin of experimental biology*. - 1976. - 5. - P.594-596.

[68] Oxenkrug G.F., Anderson G.F., Dragovic L. Circadian rhythms of human pineal melatonin, related indoles and beta adrenoreceptors: Postmortem evaluation / *J. Pineal Res.* - 1990. – Vol.9. – P.1-11.

[69] Panzer A., Lottering M.L., Bianchi P., Glencross D.K., Stark J.H., Seegers J.C. Melatonin has no effect on the growth, morphology or cell cycle of human breast cancer (MCF-7), cervical cancer (HeLa), osteosarcoma (MG-63) or lymphoblastoid (TK6) //*Cancer Lett.* – 1998. – Vol.122. – P.17-23.

[70] Panzer A., Viljoen M. The validity of melatonin os an oncostatic agent // *J. Pineal. Res.* – 1997. – Vol.22,№4. – P.184-202.

[71] Petranka J., Baldwin W., Bierman J., Jayadev S., Barrett J.C., Murphy E. The oncostatic action of melatonin in an ovarian carcinoma cell line //*J.Pine-al Res.* – 1999. – Vol.26, №3. – P.129-136.

[72] Pevet P. Physiologikal role of neuropeptide in the mammalian pineal gland // *Adv. Pineal Res.* – 1991. – Vol.6. – P.275-282.

[73] Pozdnev N.V., Etingof R.N. Biosynthesis of melatonin in pineal gland in early postnatal ontogenesis at healthy and sick of a hereditary degeneration of a retina of rats // *The Bulletin of experimental biology. And honey*. - 1997. - №24. - P.131-135.

[74] Quay W.B., Payer A.F., Parkening T.A., Banerji T.K., Collins T.J. Melatonin's inhibition of pituitary, adrenal, testicular and accessory gland growth in male golden hamsters: pineal dependence and organ differences with shielding and intracranial surgery // *J. Neural Transm.* – 1982. – Vol.53, №1. – P.59-73.

[75] Raycina S.S. Spermatogenesis and structural bases of its regulation. - M.: *the Science*, 1985. – 205p.

[76] Reinstein S., Dotan A., Avivi L., Ashkenazi I. Daily rhythms in male mice meiosis //*Chronobiol. Int.* – 1998. - Vol.15, №1. – P.13-20.

[77] Reiter R.J. *The Brain as Endocrine organ* - Vienna. – 1989. – P.91-149.

[78] Reiter R.J., Richardson B.A., Vaughan M.K., Johnson L.Y. Pineal actions and mechanisms in reproductive physiology // *Jikeikai Med. J.* – 1981. – Vol.28 (Suppl.1). – P.31-46.

[79] Rodin A.F. The growth and spread of Walker 256 carcinoma in pineal ectomized rats //*Cancer Res.* - 1963. - Vol.23, №9. - P.1545-1549.

[80] Romanov J.A., Filippovich S.S., Ostroushko E.T., Stepanova L.I., Kovalenko E.A., Kameneckaya E.A. Research of daily rhythms of a reproduction of cells at rats in conditions hypokinesia / problems of the time organization of alive systems - M.: *the Science.* - 1979. - P.121-134.

[81] Romanov J.A., Irikov O.A., Filippovich S.S. Chronobyological studying of duplication of cells in epithelium of gullet of mice at influence on them lomefloxacyne at various times day // *The Bulletin of experimental biology.* - 1994. - 5. - P.527-530.

[82] Romanov U.A. About the time organization of processes of a reproduction of cells in an organism/*daily rhythms of physiological processes of an organism.* - M.: 2-nd MOLGMI. - 1972. - P.48-51.

[83] Roosen-Range E. Spermatogenesis at animals. M.: *the World,* 1980.

[84] Sanchez de la Pena S. The feedsideward of cephalo-adrenal immune interactions //*Chronobiologia.* - 1993. - Vol.20, №1-2. - P.1-52.

[85] Sewerynek E., Lewinski A. Melatonin inhibits mitotic activity adrenocortical cells in vivo and in organ culture //*J. Pineal Res.* – 1989. –Vol.7, №1. - P.1-12.

[86] Sewerynek E., Lewinski A., Choczaj V., Szkudlinski M., Szymczykiewicz P. Melatonin-induced suppression of the pinealectomy-stimulated rat adrenal cortex mitotic activity //*Cytobios.* - 1991 - Vol.65, №261. - P.115-122.

[87] Shirama K., Furuya T., Takeo Y., Maekawa K. Direct effect on the accessory sexual organs in pinealectomized male rats kept in constant darkness // *Endocrinilogy.* – 1982. – Vol. 95, №1. – P.87-94.

[88] Shiu S.Y., Xi S.C., Xu J.N., Mei L., Pang S.F., Yao K.M., Wong J.T. Inhibition of malignant trophoblastic cell proliferation in vitro and in vivo by melatonin //*Life Sci.* – 2000. – Vol.67, №17. - P.2059-2074.

[89] Silin A.G. Daily a rhythm of reproductive activity of cells and functional morphology adenohypophysis rats (histologic and autoradiographic research) // *Archives of anatomy, histology, embryology.* - 1975. - 4. - P.80-82.

[90] Simoonet H., Thieblot L., Segal V., Melik T. Modifications du tractus genital femelle consecutives a l'epiphysectomie // *J. Physiol.* (Paris). – 1951. – Vol.43. – P.864-866.

[91] Slepushkin V.D., Mordovin V.F., Zoloev G.K. Influence of a preparation from pineal gland epithalamin on gonadotropic function of a hypophysis //. - 1983. - T.29, 6. - P.51-54.

[92] Effect of pinealectomy on prolactin, testosterone and luteinizinq hormone concentration in plasma of bull calves exposed to 8 or 16 hours of liqht per day /E.P. Stawisiewski, N.K. Ames, L.T. Chapin, C.A. Blaze, H.A. Tucer // *J. Anim. Sci.*-1989.- Vol. 66, № 2.-P.464-469.

[93] Tapp E. Rolle of pineal gland in aetiology and treatment of breast cancer //*Lancet.* – 1978. – Vol.2, №8097. – P.1001-1002.

[94] Turek F., Losee S.H. Melatonin-induced testicular growth in golden hamsters maintained on short days // *Biol. Reprod.* – 1978. – Vol.18. – P.299-305.

[95] Zaharov V.B., Berezhnova N.I., Mammoth S.G. Daily's rhythm cellular of proliferation in epithelium language of mice after sialicadenectomy // *The bulletin experimental biology*.-1993. - 7. - P.72-73.

In: Progress in Circadian Rhythm Research
Editor: Anne-Laure Léglise, pp. 123-139
ISBN: 978-1-60021-876-7

Chapter V

Control of Health and Life by a Circannual Rhythm of Hibernation: An Essential Role of Hibernation-Specific Protein Complex

Noriaki Kondo
Mitsubishi Kagaku Institute of Life Sciences
Machida, Tokyo 194-8511, Japan

Abstract

In mammals, hibernators have a unique ability to adapt organisms to an extremely decreased body temperature (Tb) and survive for several months by consuming much less energy. Many years of research it has shown that during hibernation, animals are protected from various lethal diseases and events, including bacterial infections, tumorigenesis, hypothermia and ischemia, suggesting that hibernation is an adaptive state for maintaining health by protecting organisms. If its control mechanisms were understood, hibernation would offer myriad applications in medicine and health sciences. Therefore, discovery of an endogenous factor that controls hibernation has long been dreamed of. Our early studies on cardiac muscles of chipmunks, a rodent hibernator, demonstrated that Ca^{2+} is intracellularly recycled in the hibernating state by circannually modulated Ca^{2+} source for excitation-contraction coupling and suggested that this modulation may prevent intracellular Ca^{2+} overload at a decreased Tb. Based on these findings, we have isolated a novel hibernation-specific protein (HP) complex in the blood, which is produced in the liver and downregulated during hibernation in association with the circannual cardiac modulation. Recently, this complex was identified as a promising candidate hormone that is essential for circannual control of hibernation in the brain. In this review, a significant role of cirannually modulated cardiac Ca^{2+} regulation and HP complex in the brain will be discussed in maintaining health and life, and hibernation may be redefined as a hormonally regulated state independent of Tb changes.

Introduction

One of the most mystifying life phenomena in mammals is hibernation, which seasonally allows organisms to survive potentially lethal low body temperatures (Tb) near 0 °C. In nonhibernators, such a low Tb dysfunctions every metabolic process, and then organisms soon die. However, hibernators can survive a low Tb during the hibernation season by developing a capacity for tolerating it (Lyman et al., 1982; Hochachka, 1986). During hibernation, animals are shown to be protected from lethal diseases caused by various harmful factors and events, such as bacteria (Sharapov, 1983), tumorigenesis (Kemper and Ruben, 1982), ischemia (Frerichs and Hallenbeck, 1998), cellular calcium overload (Kondo, 1986a; Kondo, 1997), irradiation (Musacchia and Barr, 1968; Jaroslow et al., 1969) and muscle disuse (Harlow et al., 2001;), and that helps animals live longer (Zivadinovic and Andjus, 1996; Kondo and Narita, 1998; Kondo, 1999). This strongly suggests that hibernation is a unique adaptive state that seasonally developed for preventing various lethal stresses and diseases for maintaining health during the severe hibernation season. If understood, hibernation would offer myriad applications in medicine and health sciences. Therefore, a control mechanism of hibernation has long been explored. Especially, an endogenous factor that controls hibernation has held interest since the beginning of the twentieth century (Wang, 1988). Recently, we discovered a novel and promising candidate hormone, hibernation-specific protein (termed HP) complex, in chipmunks, a rodent hibernator, that is essential for controlling hibernation in the brain (Kondo et al., 2006). This finding newly defined hibernation as a hormonally regulated state independent of Tb and environmental changes. This paper summarizes our studies on a seasonally regulated tolerance of hearts, one of the essential organs for maintaining life, to a potentially lethal low Tb and views an endogenously generated circannual rhythm and an essential role of HP complex for surviving hibernating state. The conceptual issue of hibernation and the contribution of hibernation and HP complex to health and life will be discussed.

1. Multiple Tolerance of Hibernation

During hibernation, Tb is dramatically decreased to near 0°C. Such a severely decreased Tb causes a profound depression in metabolism and disturbs functions of various cells and organs. At a decreased Tb during hibernation, the heart rate dramatically decreases, e.g. 6-9 beats per minutes in chipmunks (approximately 1/50 or less of the active states), resulting in a greatly reduced blood flow in the heart, brain and other all tissues and organs. However, hibernators survive these hypothermic and ischemic stresses without any damage. In nonhibernators, hypothermia causes the serious perturbations arising from an imbalance between the depressed ATP synthesis and the increased requirement for maintaining homeostasis. This imbalance disturbs maintenance of regulated metabolism and membrane functions in cells, resulting in disorder and eventual death. Thus, hibernators must develop the ability to tolerate a low temperature disturbance of metabolism during hibernation to prevent cold injuries.

Hibernation has been shown to prevent not only hypothermia but also diseases caused by various harmful factors and events. Brain slices isolated from hibernating animals tolerate hypoxia and hypoglycemia, and their tolerance is increased coincident with lowering of temperatures (Frerichs and Hallenbeck, 1998). Animals during hibernation are unaffected by bacterial infections and treatment with carcinogens (Sharapov, 1983; Kemper and Ruben, 1982), and are protected from radiation (Musacchia and Barr, 1968; Jaroslow et al., 1969). As the immune function is known to be impaired at low Tb, dysfunction of immune responses to pathogens, such as a bacterial infection, is a serious problem for maintaining life during hibernation. Therefore, hibernators are considered able to maintain or increase the ability of their immune system during hibernation for protecting the organisms from infectious pathogens. Indeed, the classical pathway of the serum complement system is active during hibernation, and the splenic macrophages from torpid squirrels bind more bacterial lipopolysaccharide than those from normothermic animals (Maniero, 2002; 2005). Modifications of the intestinal immune system are suggested to help preserve gut integrity throughout the winter fast during hibernation (Kurtz and Carey, 2007). Hibernation also prevents disuse muscle atrophy and osteoporosis in bears (Harlow et al., 2001; Donahue et al., 2006). Thus, hibernation is proposed to be a seasonally regulated physiological state that is developed specifically for maintaining health and life in organisms. Since this regulated state may protect organisms from various damage and disease induced by low Tb and a rapid rewarming when periodically aroused, ischemia, bacteria, carcinogens, radiation, and muscle disuse, understanding a mechanism of hibernation greatly contributes to a wide variety of medical applications as well as maintenance of health. In order to clarify this mechanism by which a multiple tolerance is simultaneously developed at the beginning of hibernation, it is essential to explore an orchestrated physiological system responsible for hibernation.

2. Tolerance to Low Tb in Hearts

The heart is one of the principal organs for maintaining life. At Tb below 20 °C, hearts in nonhibernators, such as the guinea pig, rabbit and human, generally develop serious ventricular fibrillations or are arrested (Duker et al., 1983; Johansson et al., 1996). However, hearts in hibernators, such as ground squirrels and chipmunks, tolerate low Tb below 10 °C during hibernation (Lyman et al., 1982; Duker et al., 1983). Therefore, hibernators are considered to develop a capacity in their hearts for adaptive hypothermia during hibernation, implying a special physiological mechanism for adapting to lethal low Tb.

2.1. Ca^{2+} Source for Cardiac Excitation-Contraction Coupling in Hibernating State

Our early electrophysiological studies on the cardiac muscle of chipmunks, a rodent hibernator, have provided the primary evidence for preventing cellular Ca^{2+} overload in the hibernating state (Kondo, 1997). In these studies, the plateau phase of membrane action potential was found to have unique properties in myocardium from hibernating animals

(Kondo and Shibata, 1984). The cardiac action potential plateau markedly depresses in hibernating animals compared to that in active animals under constant experimental conditions (figure 1). As the action potential plateau is generated mainly by Ca^{2+} current carried by Ca^{2+} influx through Ca^{2+} channels, this result implies that Ca^{2+} inflow through plasma membrane is distinctly decreased in hibernating animals. Indeed, Ca^{2+} current is remarkably reduced in hibernating animals, although this current in active animals does not differ from that in nonhibernators (Kondo, 1986a). In spite of a marked decrease in Ca^{2+} influx in hibernating animals, the contractile amplitude in is similar to that in active animals at a lower stimulation frequency (Kondo and Shibata,1984; Kondo,1986a), suggesting an unchanged cytosolic Ca^{2+} concentrations for contraction in spite of a reduced transsarcolemmal Ca^{2+} influx. Since Ca^{2+} source for cardiac excitation-contraction coupling in most nonhibernating mammals depends on both Ca^{2+} influx through Ca^{2+} channels and internal Ca^{2+} release from the sarcoplasmic reticulum (SR), it is assumed that an alteration of the SR-dependent Ca^{2+} regulation is involved in the contraction in hibernating animals. This assumption is strongly supported by the evidence that ryanodine, which inhibits Ca^{2+} release from the SR, completely abolishes the contraction in hibernating animals despite much less effect in active animals (Kondo and Shibata,1984; Kondo,1986a).

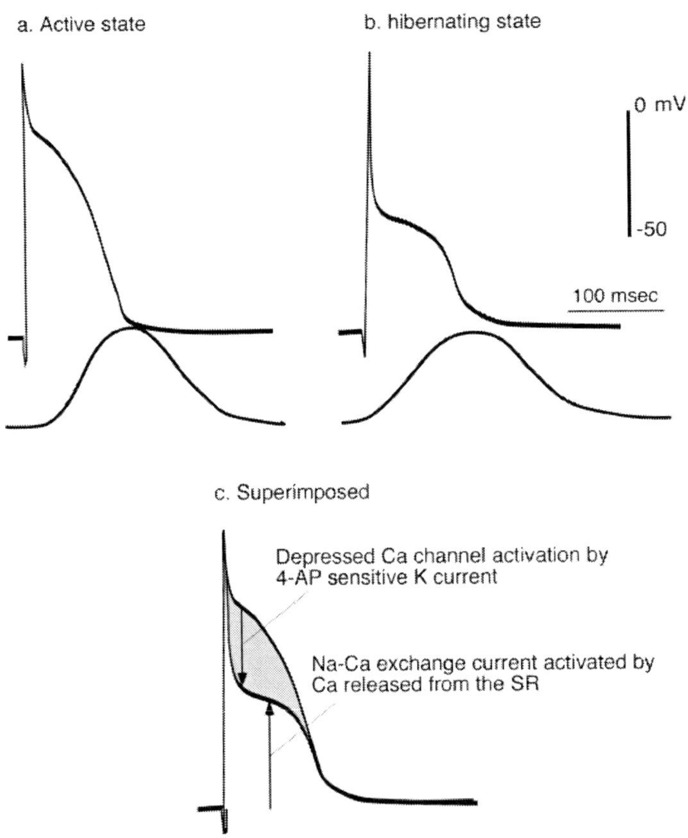

Figure 1. Cardiac membrane action potential (upper traces) and contraction (lower traces) in active and hibernating state of chipmunks. a: active state, b: hibernating state, c: superimposition of hibernating state on active. The gray area indicates difference between active and hibernating states.

The great involvement of internal Ca^{2+} released from the SR in the contraction in hibernating animals has been shown by the experiments using beta-adrenergic agonist, isoprenaline, which increases Ca^{2+} influx via facilitation of L-type Ca^{2+} channels activation, and ryanodine and caffeine, which inhibit internal Ca^{2+} release from the SR (Kondo,1986a; Kondo, 1988). While isoprenaline markedly increases the contraction in association with augmentation of the action potential plateau in active animals, the cardiac contraction in hibernating animals is not affected by the treatment with isoprenaline in spite of augmentation of the action potential plateau. This uncoupling of Ca^{2+} influx with contraction in hibernating animals is observed more clearly in cardiac muscles pretreated with ryanodine, where the contraction is abolished; in ryanodine-pretreated preparations, isoprenaline fails to cause the contraction though the action potential plateau is remarkably augmented (Kondo,1986a). These results indicate that the activator Ca^{2+} for the cardiac contraction in hibernating animals is derived mainly from the SR and suggest that increased cytosolic Ca^{2+} via Ca^{2+} channels by isoprenaline may be immediately sequestered by intracellular store sites, most likely the SR, before interacting with the contractile proteins. Such an assumed sequestration of Ca^{2+} flowed into cytoplasm was proven by cardiac muscles pretreated with caffeine, which is known to inhibit Ca^{2+} uptake by the SR (Kondo,1988). In caffeine-pretreated cardiac muscles, the increased cytosolic Ca^{2+} by isoprenaline greatly increases the contraction in hibernating animals, which is insensitive to ryanodine and sensitive to Ca^{2+} channel blocker, nifedipine. These results suggest that in cardiac muscles from hibernating animals, a large amount of cytosolic Ca^{2+} is rapidly and effectively sequestered by the SR of which ability to take up Ca^{2+} is enhanced (Kondo,1988). This is strongly supported by either study using cardiac SR vesicles from hibernating ground squirrels that a rate of Ca^{2+} uptake and a level of Ca^{2+} accumulation are increased (Belke et al., 1991) or study on expression of Ca^{2+} regulatory proteins of the SR that the level of the SR Ca^{2+}-ATPase but not Ca^{2+}-binding protein calsequestrin is increased in hibernating woodchuck (Yatani et al., 2004). Such an enhanced ability of the SR to take up Ca^{2+} may compensate the decreased Ca^{2+} influx through Ca^{2+} channels in hibernating animals' myocardium for maintaining a sufficient contraction.

Regarding mechanisms for decreased transsarcolemmal Ca^{2+} influx in hibernating animals, it is assumed that activation of Ca^{2+} channels is greatly inhibited by a large 4-aminopyridine-sensitive transient outward current that rapidly repolarizes the action potential plateau below the activation threshold of voltage-dependent Ca^{2+} channels (Kondo, 1986a; 1986b). Such less contribution of transsarcolemmal Ca^{2+} influx to the contraction is supported by the fact that either ryanodine or Ca^{2+} replacement by strontium ions, which can activate contraction by flowing into cytoplasm through Ca^{2+} channels without releasing from the SR (Kondo, 1987a), abolishes the contraction and action potential plateau in hibernating animals (Kondo, 1986b). This further suggests that Ca^{2+}-dependent inactivation processes activated by an internal Ca^{2+} released from the SR does not play a primary role in inactivation of Ca^{2+} channels. However, since Ca^{2+}-dependent inactivation of Ca^{2+} channels has bee reported to be enhanced by increased Ca^{2+} release from the SR in hibernating animals (Yatani et al., 2004), released Ca^{2+} from the SR may help to invest more effective inactivation of Ca^{2+} channels. Since kinetics of L-type Ca^{2+} channels in cardiacmyocytes from ground squirrels has also been suggested to be altered during hibernation (Alekseev et al., 1996), a possible modulation of the channel kinetics may not be excluded for explaining the decreased transsarcolemmal Ca^{2+} influx. Thus, exact mechanisms involved in the decreased Ca^{2+} influx in myocardium of hibernating animals remain to be clarified.

Although cardiac Ca^{2+} influx through Ca^{2+} channels is markedly reduced in hibernatinganimals, the low amplitude of the action potential plateau is generated (Kondo and Shibata, 1984). This shape of the action potential plateau is uniquely observed in cardiac muscles from hibernating animals (figure 1). This plateau may be generated by Na^+-Ca^{2+} exchange current that would be stimulated by augmented internal Ca^{2+} release from the SR (Kondo, 1987b). Since Na^+-Ca^{2+} exchange function has been reported to be unchanged during hibernation compared to active animals (Yatani et al., 2004), the genesis of lowered action potential plateau may not be attributed to a functional change in Na^+-Ca^{2+} exchange.

From these electrophysiological and pharmacological studies on cardiac muscles of chipmunks, it is suggested that in cardiac muscles of hibernating animals, transsarcolemmal Ca^{2+} influx is reduced or nearly abolished by much less activation of Ca^{2+} channels primarily via rapid membrane repolarization by enhanced transient outward current, while the ability to take up Ca^{2+} by the SR is augmented for sufficient supply and reuptake of Ca^{2+} in the cell for maintaining regulated contraction and relaxation (figure 1). The Na^+-Ca^{2+} exchange current instead of transsarcolemmal Ca^{2+} inward current plays a primitive role in the genesis of the action potential plateau in hibernating animals, which prolongs the refractory period of the action potential that decreases arrhythmogenic activities in hibernating state or during rewarming when periodically aroused. Similar changes in cardiac Ca^{2+} regulation are observed in hibernating state of other rodent hibernators, ground squirrels and woodchucks (Wang et al., 2002; Yatani et al., 2004), suggesting that a unique ability to modulate Ca^{2+} source for cardiac excitation-contraction coupling defined in hibernating chipmunks is commonly developed by other mammalian hibernators.

2.2. Prevention of Ca^{2+} Overload by Intracellular Recycling

In cardiac excitation-contraction coupling of nonhibernating mammals, the level of cytosolic Ca^{2+} concentration is regulated by a balance between an increase of Ca^{2+} and its removal; cytosolic Ca^{2+} is increased by both Ca^{2+} influx through Ca^{2+} channels and release from the SR and removed through the activation of plasma membrane Ca^{2+} ATPase, Na^+-Ca^{2+} exchangers, the SR Ca^{2+} ATPase, and mitochondrial uptake (Berridge et al., 2000). The inactivation process of Ca^{2+} channels, which rapidly decays Ca^{2+} influx following the channels activation, and Ca^{2+} uptake by the SR are the dominant mechanism for quickly reducing cytosolic Ca^{2+} for cardiac relaxation. However, a lowering of Tb during hibernation dramatically slows Ca^{2+} channels inactivation process, which results in abnormal increase in cytosolic Ca^{2+} by a large amount of Ca^{2+} influx, and reduces the rate of Ca^{2+} uptake by the SR through depressing Ca^{2+} ATPase activity. Simultaneously, functions of plasma membrane Ca^{2+} ATPase and temperature-sensitive Na^+-Ca^{2+} exchangers are depressed at low Tb. Such increased Ca^{2+} influx and reduced ability to remove cytosolic Ca^{2+} lead to excessive accumulation of intracellular Ca^{2+} and eventual Ca^{2+} overloaded state, which prevents relaxation. The overloaded Ca^{2+} further stimulates Ca^{2+} ATPase, which consumes more energy, and induces proteolysis and lipid peroxidation through activation of Ca^{2+}-dependent proteases and phospholipases, respectively. Mitochondria are destructed by accumulating excess Ca^{2+} in this organella. These events result in lethal damages in cells by destroying enzymes, cell membranes and mitochondria, and then induce cell death. The cardiac modulation in hibernating animals, namely inactivated Ca^{2+} channels and augmented ability to take up Ca^{2+} by the SR, may allow hearts to tolerate decreased Tb by preventing Ca^{2+} overload (figure 2). Such a prevention of Ca^{2+} overload in hibernating animals has previously been assumed (Kondo, 1986a). This modulation would maintain regulated functions in excitation-contraction coupling by repeating a cycle of Ca^{2+} release and uptake by the SR in cells, which save much energy for removing cytosolic Ca^{2+} for relaxation because of a minimum Ca^{2+} supply from the SR for the contraction (Kondo, 1997). Thus, the modulation of Ca^{2+} regulation mechanism in hibernating animals hearts may establish an ideal intracellular Ca^{2+} recycle system for surviving extremely depressed metabolic state at low Tb. More important point is that Ca^{2+} channels and the SR in hibernators, which play a critical role in maintaining heart functions during hibernation, commonly function in nonhibernators, including humans, implying that even in hearts of nonhibernators, less tolerance to low Tb is possibly improved by modulating the channels and the SR as hibernators did if understood control mechanisms of hibernation.

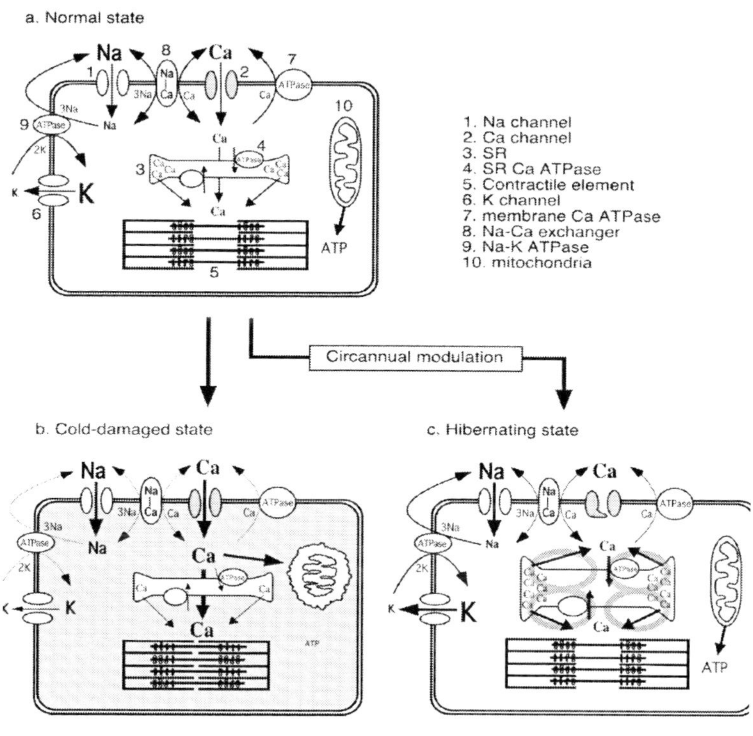

Figure 2. Schema of modulated cardiac Ca^{2+} regulation. Regulation of Na, Ca and K ions through channels, Na-Ca exchangers and ATPases is illustrated. 3Na and 2K in Na-K ATPase mean that Na-K ATPase exchanges 3Na for 2K, and 3Na and Ca in Na-Ca exchanger mean that Na-Ca exchanger exchanges 1Ca for 3Na. a and b: myocardium of active animals or nonhibernators in normal state (Tb: 37°C) and in cold-damaged state at low Tb, c: circannually modulated myocardium in hibernating state. Numbers in (a) correspond to those in inset legend. In (c), circles with gray line show intracellular recycling of Ca ions by the SR.

3. Modulation of Ca^{2+} Regulation in Cardiac Muscles Independent of Tb Changes

The modulation of cardiac Ca^{2+} regulation in hibernating animals that was discovered in chipmunks and verified in other hibernators, ground squirrels and woodchuck, are observed even in normothermic chipmunks in which a decreased Tb is prevented by keeping them under conditions of constant warmth (23 °C). In these normothermic animals, the depressed action potential plateau, which is reflected by less activated Ca^{2+} channels, and an increased dependence of contraction on Ca^{2+} released from the SR appear specifically in calendar winter but not in calendar summer (Kondo, 1987c; Kondo and Kondo,1992b). This result proves that a unique characteristics in cardiac Ca^{2+} regulation, which are observed in hibernating animals, appears seasonally even in normothermic animals, indicating that the hibernation-like cardiac changes are seasonally caused independently of changes in Tb and environment. This suggests that a cardiac Ca^{2+} regulation is switched for prevention of Ca^{2+} overload prior to the onset of lowering Tb and that a mechanism by which the Ca^{2+} regulation is seasonally modulated may be essential for surviving a decreased Tb. This leads to an idea that similar changes in cardiac Ca^{2+} regulation occur systematically in other principal organs and physiological systems, including the central nervous and immune systems, for maintaining regulated functions for health by increasing tolerance to low Tb.

These findings presented here demonstrate a seasonally and systematically controlled mechanism for hibernation. By this mechanism, a factor responsible for hibernation would develop the ability to adapt organisms to a severely depressed physiological state in hibernating animals.

4. Searching Endogenous Factors Correlated to Hibernation Rhythm

Endogenous factors that induce hibernation have long been explored in hibernators (Lyman et al., 1982; Wang, 1988). In 1930s, such factors have already been attempted to be isolated from a few tissues, including the brain. Thereafter, in 1960s, a hibernation-induction trigger has been reported to exist in the blood and induce a state similar to hibernation when injected into summer-active hibernators. However, this trigger is neither identified molecularly and biochemically nor proven to be essential for hibernation. Thus, many researchers have remained unconvinced that endogenous factors responsible for hibernation might be identified,

From the studies on cardiac excitation-contraction coupling in hibernators discussed above, seasonal modulation of cardiac Ca^{2+} regulation are suggested to be under control of a system responsible for hibernation. By using this seasonally regulated cardiac modulation as a physiological marker of hibernation, a protein complex (M.W. 140kDa) in the blood, which is regulated in association with the cardiac electromechanical changes, was isolated from chipmunks (Kondo and Kondo, 1992a; 1992b). This complex consists of four novel proteins called hibernation-specific protein (HP), three of which are structurally homologous proteins

(HP20, HP25 and HP27) with a collagen-like domain in the N-terminal regions and form a complex (HP20c) by triple-helix formation of this domain. In the blood, HP20c is further associated with the fourth protein, HP55, which is homologous to alpha 1-antitripsin, a member of the serpin supefamily (figure 3) (Kondo and Kondo, 1992a). Since HP55 inhibits elastase (Kondo and Kondo, 1996), it is assumed that HP20c is inactivated by binding to HP55 in the blood.

The complex of HP20c with HP55 (HP complex) is produced specifically in the liver and secreted into the blood (Takamatsu et al., 1993). HP complex exists abundantly in the blood in active state, while this is decreased to approximately 1/10 in hibernating state by depressed gene expression. Such a decrease in HP complex in the blood occurs prior to the onset of hibernation, and hibernation is terminated coincident with returning to the levels in active state (Kondo and Kondo, 1992a). By immunochemical analysis, HP complex is detected in the blood in other hibernators, such as ground squirrels and woodchucks, but not in tree squirrels that is nonhibernators in the same squirrel family (Kondo and Kondo, 1992a; 1993). Interestingly, although tree squirrels have HP genes, these expression is inhibited by base substitution in promoter regions (Kojima, 2001). From these results, it was indicated that HP complex is downregulated in association with hibernation rhythm and a highly specific for hibernation.

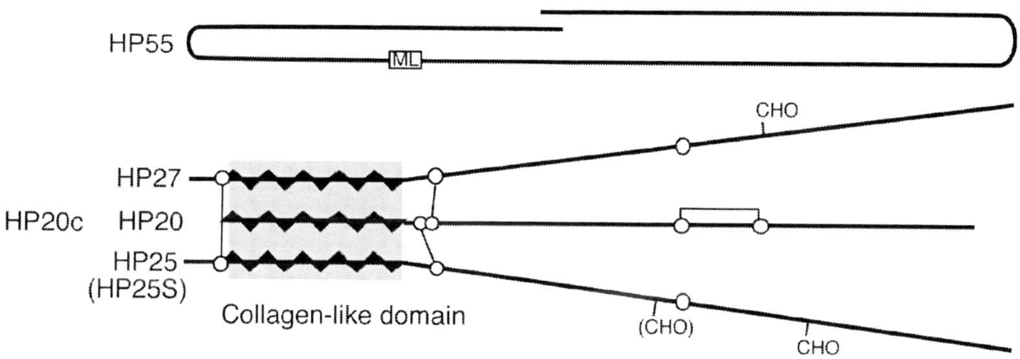

Figure 3. Structural characteristics of HP complex. Amino acid sequences are drown as lines, and the length of each line corresponds to the number of amino acid in each subunit of HP complex, respectively. The gray area in HP20c indicates collagenous domain. Open circles and CHOs indicate cysteine residues and glycosylation sites, respectively. The bars between open circles mean formations of disulfide bond. HP25S indicates additional glycosylated HP25. ML in HP55 shows sequence of active center (Met and Leu)

5. Circannual Regulation of HP Complex

An association of downregulation of HP complex with hibernation is observed throughout life in hibernating chipmunks which were kept under constant conditions of cold (5°C) and darkness (Kondo et al., 2006). Surprisingly, maximum life span in animals tested is approximately 11 years, near 4-times longer than that in rats. The animals exhibit an individual hibernation rhythm between 5 and 13 months, and the duration of hibernation period also differs among animals (between 3 and 7 months). However, in each individual, the rhythm and duration are relatively constant throughout life, suggesting that an inherent circannual timing mechanism functions in individual organisms (Kondo et al., 2006). In association with such an individually distinct hibernation timing and duration, HP complex in the blood is downregulated throughout life, suggesting that HP complex is an important factor that is strictly controlled by an endogenous circannual timing mechanism.

An evidence that HP complex is controlled by a circannual rhythm is obtained by the experiments in chipmunks in which a decreased Tb is prevented by keeping them under conditions of constant warmth (23 °C) and 12 hr light-dark cycle (Kondo et al., 2006). Even in these normothermic animals, HP complex in the blood is circannually regulated as observed in hibernating animals in the cold, and Tb is lowered immediately after the cold exposure only when HP complex is downregulated. This demonstrated that HP regulation in the blood is controlled by a circannual rhythm independently of Tb changes and implied that regulated HP complex is essential for the timing and maintenance of physiological adaptation for hibernation. This implication is further supported by the fact that in a few chipmunks, which are unable to hibernate throughout life and died within a few years under constant cold and dark conditions, HP complex in the blood is never downregulated (Kondo et al., 2006).

These results suggest that regulated HP complex by an endogenously generated circannual rhythm may play a critical role in adaptation for hibernation.

6. HP Complex in the Brain

In the cerebrospinal fluid (CSF), HP is detected by immunochemical analysis (Kondo et al., 2006). Although HP levels in CSF in active state is less than 1/1000 of those in the blood, HP levels in CSF increase coincident with HP reduction in the blood and then hibernation begins. Thereafter, HP levels in CSF further increase and reach a peak at the middle stage of hibernation (approximately 20 times higher than in active state). At the late stage, HP levels in CSF abruptly decrease coinciding with termination of hibernation though HP in the blood is increased. By the evidence that HP signal is immunohistochemically detected during hibernation in the choroidal epithelial cells of the blood-CSF barrier, the major producer of CSF, and HP mRNA expression is not detected in the brain (Kondo et al., 2006), HP was shown to increase by transporting into the brain through the blood-CSF barrier in hibernating state. Similar upregulation in HP in CSF is circannully caused in animals in which a decreased Tb is prevented by keeping them under constant warm conditions, indicating that an increase in HP in the brain as well as a decrease in HP in the blood is controlled by a circannul rhythm independently of Tb and environmental changes.

Further biochemical analysis of HP in CSF reveals dissociation of HP complex to HP20c and HP55 in CSF though HP20c is associated with HP55 in the blood (Kondo et al., 2006). This suggests that HP complex in the blood is activated by the dissociation through transport into CSF and that HP20c dissociated from HP55 may function in the brain. Thus, HP complex is considered to control hibernation in the brain by transport and activation, and the blood-CSF barrier may be a key in HP signaling.

These results show a unique property of the blood-CSF barrier, the choroidal epithelium. A high molecular protein complex of HP (M.W. 140kDa) is not only transported but also activated (Kondo et al., 2006). Since it has been reported that leptin, a regulator suppressing food intake, and insulin-like growth factors, a regulator of cell growth and development, are transported through the choroid plexus, the blood-CSF barrier may play a significant role as a signaling pathway of hormones for regulating functions of the central nervous system. Furthermore, the finding that HP complex in the blood (inactivated HP20c with HP55) is transported into the brain with its activation (dissociated HP20c) may help to develop HP complex as a unique carrier of drug delivery system into the brain. A role of the blood-CSF barrier, including the blood-brain barrier, should be reconsidered as a signaling system and a signal activation machinery.

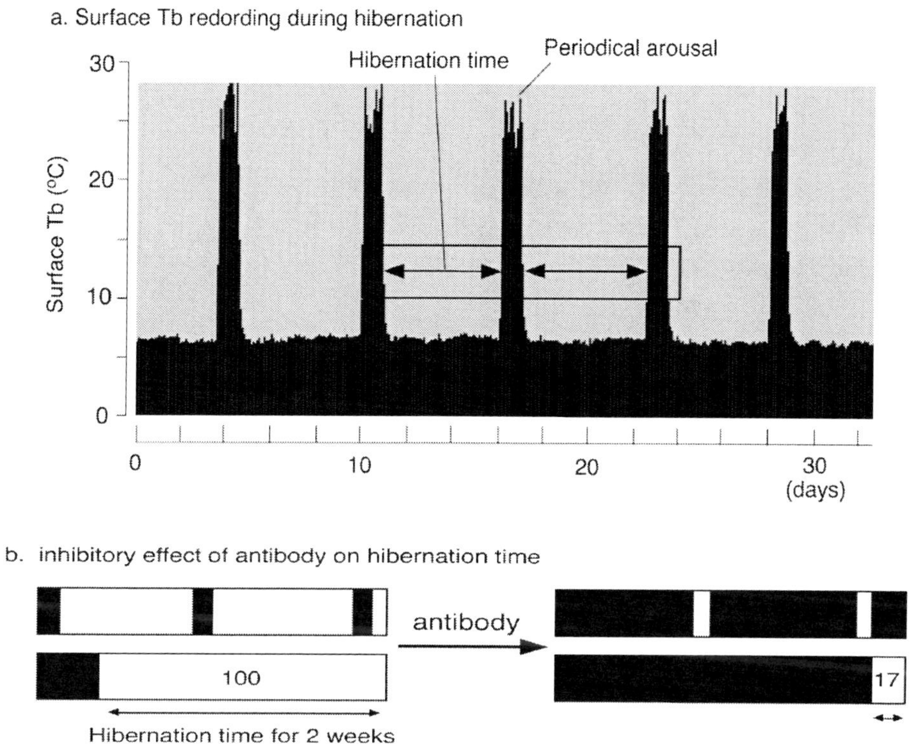

Figure 4. Tb changes during hibernation season and effects of anti HP20c antibody on hibernation time. a: surface Tb changes measured by thermal camera at 0.5 hr intervals (Kondo et al, 2006). Each periodical arousal is detected by periodically increased Tb. The rectangle column indicates the period of administration of anti HP20c antibody into lateral ventricle for two weeks by osmotic pump. b: inhibitory effects of the antibody administration on hibernation time for two weeks. The rectangle column corresponds to that shown in (a). The open and closed areas in the upper columns show the periods of decreased Tb (hibernation time) and periodical arousal, respectively. In the lower columns, each period (open or closed column) is shown as the period combined in each. The left and right columns show before and after antibody administration, respectively. In this case, the antibody decreased the hibernation time to 17% (the lower column in the right).

7. HP Complex in the Brain Controls Hibernation

From the studies on HP complex in CSF, it is proposed that HP complex in the blood targets the brain for hibernation. Since animals begin to hibernate when dissociated HP20c from HP55 increased in the brain, the brain HP20c is assumed to be an important factor in adapting organisms to hibernating state. Then, effects of the antibody against HP20c on hibernation have been investigated (Kondo et al., 2006). The blocking brain HP20c by the administration of anti-HP20c antibody into the lateral ventricles in hibernating animals using osmotic pumps which function for two weeks dramatically decreases hibernation time, which is defined by the amount of time spent in low Tb (figure 4). The decreased hibernation time depends on administered doses of the antibody, and in some animals, a lowering of Tb is prevented by the antibody administration (Kondo et al., 2006). After finishing the administration, decreased hibernation time recovers to normal levels. Thus, a time capable for hibernation is reversibly reduced or lost by blocking the brain HP20c, demonstrating that the brain HP20c is an essential molecule for developing a capacity for hibernation (Kondo et al., 2006).

8. What Does a Decreased Hibernation Time Mean?

The duration of hibernation, a period from the onset to the termination of hibernation, is composed of periodical arousals and periods of continuously decreased Tb between periodical arousals (figure 4). The periodical arousal repeatedly occurs with constant intervals and is critical for surviving a decreased Tb during the hibernation season because animals never survive hypothermic state without periodically arousing (Lyman et al., 1982). Since a decreased Tb generally causes disorder in various physiological functions arising from an imbalance between ATP synthesis and requirement, periodical arousals are thought to be of prime importance in prevention of the lethal hypothermic disorders by recovering the energy balance and excreting metabolic waste in normothermic arousal state. In this regard, a hibernation time between periodical arousals shows the limit of ability to survive continuous low Tb in hibernating state. Although even nonhibernators, including humans, can survive a decreased Tb, a survival time is much less. In chipmunks in hibernating state, a typical hibernation time between periodical arousals is about 7days (168 hours), while in humans, a time for tolerating a similarly decreased Tb should be less than 1 hour. The difference in a time for tolerating low Tb between hibernating and nonhibernating animals may reflect an intensity of the ability to hibernate. Thus, hibernation can be redefined by the ability to modulate cellular functions for tolerating severely depressed metabolic state by a decreased Tb rather than the ability to decrease Tb. Therefore, a factor that augments the ability to hibernate would increase a time capable for tolerating decreased Tb, namely a hibernation time. The longer a hibernation time is, the greater the ability to hibernate is.

Based on the above discussion, it is reasonable to conclude that a factor that controls hibernation increases hibernation time. The evidence that anti-HP20c antibody in the brain decreases hibernation time or interrupts hibernation only during the antibody administration demonstrates that HP20c in the brain is essential for enabling organisms to hibernate. The brain HP20c is suggested to develop a capacity for tolerating decreased Tb.

9. Circannual Rhythm and Health

From the studies on seasonal modulation of cardiac Ca^{2+} source for excitation-contraction coupling and hormonal role of HP complex, it is evident that hibernation is a circannually controlled adaptive state, in which cells enhance tolerance to low Tb for adapting to severe physiological state during hibernation. In animals during hibernation, it has previously been reported that organisms are prevented from potentially lethal irradiation, bacterial infection, tumorigenesis, cellular Ca^{2+} overload, ischemia, and muscle disuse. These suggest that hibernation is based on a physiological system controlled by an endogenous circannual timing mechanism by which organisms are protected not only from hypothermia but also from serious diseases. Since a decreased Tb may cause necrosis and apoptosis in cells by disturbing maintenance of ion regulation and energy metabolism and reduce resistance to bacteria and viruses by depressing immune system, the fact that the brain HP20c is essential for developing a capacity for tolerating decreased Tb allows us to assume its potent pharmacological effects on the prevention of diseases.

Importantly, in nonhibernators, including humans, there are proteins structurally homologous to HP20c, such as complement protein C1q, precerebellin and ACRP30/adiponectin (Kishore et al., 2004). C1q is involved in several immunological processes, including maintenance of immune tolerance, phagocytosis of bacteria, neutralization of retroviruses, and modulation of dendritic cells, B cells and fibroblasts. Precerebellin may play a role in the development and stability of the purkinje cell synapse, and ACRP30 is an antidiabetic and antiatherogenic adipokine, which is involved in carbohydrate and lipid catabolism and resolution of inflammation. Recently, the members of the C1q superfamily, including HP20c, is suggested to have a structural and evolutionary similarity to those of tumor necrosis factor family, which are involved in adaptive immunity, apoptosis selectively in tumor cells, inflammation, cell survival, and bone and energy homeostasis (Kishore et al., 2004). These lead to an idea that HP complex is involved in immune mechanisms, cell survival and energy homeostasis that is critical for surviving severely decreased Tb. Furthermore, during the hibernation season, fatness is cancelled by suppressed food intake and increased lipid catabolism, and paired helical neurofilaments by hyperphosphorylated microtuble-associated protein tau, which is known to be a course of Alzheimer's disease, are reversibly formed and degraded (Arendt ea al., 2003). These suggest that hibernation is an adaptive state for maintaining health, and the finding of HP complex will stimulate promising pharmacological applications in humans to the prevention and therapy of various diseases. Also, studies on HP complex showing that switching to the physiological states that enable organisms to hibernate is caused independently of Tb changes indicate that a unique ability to protect organisms is developed in normothermic state.

This concept of hibernation independent of decreased Tb is strongly supported by our preliminary study on longevity of chipmunks (Kondo and Narita, 1998; Kondo, 1999). Animals with circannual rhythm, in which the brain HP20c is circannually increased then Tb decreases, can survive more than 10 years, whereas animals without circannual rhythm, in which the brain HP20c is not increased and Tb never decreases, survive only 3 years or less. Such an exceptionally long life span observed in hibernating animals is realized in animals with circannual rhythm of the brain HP20c even though these animals are prevented from a decreased Tb by keeping them under constant warm conditions (23°C). This indicates that the surprisingly long life span is not due to slowed metabolic rate by a low Tb and suggests an involvement of circannually regulated HP complex in the brain in this life span. Therefore, it is excitingly implied that HP complex in the brain is involved in not only maintaining health by protecting organisms from diseases but also controlling prolongation of life span by preventing aging. The exact mechanisms of HP action, and a signaling pathway and network for circannual control of HP are the attractive future subjects.

10. Conclusion

In the hearts of hibernators, Ca^{2+} source for excitation-contraction coupling is modulated during the hibernation season; Ca^{2+} influx through Ca^{2+} channels is decreased and the ability of the SR to release and uptake Ca^{2+} is augmented. This modulation establishes an intracellular Ca^{2+} recycling system ideal for maintaining regulated cardiac function with consuming less energy and plays a primary role in protecting from intracellular Ca^{2+} overload in depressed metabolic state by low Tb. The cardiac modulation is controlled by an endogenously generated circannual rhythm independently of Tb and environmental changes, suggesting endogenous factors controlled by this rhythm that seasonally develop the ability to tolerate an extremely decreased Tb in organisms. Along this suggestion, the liver-derived HP complex is isolated from the blood in chipmunks. HP complex is downregulated in the blood through depressed expression of HP genes in the liver in correlation with hibernation, while coincident with this downregulation, HP complex is upregulated in the brain through facilitated transcytosis of choroidal epithelium, the blood-CSF barrier. The transported HP complex dissociates to HP20c and HP55, and then HP20c may modulate brain functions in order to develop a capacity for tolerating low Tb because blocking HP20c in the brain by the antibody remarkably decreases the hibernation time, during which low Tb is tolerated, or interrupts hibernation. The regulation of HP complex in both the blood and CSF is controlled by an endogenous circannual rhythm independently of Tb changes like a circannual modulation of Ca^{2+} source shown in cardiac muscles. The brain HP20c might be involved in an exceptionally long life span in hibernators. Thus, HP complex is a promising candidate hormone that modulates brain functions for maintaining health and life during severe hibernation season by developing a capacity for protecting organisms from various lethal diseases and aging.

References

Alekseev, A. E., Markevich, N. I., Korystova, A. F., Terzic, A. and Kokoz, Y. M. (1996). Comparative analysis of the kinetic characteristics of L-type calcium channels in cardiac cells of hibernators. *Biophys. J., 70,* 786-797.

Arendt, T., Stieler, J., Strijkstra, A. M., Hut ,R. A., Rudiger, J., Van der Zee, E. A., Harkany, T., Holzer, M. and Hartig, W. (2003). Reversible paired helical filament-like phosphorylation of tau is an adaptive process associated with neuronal plasticity in hibernating animals. *J. Neurosci., 23,* 6972-6981.

Belke, D. D., Milner, R. E. and Wang, L. C. (1991). Seasonal variations in the rate and capacity of cardiac SR calcium accumulation in a hibernating species. *Cryobioloby, 28,* 354-363.

Berridge, M. J., Lipp, P. and Bootman, M. D. (2000). The versatility and universality of calcium signalling. *Nat. Rev. Mol. Cell. Biol., 1,* 11-21.

Donahue, S. W., Galley, S. A., Vaughan, M. R., Patterson-Buckendahl, P., Demers, L. M., Vance, J. L. and McGee, M. E. (2006). Parathyroid hormone may maintain bone formation in hibernating black bears (Ursus americanus) to prevent disuse osteoporosis. *J. Exp. Biol., 209,* 1630-8.

Duker, G. D., Olsson, S. -O., Hecht, N. H., Senturia, J. B. and Johansson, B. W. (1983). Ventricular fibrillation in hibernators and nonhibernators. *Cryobiol., 20,* 407-420.

Johansson, B. W. (1996). The hibernator heart-nature's model of resistance to ventricular fibrillation. *Cardiovasc. Res., 31,* 826-832.

Frerichs, K. U. and Hallenbeck, J. M. (1998). Hibernation in ground squirrels induces state and species-specific tolerance to hypoxia and aglycemia: an in vitro study in hippocampal slices. *J. Cereb. Blood Flow Metab., 18,* 168-175.

Harlow, H. J., Lohuis, T., Beck, T. D. I. and Iaizzo, P. A. (2001). Muscle strength in overwintering bears. *Nature, 409,* 997.

Hochachka, P. W. (1986). Defense strategies against hypoxia and hypothermia. *Science, 231,* 234-241.

Jaroslow, B. N., Smith, D. E., Willians, M. and Tyler, S. A. (1969). Survival of hibernating ground squirrels (Citellus tridecemlineatus) after single and fractionated doses of cobalt-60 gamma radiation. *Radiation Res., 38,* 379-388.

Kemper, G. B. and Ruben, R. L. (1982). Effect of 7,12-dimethylbenz(a)anthracene on the integument of the hibernating and nonhibernating 13-lined ground squirrel. *Comp. Biochem. Biophys., 73C,* 445-450.

Kishore, U., Gaboriaud, C., Waters, P., Shrive, A. K., Greenhough, T. J., Reid, K. B. M., Sim, R. B. and Arlaud, G. J. (2004). C1q and tumor necrosis factor superfamily: modularity and versatility. *Trends Immunol., 25,* 551-561.

Kojima, M., Shiba, T., Kondo, N. and Takamatsu, N. (2001). The tree squirrel HP-25 gene is a pseudogene. *Eur. J. Biochem., 268,* 5997-6002.

Kondo, I. and Kondo, N. (1996). Structural aspects of complex of hibernation-specific proteins. In F. Geiser, A.J. Hulbert and S.C. Nicol (Eds.), Adaptation to the cold: Tenth International Hibernation Symposium (pp.351-355). Armidale: University of New England Press.

Kondo, N. (1986a). Excitation-contraction coupling in myocardium of nonhibernating and hibernating chipmunks: Effects of isoprenaline, a high calcium medium and ryanodine. *Circ. Res., 59,* 221-228.

Kondo, N. (1986b). Excitation-contraction coupling in the myocardium of hibernating chipmunks. *Experientia, 42,* 1220-1222.

Kondo, N. (1987a). Can strontium replace calcium as an activator of internal calcium release in cardiac muslces?: Study on a dual action of A23187. *J. Mol. Cell. Cardiol., 19,* 391-399.

Kondo, N. (1987b). Electrophysiological effects of Ca antagonists, tetrodotoxin, [Ca]o and [Na]o on hibernating chipmunks myocardium: Possible involvement of Na-Ca exchange mechanisms. *Br. J. Pharmacol., 91,* 315-319.

Kondo, N. (1987c). Identification of a pre-hibernating state in myocardium from nonhibernating chipmunks. *Experientia, 43,* 873-875.

Kondo, N. (1988). Comparison between effects of caffeine and ryanodine on electromechanical coupling in myocardium of hibernating chipmunks: Role of internal Ca stores. *Br. J. Pharmacol., 95,* 1287-1291.

Kondo, N. (1997). Physiological and biochemical studies on hibernation control mechanism in mammalian hibernation. In O. Hayaishi and S. Inoue (Eds), Sleep and Sleep Disorders: From Molecular to Behavior (pp. 129-143). Tokyo: Academic Press.

Kondo, N. (1999). Hibernation control mechanism and possible applications to humans. *J. Br. Interplanet. Soc., 58,* 343-347.

Kondo, N. and Kondo, J. (1992a). Identification of novel blood proteins specific for mammalian hibernation. *J. Biol. Chem., 267,* 473-478.

Kondo, N. and Kondo, J. (1992b). Identification of novel types of protein specific for mammalian hibernation with circannual rhythm. In T. Hiroshige and K. Honma (Eds.), Circadian Clocks from Cell to Human (pp. 89-96). Sapporo: Hokkaido University Press

Kondo, N. and Kondo, J. (1993). Identification and characterization of novel types of plasma protein specific for hibernation in rodents. In C. Carey, G. L. Florant, B. A. Wunder and B. Horwitz (Eds.), Life in the Cold: Ecological, Physiological and Molecular Mechanisms (pp. 467-473). Boulder: Westview Press Inc.

Kondo, N. and Narita, A. (1998). Dose hibernation increase survival rate in chipmunks? *Zool. Sci. 15,* Suppl. 106.

Kondo, N., Sekijima, T., Kondo, J., Takamatsu, N., Tohya, K. and Ohtsu, T. (2006). Circannual control of hibernation by HP complex in the brain. *Cell, 125,* 161-172.

Kondo, N. and Shibata, S. (1984). Calcium source for excitation-contraction coupling in myocardium of nonhibernating and hibernating chipmunks. *Science, 225,* 641-643.

Kurtz, C. C. and Carey, H. V. (2007). Seasonal changes in the intestinal immune system of hibernating ground squirrels. *Dev. Comp. Immunol., 31,*1415-28.

Lyman, C. P., Willis, J. S., Malan, A., and Wang, L. C. H. (1982) *Hibernation and torpor in mammals and birds.* New York: Academic Press.

Maniero, G. D. (2002). Classical pathway serum complement activity throughout various stages of the annual cycle of a mammalian hibernator, the golden-mantled ground squirrel, Spermophilus lateralis. *Dev. Comp. Immunol., 26,* 563-74.

Maniero, G. D. (2005). Ground squirrel splenic macrophages bind lipopolysaccharide over a wide range of temperatures at all phases of their annual hibernation cycle. *Comp. Immunol. Microbiol. Infect. Dis., 28,* 297-309.

Musacchia, X. J. and Barr, R. E. (1968). Survival of whole-body-irradiated hibernating and active ground squirrels; citellus tridecemlineatus. *Radiation Res., 33,* 348-356.

Sharapov, V. M. (1983). Influence of animal hibernation on the development of mycoses. *Mycopathologia, 84,* 77-80.

Takamatsu, N., Ohba, K. -I., Kondo, J., Kondo, N. and Shiba, T. (1993). Hibernation-associated gene regulation of plasma proteins with a collagen-like domain in mammalian hibernators. *Mol. Cell. Biol., 13,* 1516-1521.

Wang, L. C. H. (1988). Mammalian hibernation: An escape from the cold. In R. Gilles (Ed), *Advances in Comparative and Environmental Physiology.* (Vol. 2, pp. 1-45). Berlin: Springer-Verlag.

Wang, S. Q., Lakatta, E. G., Cheng, H., Zhou, Z. Q. (2002). Adaptive mechanisms of intracellular calcium homeostasis in mammalian hibernators. *J. Exp. Biol., 205,* 2957-2962.

Yatani, A., Kim, S. J., Kudej, R. K., Wang, Q., Depre, C., Irie, K., Kranias, E. G., Vatner, S. F., Vatner, D. E. (2004). Insights into cardioprotection obtained from study of cellular Ca2+ handling in myocardium of true hibernating mammals. *Am. J. Physiol. Heart Circ. Physiol., 286,* H2219-H2228

Zivadinovic, D. and Andjus, R. K. (1996). Life span of the European ground squirrel Spermophilus citellus under free-running conditions and entrainment. In F. Geiser, A. J. Hulbert and S. C. Nicol (Eds), *Adaptation to the cold: Tenth International Hibernation Symposium.* (pp. 103-108). Armidale: University of New England Press.

In: Progress in Circadian Rhythm Research
Editor: Anne-Laure Léglise, pp. 141-161
ISBN: 978-1-60021-876-7

Chapter VI

Cognitive Strategies and Circadian Typology

Ana Adan[1*]***, Vincenzo Natale***[2] ***and Hervé Caci***[3]
[1] Department of Psychiatry and Clinical Psychobiology, School of Psychology, University of Barcelona. Barcelona, Spain.
[2] Department of Psychology, University of Bologna. Bologna, Italy.
[3] Service de Pédiatrie. CHU de Nice. Hôpital Archet 2. Nice, France.

Abstract

In recent years many studies have been published trying to establish the differences among individuals depending on their circadian typology (morning-, neither-, and evening-type). This is an individual difference underlying differences in the rhythmic expression of both biological and behavioural parameters, especially the circadian period. In general, morning-type subjects show phase advances with respect to evening-type subjects, which vary from 2 to 12 hours, depending on the variable studied. The neither-type group is usually in an intermediate position regarding the extreme groups (morning- and evening-type), although many studies do not contemplate this group despite the fact that it is predominant in the general population (around 60 %). Our study exposes the optimal moments of efficiency according to the type of cognitive ability that healthy subjects must perform depending on their circadian typology. This aspect is of vital importance for the subjects' performance at school and work, and also when the environmental conditions of the light-dark cycle (jet-lag and shift-work) are modified. We also comment on the observed differences in eating habits, consumption of psychoactive substances, and preferred spare-time activities among chronotypes, which in general are healthier in the morning-type than in the evening-type. Finally, we present data on the associations found between other personality characteristics, such as extraversion or impulsivity, and with the vulnerability to suffer certain

* Correspondence to: Dr. Ana Adan. Department of Psychiatry and Clinical Psychobiology. School of Psychology. University of Barcelona. Passeig Vall d'Hebron 171. 08035 Barcelona. Spain. e-mail: aadan@ub.edu

psychopathologies such as Seasonal Affective Disorder (SAD), bulimia nervosa or Attention Deficit Hyperactivity Disorder. The relevance of this individual difference makes it advisable that research which is not directly interested in the study of this variable, but which evaluates the parameters involved in the circadian typology, should have an adequate control of it.

Introduction

The pioneering works on chronobiology had already pointed out the existence of individuals who, in relation to sleep habits, present a morning preference (getting up and going to bed early) while others present an evening preference (getting up and going to bed late). However, it was not until the first self-assessment questionnaire was designed in the 1970s to classify subjects according to this variable [1] that its study was systematically carried out in the studies on rhythmicity. Subjects are classified in three groups according to their answers to the morningness-eveningness questionnaire. These groups are called circadian typology or chronotypes: morning-, neither-, and evening-type. Data show that the extreme groups hold 20% of the population each, whereas the remaining 60% is classified as neither-type. In the chapter by Caci, Natale and Adan the interested reader may find more information on the psychometric properties and the transcultural implications of the questionnaires for the measurement of circadian typology.

Circadian typology is an individual difference that currently holds the interest of many research groups, since it provides a more robust explanation of the differences in individual functioning than other subject characteristics. It has a special relevance in the diurnal variations in cognitive and physical performance, as well as in the subjective state perceived, which show clear phase mismatches regarding circadian typology. The implications of such differences in the academic and work environments are very important, although our society is not much sensitised about it. Moreover, this individual difference is appearing as a key explicative variable of behavioural aspects related to health and the risk of suffering psychopathological disorders.

Circadian typology is determined by a different organization of the endogenous biological clock [2]. This has been interpreted as a lesser dependency on the environmental synchronizers in the evening-type with respect to the morning-type, and consequently, on a more powerful hypothalamic circadian control. In the expression of this individual difference there are underlying genetic differences [3,4,5], but also environmental factors. Among these, we highlight the photoperiod to which the individual is exposed during the first months of life [6,7,8]. The circadian organization of the individuals born with a short photoperiod (autumn-winter) tends to be more morning-like, while those with a long photoperiod (spring-summer) tend to be more evening-like. This is more clearly seen in men than in women, as figure 1 shows, and in adolescents too.

Belonging to one or the other typology is not a problem if the schedule preference marked by the endogenous biological clock can be followed, although especially among extreme groups there can be observed important phase mismatches in the circadian functions of biological parameters (biochemical, endocrine, psychophysiological) and of performance. However, if socio-environmental requirements force the subject to follow a rhythm different than that determined by his circadian typology, then temporary or permanent problems may appear that, in certain cases, will require the application of chronobiological therapeutic treatments.

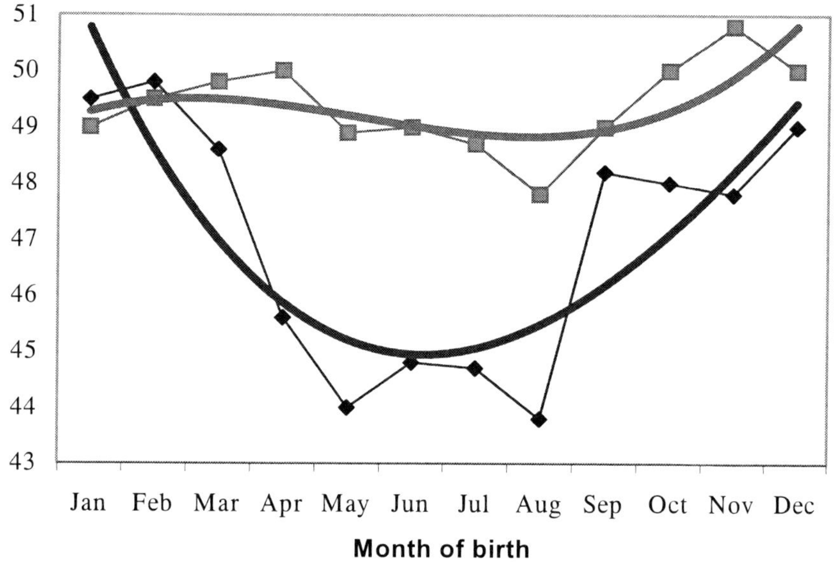

Figure 1. Curves for the mean values of the morningness-eveningness score MEQ according to the month of birth, separately for females (■) and for males (◆), and the corresponding cosine curves derived from nonlinear regression. Taken from Natale et al. (2002).

Finally, circadian typology has been related to several individual variables, such as gender, age, and certain personality traits. Since it has been found associated with behaviours like psychoactive substances abuse, or disorders like depression, the chronotype is nowadays more and more considered either as a protection factor (morning-type) or as a risk factor (evening-type) in the development of such problems that have non-negligible socio-economic issues. The present chapter exposes the most important findings in circadian typology and suggests future lines of research to develop in this field of research in order to elucidate some still controversial aspects.

Cognitive Performance

There are differences between the chronotype groups in the rhythmic expression of circadian periodicity both in biological and in behavioural parameters. The morning-types present an advancement of phase of 2 to 4 hours in relation to evening-types in most of the biological parameters [9,10,11,12]. The main circadian rhythm is the sleep-awake cycle, which determines the times of rest and functioning as well as those in which the subjects find themselves in better or worse conditions to face the socio-environmental requirements around them. Thus, the sleep habits of the morning-type subjects present gaps with an advancement of about 2-3 hours from the evening-type subjects regarding the time of going to bed and getting up, if the subjects are free to follow their own rhythm. The morning subjects also show more individual regularity and less group dispersion, while the evening subjects are more flexible and individual variability is larger [13,14,15,16]. Polisomnographic results (sleep quality) show a longer sleep latency and the delayed appearance of the first REM episode in the evening subjects, who also overestimate the latency to fall asleep with subjective assessments. There were no differences comparing extreme chronotypes in the relation between slow-wave-activity decay and dim light melatonin onset [17].

Table 1. Description of the circadian typology groups according to their principal rhythmic characteristics

Morning-type
Wake up early, feeling very refreshed Best time of day between 6:00-11:00 h. Go to bed between 22:00-0:30 h. Define themselves as clearly morning-type or more morning-than evening-type High regularity of their sleep schedule between working and leisure days Sleep-awake rhythm very similar to the light-darkness cycle More difficult adaptation to shift work and jet-lag
Neither-type or intermediate-type
Wake up between 7:00-10:00 h., feeling quite refreshed Best time of day between 12:00-18:00 h. Go to bed between 23:30-1:30 h. Define themselves as more morning-type than evening-type Medium regularity of their sleep schedule between working and leisure days Sleep-awake rhythm quite similar to the light-darkness cycle Intermediate capacity for adaptation to shift work and jet-lag
Evening-type
Wake up late, feeling rather or very tired Best time of day between 16:00-0:00 h. Go to bed between 0:00-4:00 h. Define themselves as clearly evening-type or more evening- than morning-type Low regularity of their sleep schedule between working and leisure days Sleep-awake rhythm not very similar to the light-darkness cycle Less difficult adaptation to shift work and jet-lag

Evening subjects show a greater tolerance to the sudden modification of the environmental cues (transmeridian travel and work-shifts) and to changes in activity and sleep schedules, depending on whether these occur on working days or leisure days [18,19]. The extreme evening-type subjects with a early-morning onset schedule present an accumulation of sleep restriction over the week, and they usually show an increase in sleep time in leisure days compared to work days [20,21]. In contrast, morning- and neither-types are less susceptible to this phenomenon, since the socio-environmental activity demands are closer to their preferred schedule and have a lesser impact in terms of sleep deprivation. Table 1 summarizes the principal rhythmic characteristics of the circadian typology groups.

Next we describe the main differences in performance between chronotype groups in objective tasks, which initiate a skill or ability of the subject to complete them. The extreme groups present important gaps in diurnal performance, ranging from 2h. for complex tasks, to even 12h. in simple tasks of surveillance and subjective assessments [11,22]. We make a distinction between attention tasks, according to the weight on sensorial or motor requirements, and memory tasks, according to the time passed between the presentation and the recovery of the material presented. An additional aspect which must be considered in certain tasks is what estimation should be selected as the best indicator of performance. The chronobiological works report differences depending on whether they make a quantitative assessment –usually of speed, the time used to perform the task—or qualitative (mistakes, omissions, etc.). In certain cases, we observe that more speed implies worse quality or less precision in the task. Moreover, during the awake period there is the post-lunch phenomenon, caused by a semicircadian rhythmicity related to the sleep episode [23,24,25], that is not suppressed in the absence of food intake. Finally, we comment on the diurnal variations of subjective assessments frequently used in chronobiology and which provide very sensitive and valid information. To read more on the normative data of the diurnal variations in performance, several reviews can be consulted [26,27,28,29,30].

Attention Tasks

We regard attention tasks as repetitive and monotonous tasks that involve perceptual and motor skills, and whose common characteristic is to prioritise the reaction time. We further distinguish between those of surveillance and manual dexterity since they present different temporal patterns and important phase mismatches in the optimal moments of execution.

In surveillance tasks, the execution speed in the early hours is low and increases along the day up to a peak around 19.00-21.00 h. In contrast, the precision is higher in the early hours (08:00-10:00 h.), decreases along the day to minimum values that correspond with moments of maximum speed. The diurnal variation in speed is in line with the fact that the sensorial resolution is optimal between 19:00-21:00 h., while the loss of precision may evidence a change of strategy depending on the hour of the day together with the accumulation of fatigue. The post-pandrial depression usually causes a decrease in surveillance tasks, especially in the speed of resolution. The extreme chronotypes differ in the speed of these tasks with phase mismatches of around 7 to 12 hours. The optimum is reached in the early hours of the day for morning-types and at the end of the day for evening-types.

The post-lunch drop, is quite marked and of similar magnitude in the morning- and neither-types, while it has little effect on the evening-types. The precision for these tasks reaches a maximum at the beginning of the day in morning-types, at mid-morning in neither-types, and in the first hours of the evening in evening-types.

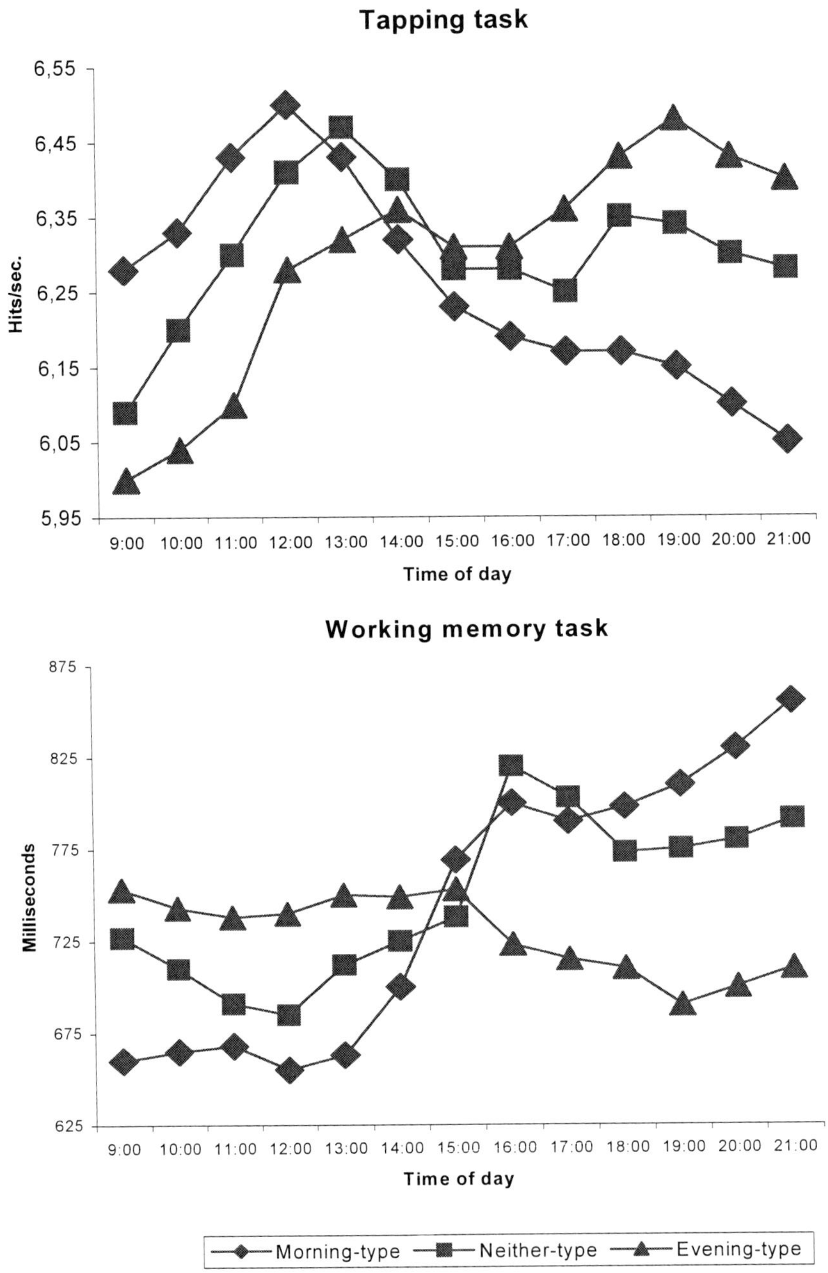

Figure 2. Diurnal variations of speed of execution in a motor execution task (tapping) and a task of working memory according to the circadian typology groups. Better performance: more hits/sec. and less milliseconds.

The diurnal function for tasks of manual skill, such as the Purdue pegboard or tapping, presents the best executions around midday (12.00-14.00 h.) and neither the preceding increase nor the following decrease are so marked as in the surveillance tasks. However, taking into account circadian typology allows to state that this is the evolution in the morning- and neither-type groups, while for the evening-type the peak of performance happens in the evening (18:00-20:00 h). In the hand skill tasks there is also a significant decrease during the post-lunch, less marked if the typology of the subjects is evening-type. Figure 2 presents the diurnal function of speed of execution in the hand skill task of tapping for the three groups of circadian typology.

Memory and Cognitive Tasks

Memory tasks involve the presentation of a material that the subject must recall after a certain amount of time, or use it to reply to an exercise that he is presented, or freely narrate its content. These tasks are differenciated according to the capacity and duration of information storage that are parameters of the diurnal functions. In the immediate or short-term memory tasks, the interval between presentation-retrieval is short, and the capacity of storage is limited (e.g., retrieving lists of digits or words). The optimal moment of execution is during the first hours of the morning (08.00-11.00 h.), and from this moment on they begin a slow but constant decline. Working memory tasks are those which require a longer duration of the information storage and a limited but larger capacity than in the short-term tasks (arithmetic, verbal reasoning and comprehension). They present a diurnal evolution, in relation to the speed of increase in the morning until they reach their peak at midday (12.00-14.00 h.), and then decrease significantly until the last hours of the day. In contrast, precision tends to show an advanced schedule with respect to speed, being superior in the first hours of the morning (09.00-11.00 h.), and then present a continued decrease until the end of the day.

In all the cases, taking into account circadian typology explains the phase mismatches among the groups, with optimal schedule advances in morning-types with respect to evening-types, and neither-types being in an intermediate position, although there should be more studies which include subjects from the latter group. In figure 2 the diurnal function of performance for the circadian typology groups in a working memory task, extreme groups differ by 7 hours, the effect of the post-lunch also being important for the morning and neither-types, and insignificant for evening-types.

Finally, we have the long-term memory tasks, whose interval between presentation and retrieval ranges from several hours to days or even months; the duration of the storage of the information is long and of unlimited capacity. The material is more and better remembered if it has been presented towards the end of the afternoon (18.00-20.00 h.), when there is more attention on the central information or on a processing based more on the meaning and less on the physical characteristics. This seems to question the validity of the traditional school timetables and the study habits that are usually advertised, since normally these tasks take place during the first hours of the day. Post-pandrial depression has influence on all types of memory, but it affects more the reaction time than the precision. The best moments to carry out these tasks in young adults are in the first hours of the afternoon if they are morning-type, and during the first hours of the evening if they are evening-type. The intermediate subjects are those who meet with the optimal normative interval.

Morning- and evening-type individuals differ in their performance to memory and more cognitive tasks, but only if motivation is controlled. Subjects with little motivation meet the traditional hypothesis that the morning subjects obtain better results in the morning and the evening subjects do so in the evening, while these differences disappear in highly-motivated subjects. The post-lunch effect and the midday sleep pressure are lower in evening-type than in morning- and neither-type individuals also in the complex tasks [9,31].

Subjective State

The measurement of the subjective state of the individuals in chronobiology is usually carried out by means of visual-analogue scales (VAS), such as the eight proposed by Monk [32]. Each scale is 100 mm long and allows the subject to put a cross at points between "very little" (situated at the left) and "very much" (situated at the right) based on perception at the moment of response (scores range from 0 to 100). These are faster to answer and facilitate the collection of information in cases of several measures during the day. The VAS demonstrate their validity, reliability, and sensitivity in both healthy and psychiatric subjects, circadian rhythms, and shift work or temporal isolation settings [25,32,33,34]. Four scales refer to dimension of subjective activation (alertness, vigour, weariness and sleepiness) and four refer to mood or affect (happiness, calmness, sadness and tension). A score of zero corresponds to the lowest rating in reference to positive scales (e.g. alertness) or the optimal response in reference to negative scales (e.g. sleepiness). Each group of four scales can then be added algebraically to give a simple global value of activation and mood. The formulas used are [(alert) + (vigorous) + 200 - (weariness) - (sleepiness)/4] for activation, and [(happiness) + (calm) + 200 - (sadness) - (tense)/4] for mood. Each scale can also be analysed independently to determine whether they differ in sensitivity to the variable studied.

Subjective activation presents a pattern that increases rapidly during the first hours in the morning, reaching its maximum between 11.00-13.00 h., at which time they begin a slow but continued decrease. This is also the evolution of the "alertness" scale, which itself is an extremely sensitive variable to the effects of hour of the day. The diurnal variations of the assessments of subjective affect or mood are less marked than those of activation, although they are superior in the morning with respect to the afternoon / evening, thus being very sensitive markers when the subjects suffer some psychopathological disorder [23,35]. The post-lunch drop interferes more markedly in the assessments of activation than of affect, and correlates with the decreases in the execution of performance tasks.

The consideration of circadian typology in subjective assessment yields highly significant differences. The maximum peak of subjective activation happens during the first hours in the morning for morning-types, during midday for neither-types and towards the end of the day for evening-types. Post-lunch produces a marked decrease both in morning- and neither-types, while this is almost insignificant in the evening-type [25,31]. Figure 3 presents the diurnal functions in the VAS for the alertness and weariness scales in relation to the three chronotypes. The mood ratings are superior during the first half of the day in morning types and during the second half of the day in the evening-types, the medium diurnal level being higher in the first group than in the second.

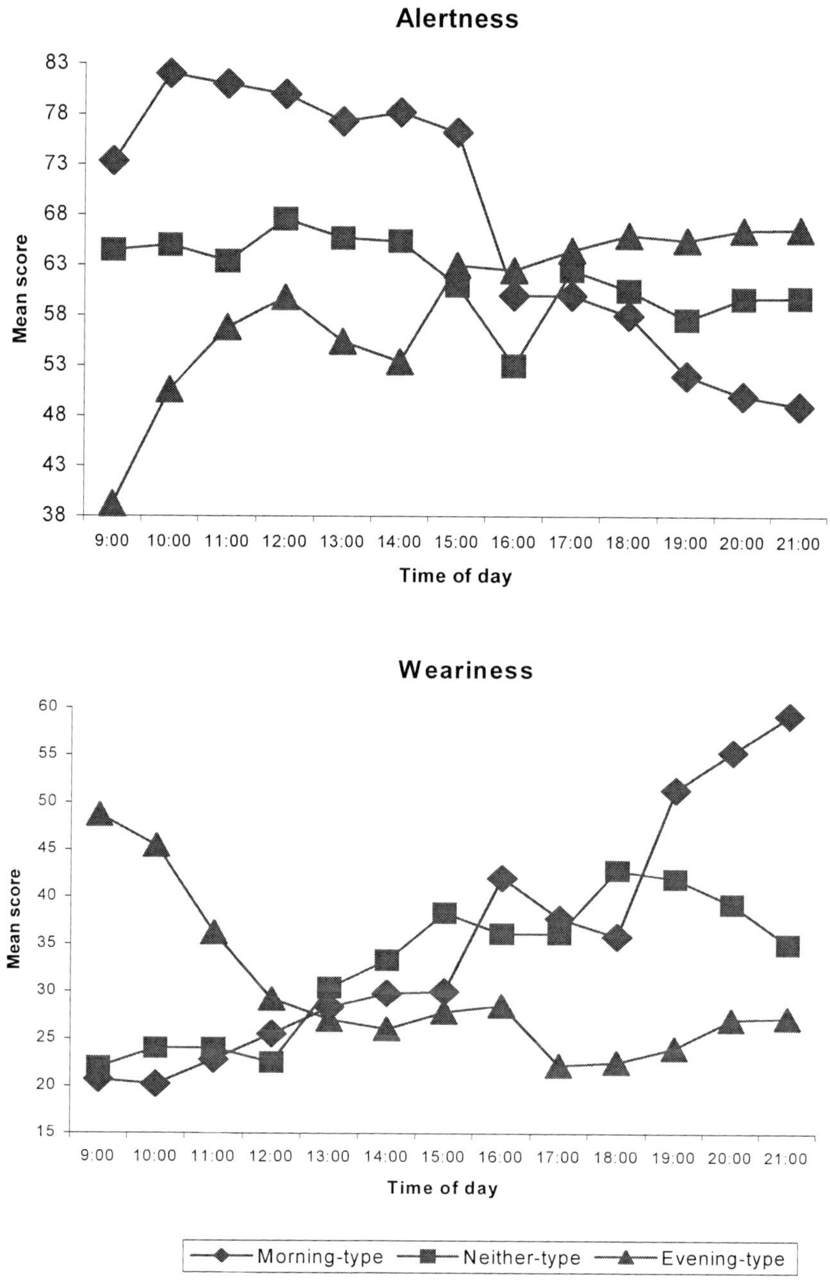

Figure 3. Diurnal variations of alertness (positive activation scale) and weariness (negative activation scale) according to the circadian typology groups. Scales range from 0 to 100.

It should also be noted that most of the works which study the performance in relation with circadian typology only select extreme subjects (morning- and evening-type). This reduces the possibility of generalising their results to the majority of the population. However, in addition to this, they take for granted that the rhythmic pattern of the neither-type typology provides simple intermediate values. The data commented on this section show and alert us that the diurnal pattern of performance of the neither-type is mixed, resembling more one or the other extreme group depending on the type of task evaluated. Thus, in the future, chronopsychological research must design studies which incorporate individuals from the three typologies. Another weak point currently is the lack of concern for the diurnal variations in the execution of complex tasks that involve superior cognitive skills. Without ignoring the problem of test-retest for repeated administrations of such tasks, intra-subject designs with a strong control of the individual characteristics may be an option for collecting new data whose applied interest is indubitable.

Constitutional Factors and Personality Characteristics

Circadian typology has been associated to constitutional factors, among which is age. As the adults grow older, there is a tendency to develop a morning-like rhythmic pattern [13,21,36,37,38,39], evident in the parameters which determine the cycle sleep-wake [5,14,16,40,41,42]. There is a tendency to go to bed and wake up earlier, and to present the highest levels of activation at an earlier time than when they were younger. Moreover, the quality of both the wake and the sleep periods decrease. The maintenance of the circadian rhythm and, very especially, that of melatonin production, is considered currently a biological marker associated to successful ageing, while its abolition is related to the appearance of neurodegenerative diseases and the concomitant presence of psychiatric pathology [43,44,45,46,47].

Adolescence (12 – 17 years) is a critical period, in which there is a change in the rhythmical patterns with a phase delay or the presence of a more pronounced tendency to the evening pattern, different from that which the young adult will adopt [48,49,50]. It is precisely at these ages that the development of an extreme evening pattern should be controlled, which can promote an individual functioning with not very healthy habits and impede the adaptation to the socio-environmental demands of activity [15]. Thus, the incorporation to the work setting may require an adjustment of the expression of the circadian rhythm which may not be successful in many extreme cases without a chronotherapeutic approach to advance the phase of the rhythmicity of the circadian period. With regard to this, it has also been observed that the work schedules are an important predictor of circadian typology [21,36]. Among the evening-type subjects there are more night-shift workers, several-shift workers and more unemployment, while among the morning types there is more diurnal and regular work shifts. An extreme morning pattern can also pose a problem because of its excessive rigidity, both in order to assume in an efficient way change in activity schedule, transmeridian travel or social activities which involve the late evening hours.

The possibility that age is related to circadian typology has been approached in several studies but there is no agreement on it currently. Some works have not found any differences between men and women [21,51], while others obtain a larger proportion of evening-type subjects among men and more morning-type among women [4,6,52,53]. It should be noted that the positive results are characteristic in that they use the Horne and Östberg questionnaire and select large samples of subjects. Hence, a small difference may appears significant just because of an excessive power in the statistical test used. Nevertheless, the difference in favour of morningness in women is in accordance with the empirical observation that the diurnal variations in women produce advances of phase (more morningness) with respect to men [5,35,38,54], with magnitude depending of the considered parameter. Such difference can be explained by the control of the circamensual rhythmicity associated to the menstrual cycle in women, which would act against the intensity of the rhythmic control of circadian periodicity.

With respect to personality traits, several works in the last decades have related the dimensions in the model proposed by Eysenck with circadian typology. There is a tendency in the morning-type subjects to score towards the introversion pole, while the evening types score towards the extraversion pole [22]. However, the observation that morning types were more conscientious and higher in self-esteem than the evening types suggest that conscientiousness rather than the extraversion dimension may be the personality factor that best distinguishes circadian typologies [55]. The association with the neuroticism construct or anxiety-trait is less obvious [22], although some works obtain higher scores in the evening-type subjects [56,57,58,59]. There are also data suggesting an association by watching late night TV, some kind delayed sleep syndrome and a growing disorder in children and adolescents, namely anxious school refusal [60]. This anxiety disorder is related to separation anxiety and/or social phobia, and can lead to anxious and/or depressive disorders with social impairment at adulthood.

Moreover, emotional reactivity or stress have less impact in morning-types on the performance of cognitive tasks [57]. Regarding this, it has been showed using the Millon Inventory of Personality Types that the morning-types prefer obtaining information from their surroundings based on concrete sources and direct experience, whereas evening-types prefer information based on intangible and abstract data. In addition, the evening types interpret the information using subjective and emotional parameters, and tend to elaborate it with knowledge schemata which are newer and more original [61,62].

More recently it has been found that the evening-type subject present higher scores in impulsivity and sensation-seeking [31,58,63], also expressing less affinity in their relationships with other people (acquiescence) and less respect for customs and social norms [61,62] in comparison with the morning- and neither- types. Although none of these variables is in itself a disorder, it is true that all of them gathered in extreme cases may constitute risk factors for the appearance of several problematic situations. Recently, we have reported that adults suspected of having Attention Deficit/Hyperactivity Disorder scored lower on morningness, and that this relationship was stronger for the Inattention factor than the Impulsive-Hyperactive factor. The non-significant correlation between morningness and this impulsivity factor was discussed with regard to item content [64] .

Some works have found differences in the preference for leisure activities and sport practices according to the optimal schedule of activation of the subjects. Morning- types prefer outdoor activities (e.g., trekking), and practice more precision and manual skill sports, and even if they prefer other types of physical exercise, they tend to do it in the morning [38,65,66]. The evening types prefer evening activities of social interaction and if they do some exercise, they choose anaerobic activities which are carried out towards the end of the day (soccer, swimming, etc.). In the same way, the morning subjects tend to have their meals in more regular schedules and usually follow healthier diets [14]. The evening types, in addition to having more variable time patterns, also tend to "snack" between meals and express a lesser preference to follow a "Mediterranean" diet and to control the use of normal and spicy seasonings. With this regard, there is a field of growing interest concerned with circadian rhythms in eating disorders such as night-eating syndrome, bulimia and anorexia nervosa. Recently, we have speculated that the isolation with bed-rest proposed in the treatment of anorectics for more than one and half century by French psychiatrists is an unrecognized chronotherapeutic treatment [67].

Associations with Health Parameters

Another field of research yielding highly relevant results regards the possible relations between circadian typology and health habits, as well as the risk or vulnerability to develop psychopathological disorders. The sleep beliefs and habits, drug consumption, the general state of health and the incidence of depressive and bipolar I disorders are some of the aspects developed in the present section.

Evening-types present a higher level of wrong beliefs and a lesser knowledge and practice of sleep hygiene, whereas morning-types show a more correct level of beliefs and practice of sleep hygiene [68,69]. This explains the differences observed in the quality of the wake period, both subjectively and objectively. This observation has clinical interest in the area of sleep disorders and their prevention, since the correct sleep beliefs and attitudes are related with the sleep hygiene practice and with a greater efficacy in the treatment of insomnia [70,71,72]. However, in younger population without sleep problems the wrong sleep beliefs did not seem to negatively affect their sleep, although it could be considered a risk factor for the future appearance of insomnia [73]. For this reason, the morning-type typology could be considered as a protective factor for sleep disorders, and these individual differences may be taken into account in order to adopt preventive measures [69].

The pioneering work by Adan [74], who observed differences in the consumption habits, both daily and sporadic, of psychoactive substances among chronotypes, has been confirmed both in adolescents and adult samples [59,75]. The results are totally clear: the largest proportion of consumers (alcohol, coffee and tobacco) is found among evening-types, the smallest among morning- types, while neither-types hold an intermediate position. Tea consumption is the only exception since, more morning-types take it daily (see figure 4). This can be explained taking into account the differences in diurnal activation among chronotypes, and the strategies to adjust these differences to the socio-laboural requirements of the environment. Thus, while morning-types wake up with a level of activation adjusted to begin their daily activities, the evening types present defective levels of activation and need a considerable amount of time to activate themselves. The consumption of psychostimulant substances such as coffee or nicotine helps them to boost their functioning, but the beneficial effects turns out to reinforce the consuming behaviour. In Spain tea, which is also a stimulant beverage, is usually taken during the first hours of the evening, when morning-types perceive an insufficient activation and consumed this drink and in larger quantities. In contrast, the consumption of alcohol, a depressive substance, usually happens towards the end of the day, when the evening-type subjects are highly activated, maybe even excessively, depending on the activity they have to carry out. The initial contact with alcohol, which promotes the establishment of a habitual pattern of consumption, is closely related to going out socially during the night hours, an activity which evening-types are more predisposed to than morning-types. During adolescence, loitering in the evening is a risk factor for antisocial behaviours, drugs and alcohol consumption [76].

Evidently, there are many other factors that may explain addictive behaviours. Being one of them, circadian typology should be taken into consideration by professionals in this field should have it into account. It should be added that the consumption of substances as a regulating mechanism of activation is valid only if the consumption is eventual, since with regular consumption there are neuroadaptations and what is know as the spiral of stress which explains the dependence on psychoactive substances [34,77,78]. Thus, individuals continue consuming in order to avoid the negative consequences of the absence of substance in their organism, which is the symptomatology of withdrawal.

In comparison with morning-types, evening-types are 2.5 times more likely to report that their general health is poor or fair [21]. Among evening-types, there is also a higher proportion of individuals with somatic and psychological problems of superior intensity when present [79,80]. It has also been observed that among young evening-type women there is a higher frequency of unstable menstrual cycles and more occurrence and severity of premenstrual mental and physical symptoms compared to the morning-type women [81]. As mentioned early, the incidence of bulimia nervosa is higher in evening-types [80]. A factor that could be considered as a possible trigger of all these problems is the sleep restriction, since it can suppress immune function, negatively affect mood and induce glucose intolerance and increase appetite [21]. The clinical management of these disorders which includes a chronobiological perspective can increase the symptom remission and achieve periods without durable symptoms.

It has been observed recently that phase retardation in the circadian rhythmic expression (eveningness) is associated with vulnerability to develop emotional problems. The presence of depressive symptoms which can meet the diagnostic criteria for major depression is higher in the evening-type and lower in the morning-type [49,66,82], and specifically with the seasonal affective disorder [53,83], although this needs to be confirmed in larger samples and with subjects who live in different latitudes. Similarly, in patients diagnosed of bipolar I disorder there is a high proportion of evening-types, and there seems to be a direct relationship between greater eveningness and earlier age of onset, as well as larger duration of the most severe depressive episodes [84]. In the future, more attention should be drawn to these lines, since although it cannot be discarded that the results are influenced by the impact of medication, the mood state, and chronicity also suggest practical therapeutic avenues for the treatment of these patients.

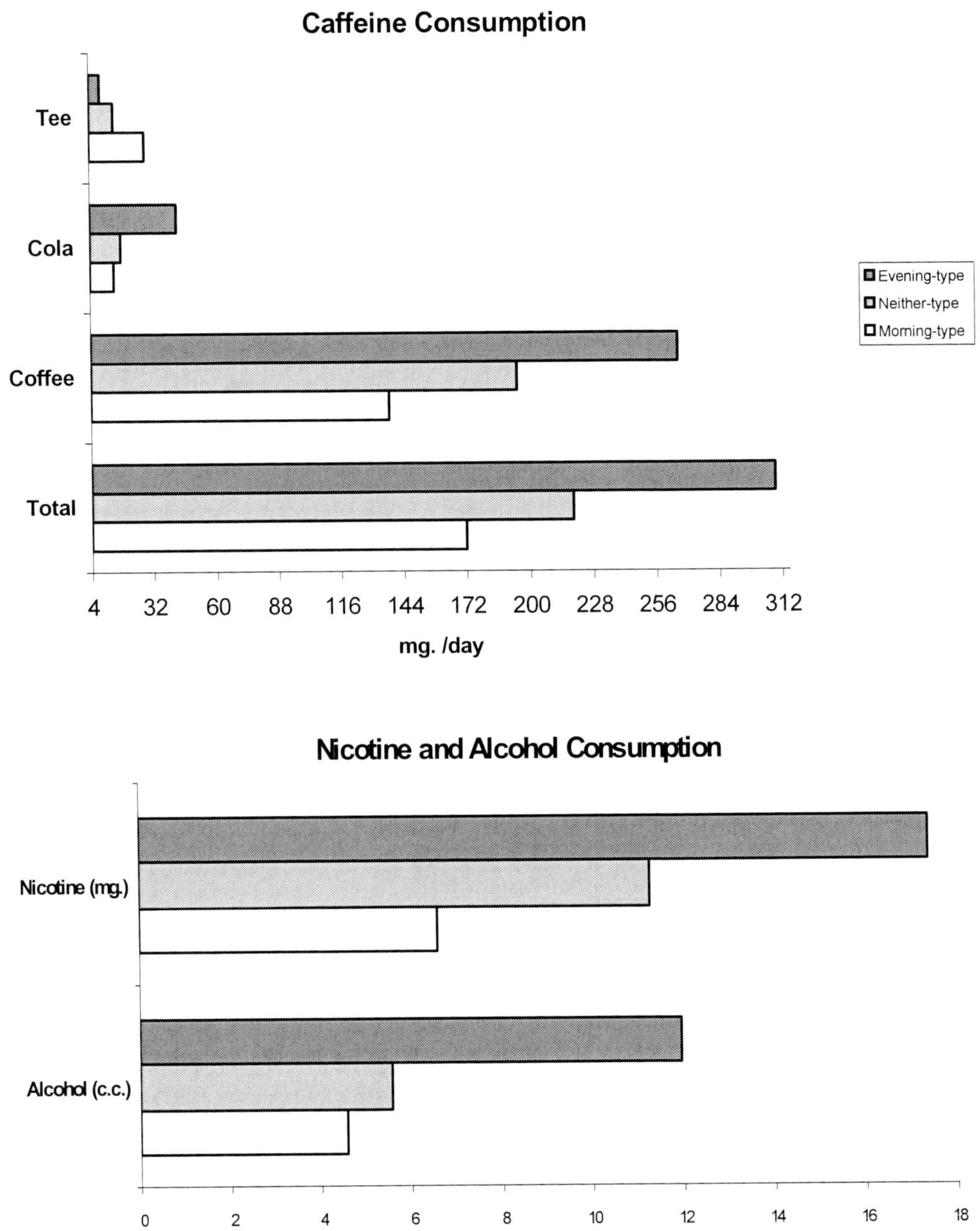

Figure 4. Habitual daily consumption of caffeine, nicotine and alcohol according to the circadian typology of the subjects. Data revised from Adan (1994).

Conclusions

The differences among chronotypes should be understood from a wide perspective that would lead us to state that they are associated to different life-styles. Although one should avoid a simplistic short cut of associating the evening typology to some negative aspects, data point to the idea that an evening pattern is a risk factor for some disorders and morningness is a protection factor. Some studies have found differences in the consumption habits of psychoactive substances, the patterns and content of meals, the preference for certain leisure activities and patterns of social interaction [65]. The habits of the morning-types are healthier in general than those found in evening-types. Hence, it may be of interest to consider this variable in a bio-psycho-social health model, in order to design both preventive and therapeutic approaches which are more efficient. This may be extremely useful in the field of drug addictions, where prevention programmes and also treatments for detoxification and rehabilitation should consider this individual difference. However, it is not advisable to consolidate at adulthood an extreme tendency for morningness, since it will probably lead to important difficulties in adapting to temporary changes in schedule (e.g. jet-lag) or permanent (e.g. shift-work), and also during the ageing period a higher probability of developing a disorder of advanced-phase sleep.

So far collected data suggest some preventive measures in order to avoid the consolidation or maintenance of an extreme circadian typology. Taking into account the effect of the photoperiod in the endogenous rhythmic expression, it should be advised that children born in autumn / winter be exposed to the light despite the weather conditions, while avoiding that those born in the spring / summer have an excess of light exposure. Around two or three hours of exposure is the ideal time, and it does not make a difference whether the baby is sleeping or not, given that the simple incidence of light on the eyelids activates the transmission of the retino-hypothalamic tract. During the period of adolescence, habits must be promoted concerning sleep-wake, alimentary, leisure time activities, etc. that do not happen in the late afternoon, or regarding the temporal socio-laboural organisation, in order to avoid the development of an extreme evening pattern. An aspect which should also be educated is the habit seen in some youth to change the sleep-wake cycle between days with obligations and leisure days, since in some cases this is inverted and they tend to sleep during the day and be active all night.

Finally, if an extreme pattern of circadian typology has installed causing problems for the socio-laboural functioning of the individual, the most effective and secure treatment is to apply a behavioural approach to resynchronise the rhythmic expression of the circadian period. Those treatments based on chronobiology, such as phototherapy [85,86,87], or the administration of melatonin, have also proved their effectiveness in subjects resistant to behavioural treatment of rhythmic adjustment, and in several disorders. This is the case in several sleep pathologies, major seasonal or non-seasonal depression, or the presence of neurodegeneration due to excitotoxicity and psychiatric symptomatology in the elderly. Phototherapy must be chosen as the first option if it is clear that there is an insufficient exposure to light in the individual and the non-existence of deficits that prevent the transduction of the light signal to the central nervous system. Melatonin, the hormone secreted by the pineal gland, is the signal of darkness for our organism, and the most potent antioxidant agent known to date [88]. Its major inconvenient is that in most countries it is sold as a nutritional complement rather than as a medical treatment, and therefore it is not subject to any health regulation nor are the manufacturers subjected to certify its purity. The synthesis of new pharmacological drugs which are agonistic to melatonin is rapidly increasing. Recently the chronohipnotic Ramelteon has been commercialized, and Agomelatin is phase III of clinical investigation with excellent results in the treatment of severe major depression. Although there is still much work to be done, the findings in chronobiology are already entering the clinical settings, and their consideration is helping to design better strategies of therapeutic treatment and to increase the quality of life of the patients.

References

[1] Horne, J,A, and Östberg, O. (1976). A self-assessment questionnaire to determine morningness-eveningness in human circadian rhythms. *International Journal of Chronobiology,* 4, 97-110.

[2] Grandin, L.D., Alloy, L.B and Abramson L.Y (2006).The social zeitgeber theory, circadian rhythms, and mood disorders: review and evaluation. *Clinical Psychology Review,* 26, 679-694.

[3] Katzenberg, D., Young, T., Finn, L., Lin, L., King, D.P., Takahashi, J.S. and Mignot, E. (1998). A clock polymorphism associated with human diurnal preference. *Sleep,* 21, 569-576.

[4] Vink, J.M., Groot, A.S., Kerkhof, G.A. and Boomsma, D.I. (2001). Genetic analysis of morningness and eveningness. *Chronobiology International,* 18, 809-822.

[5] Klei, L., Reitz, P., Miller, M., Wood, J., Maendel, S., Gross, D., Waldner, T., Eaton, J., Monk, T.H. and Nimgoankor, V.L. (2005). Heritability of morningness-eveningness and self-report sleep measures in a family-based sample of 521 hutterites. *Chronobiology International,* 22, 1041-1054.

[6] Natale, V., Adan, A. (1999). Season of birth modulates the morningness-eveningness preference. *Neuroscience Letters,* 274, 139-141.

[7] Natale, V., Adan, A. and Chotai, J. (2002). Further results on the association between morningness-evenigness preference and the season of birth in human adults. *Neuropsychobiology,* 46, 209-214.

[8] Mongrain, V., Paquet J. and Dumont, M. (2006). Contribution of the photoperiod at birth to the association between season of birth and diurnal preference. *Neuroscience Letters,* 406, 113-116.

[9] Taillard, J., Philip, P., Coste, O., Sagaspe, P. and Bioulac, B. (2003). The circadian and homeostatic modulation of sleep pressure during wakefulness differs between morning and evening chronotypes. *Journal of Sleep Research,* 12, 275-282.

[10] Rosenthal, L., Day, R., Gerhardstein, R., Meixner, R. Roth, T., Guido, P. and Fortier, J. (2001). Sleepiness/alertness among healthy evening and morning type individuals. *Sleep Medicine,* 2, 243-248.

[11] Natale, V., Alzani, A. and Cicogna, P.C. (2003). Cognitive efficiency and circadian typologies: a diurnal study. *Personality and Individual Differences,* 35, 1089-1105.

[12] Natale, V., Esposito, M.J., Martoni, M. and Fabbri, M. (2006). Validity of the reduced version of the morningness-eveningness questionnaire. *Sleep and Biological Rhythms,* 4, 72-74.

[13] Monk, T.H., Frank, E., Potts, J.M. and Kupher, D.J. (2002). A simple way to measure daily lifestyle regularity. *Journal of Sleep Research,* 11, 183-190.

[14] Monk, T.H., Buysse, D.J., Potts, J.M., DeGrazia, J.M., Kupher, D.J. (2004). Morningness-eveningness and lifestyle regularity. *Chronobiology International,* 21, 435-443.

[15] Waterhouse, J., Folkard, S,, Van Dongen, H., Minors, D., Osen, D., Kerkhof, G., et al. (2001). Temperature profiles, and the effect of sleep on them, in relation to morningness-eveningness in healthy female subjects. *Chronobiology International,* 18, 227-247.

[16] Mongrain, V., Lavoie, S., Selmaqui, B., Paquet, J. and Dumont, M. (2004) Phase relationships between sleep-wake cycle and underlying circadian rhythms in morningness-eveningness. *Journal of Biological Rhythms,* 19, 248-257.

[17] Mongrain, V., Carrier, J. and Dumont, M. (2006). Circadian and homeostatic sleep regulation in morningness-eveningness. *Journal of Sleep Research,* 15, 162-166.

[18] Seo, Y., Matsumoto, K., Park, Y., Shinkoda, H. and Noh, T. (2000). The relationship between sleep and shift system, age and chronotype in shift workers. *Biological Rhythm Research,* 31, 559-579.

[19] Natale, V., Martoni, M. and Cicogna, P.C. (2003). Effects of circadian typology on sleep-wake behavior of air traffic controllers. *Psychiatric Clinic Neuroscience,* 57, 539-541.

[20] Roenneberg, T., Wirz-Justice, A. and Merrow, M. (2003). Life between clocks: daily temporal patterns of human chronotypes. *Journal of Biological Rhythms,* 18, 80-90.

[21] Paine, S.J., Gander, P.H. and Travier, N. (2006). The epidemiology of morningness/eveningness influence of age, gender, ethnicity and socioeconomic factors. In adults (30-49 years). *Journal of Biological Rhythms,* 21, 68-76.

[22] Tankova, I., Adan, A. and Buela-Casal, G. (1994). Circadian typology and individual differences. A review. *Personality and Individual Differences,* 16, 671-684.

[23] Monk, T.H., Buysse, D.J., Reynolds, C.F. and Kupher, D.J. (1996). Circadian determinants of the postlunch dip in performance. *Chronobiology International,* 13, 123-133.

[24] Tietzel, A.J. and Lack, L.C. (2002). The recuperative value of brief and ultra-brief naps on alertness and cognitive performance. Journal of Sleep Research, 11, 213-218.

[25] Adan, A. (2005). Mood rhythmicity and individual differences. In: Clark, V.A. (Ed.). *Causes, Role and Influence of Mood States.* Pp: 33-56. New York: Nova Science Publishers.

[26] Adan, A. (1993). Circadian variations in psychological measures. A new classification. *Chronobiologia,* 20, 145-162.

[27] Adan, A. (1997). Individual differences in physiological and behavioral diurnal variations. *Revista Latinoamericana de Psicología,* 29, 81-114.

[28] Van Dongen, H.P.A. (1998). *Inter- and intra-individual differences in circadian phase.* Leiden: Leiden University, Department of Physiology.

[29] Carrier, J. and Monk, T.H. (2000). Circadian rhythms of performance: new trends. *Chronobiology International,* 17, 719-732.

[30] Owens, D.S., Macdonald, I., Tucker, P., Sytnik, N., Minors, D., Waterhouse, J. and Folkard, S. (2000). Diurnal variations in the mood and performance of highly practiced young women living under strictly controlled conditions. *British Journal of Psychology,* 81, 63-72.

[31] Adan, A. (2006). Cronobiología del rendimiento cognitivo y físico. In: Madrid, J.A. and Rol de Lama, M.A. (Eds.). *Cronobiología Básica y Clínica.* Pp: 713-739. Madrid: Editec@red.

[32] Monk, T.H. (1989). A visual analogue scale technique to measure global vigor and affect. *Psychiatry Research,* 27, 89-99.

[33] Monk, T.H. (1989). Circadian rhythms in subjective activation, mood and performance efficiency. In: Kryger, M.H., Roth, T. and Dement, W.C. (Eds). *Principles and Practice of Sleep Medicine.* Pp. 163-172. Philadelphia: W.B. Saunders Company.

[34] Adan, A., Prat, G., Sánchez-Turet, M. (2004). Effects of nicotine dependence on diurnal variations of subjective activation and mood. *Addiction,* 98, 1599-1607.

[35] Adan, A. and Sánchez-Turet, M. (2001). Gender differences in diurnal variations of subjective activation and mood. *Chronobiology International,* 18, 491-502.

[36] Adan, A. and Almirall, H. (1992). The influence of age, work schedule and personality on morningness dimension. *International Journal of Psychophysiology,* 12, 95-99.

[37] Monk, T.H., Kupher, D.J. (2000). Circadian rhythms in healthy aging – Effects downstream from the pacemaker. *Chronobiology International,* 17, 355-368.

[38] Park, Y.M., Matzumoto, K., Seo, Y.J., Kang, M.J. and Nagashima, H. (2002). Effects of age and gender on sleep habits and sleep trouble for aged people. *Biological Rhythms Research,* 33, 39-51.

[39] Taillard, J., Philip, P., Chastang, J.-F. and Bioulac, B. (2004). Validation of Horne and Östberg Morningness-Eveningness questionnaire in a middle-aged population of French workers. *Journal of Biological Rhythms,* 19, 76-86.

[40] Carrier, J., Monk, T.H., Buysse, D.J. and Kupfer, D.J (1997). Sleep and morningness-eveningness in the 'middle' years of life (20-59y). *Journal of Sleep Research,* 6, 230 237.

[41] Park, Y.M., Matzumoto, K., Seo, Y.J., Shinkoda, H. and Park, K.P. (1997). Scores on morningness-eveningness and sleep habits of Korean students, Japanese students, and Japanese workers. *Perceptual and Motor Skills,* 85, 143-154.

[42] Monk, T.H., Reynolds, C.F., Kupher, D.J., Hoch, C.C., Carrier, J. and Houck, P.R. (1997). Differences over the life span in daily life-style regularity. *Chronobiology International,* 14, 295-306.

[43] Richardson, G. and Tate, B. (2000). Hormonal and pharmacological manipulation of the circadian clock: recent developments and future strategies. *Sleep,* 23 (Suppl.), S77-S85.

[44] Mattson, M.P., Duan, W., Chan, S.L., Cheng, A., Haughey, N., Gary, D.S., Lee, J. and Furukawa, K. (2002). Neuroprotective and neurorestorative signal transduction

mechanisms in brain aging: modification by genes, diet and behavior. *Neurobiology of Aging,* 23, 695-705.

[45] Magri, F., Sarra, S., Cinchetti, W., Guazzoni, V., Fioravanti, M., Cravello, L. and Ferrari, E. (2005). Qualitative and quantitative changes of melatonine levels in physiological and pathological aging and in centenarians. *Journal Pineal Research,* 37, 256-261.

[46] Wu, Y-H. and Swaab, D.F. (2005). The human pineal gland and melatonin in aging and Alzheimer's disease. *Journal Pineal Research,* 38, 145-152.

[47] Tractenberg, R.E., Singer, C.M. and Kaye, J.A. (2006). Characterising sleep problems in persons with Alzheimer's disease and normal eldery. *Journal of Sleep Research,* 15, 97-103.

[48] Kim, S., Dueker, G.L., Hasher, L. and Goldstein, D. (2002). Children's time of day preference: Age, gender, and ethnic differences. *Personality and Individual Differences,* 33, 1083-1090.

[49] Gau SF and Merikangas KR (2004) Similarities and difference in sleep-wake patterns among adults and their children. *Sleep,* 27, 299-304.

[50] Roenneberg, T., Kuehnle, T., Pramstaller, P. P., Ricken, J., Havel, M., Guth, A. and Merrow, M. (2004). A marker for the end of adolescence. *Current Biology,* 14, R1038-R1039.

[51] Caci, H., Adan, A., Bohle, P., Natale, V., Pornpitakpan, C. and Tilley, A.J. (2005). Transcultural properties of the Composite Scale of Morningness: the relevance of the "morning affect" factor. *Chronobiology International,* 22, 523-540.

[52] Adan, A. and Natale, V. (2002). Gender differences in morningness-eveningness preference. *Chronobiology International,* 19, 709-720.

[53] Natale, V., Adan, A. and Scapellato, P. (2005). Are seasonality of mood and eveningness closely associated?. *Psychiatry Research,* 136, 51-60.

[54] Natale, V. and Danesi, E. (2002). Gender and circadian typology. *Biological Rhythm Research,* 33, 261-269.

[55] Jackon, L.A. and Gerard, D.A. (1996). Diurnal type, the "big five" personality factors, and other characteristics. *Journal of Social Behavior and Personality,* 11, 275-284.

[56] Mecacci, L,, Scaglione, M.R. and Vitrano, I. (1991). Diurnal and monthly variations of temperature and self-reported activation in relation to sex and circadian typology. *Personality and Individual Differences,* 12, 819-824.

[57] Mecacci, L., Righi, S. and Rocchetti, G., (2004). Cognitive failures and circadian typology. *Personality and Individual Differences,* 37, 107–113.

[58] Neubauer, A.C. (1992). Psychometric comparison of two circadian rhythms questionnaires and their relationship with personality. *Personality and Individual Differences,* 13, 125-131.

[59] Gau SS, Soong WT and Merikangas KR (2004) Correlates of sleep-wake patterns among children and young adolescents in Taiwan. *Sleep,* 27, 512-519.

[60] Tomoda, A., Miike, T. Uezone, K. and Kawasaki, T. (1994). A school refusal case with biological rhythm disturbance and melatonin therapy. *Brain Development,* 16, 71-76.

[61] Díaz-Morales, J.F. and Aparicio, M. (2003). Relaciones entre matutinidad-vespertinidad y estilos de personalidad. *Anales de Psicología, 19,* 247-256.

[62] Díaz-Morales, J.F. (in press). Morning and evening-types: exploring their personality styles. *Personality and Individual Differences*. Doi: 10.1016/j.paid.2007.02.002.

[63] Caci, H., Robert, P. and Boyer, P. (2004). Novelty seekers and impulsive subjects are low in morningness. *European Psychiatry* ,19, 79-84.

[64] Caci, H., Bouchez, J. and Boile, F.J. (submitted). Adults suspect of attention-deficit disorder are evening oriented.

[65] Cofer, L.F., Grice, J.W., Sethre-Hofstad, L., Radi, C.J., Zimmermann, L.K., Palmer-Seal. D. and Santa-María, G. (1999). Developmental perspectives on morningness-eveningness and social interactions. *Human Development,* 42, 169-198.

[66] Giannotti, F., Cortesi, F., Sebastiani, T. and Ottavino, S. (2002). Circadian preference, sleep and daytime behavior in adolescence. *Journal of Sleep Research,* 11, 191-199.

[67] Caci, H. "Is anorexia nervosa a chronobiological disorder?', AEP Congress, *European Psychiatry*, 2006, 21(suppl.1), S80

[68] Gau SSF, Chong MY, Chen TH and Cheng ATA (2005) A three-year panel study of mental disorders among adolescents in Taiwan. *American Journal of Psychiatry,* 162, 1344-1350.

[69] Adan, A., Fabbri, M., Natale, V. and Prat, G. (2006). Sleep Beliefs Scale (SBS) and circadian typology. *Journal of Sleep Research,* 15, 125-132.

[70] Morin, C.M., Blais F. and Saverd, J. (2002). Are changes in beliefs and attitudes about sleep related to sleep improvements in the treatment of insomnia? *Behaviour Research and Therapy,* 40, 741-752.

[71] Ree, M.J. and Harvey, A,G. (2004). Investigating safety behaviours in insomnia: the development of the sleep-related behaviours questionnaire (SRBQ). *Behaviour Change,* 21, 26-36.

[72] Cervena, K., Dauvilliers, Y., Espa, F., Touchon, J., Matousek, M., Billiard, M. and Besset, A. (2004). Effect of cognitive behavioural therapy for insomnia on sleep architecture and sleep EEG power spectra in psychophysiological insomnia. *Journal of Sleep Research,* 13, 385-393.

[73] Morgan, K. (2003). Day time activity and risk factors for late-life insomnia. *Journal Sleep Research,* 12, 231-238.

[74] Adan, A. (1994). Chronotype and personality factors in the daily consumption of alcohol and psychostimulants. *Addiction,* 89, 455-462.

[75] Gau, S.F. and Soong, W.T. (2003) The transition of sleep-wake patterns in early adolescence. *Sleep,* 26, 449-454.

[76] Andershed, K-A. (2005). In sync with adolescence: the role of morningness-eveningness in adolescence. Springer: New York.

[77] Koob, G.F. and Le Moal, M. (1997). Drug abuse: hedonic homeostatic dysregulation. *Science,* 278, 52-58.

[78] Koob, G.F. (2006). The neurobiology of addiction: a neuroadaptational view relevant for diagnosis. *Addiction,* 101, 23-30.

[79] Mecacci, L. and Rocchetti, G. (1998) Morning and evening types: stress-related personality aspects. *Personality and Individual Differences,* 25, 537-542.

[80] Kasov (2001). Eveningness and bulimic behaviour. *Personality and Individual Differences*, 31, 361-369.

[81] Takeuchi, H., Oishi, T. and Harada, T. (2005). Association between morningness-eveningness preference and mental/physical premenstrual symptoms in Japanese females 12 to 31 years of age. *Chronobiology International,* 22, 1055-1068.

[82] Chelminski, I., Ferraro, F.R., Petros, T.V. and Plaud, J.J. (1999). An analysis of the "morningness-eveningness" dimension in "depressive" college students. *Journal of Affective Disorders,* 52, 19-29.

[83] Murray, G., Allen, N.B. and Trinder, J. (2003). Seasonality and circadian phase delay: prospective evidence that winter lowering of mood is associated with a shift towards eveningness. *Journal of Affective Disorders,* 76, 15-22.

[84] Mansour, H.A., Wood, J., Chowdari, K.V., Dayal, M., Thase, M.E., Kupher, D.J., Monk, T.H., Devlin, B. and Nimgaourkar, V.L. (2005). Circadian phase variation in bipolar I disorder. *Chronobiology International,* 22, 571-584.

[85] Skjerve, A., Holsten, F., Aarsland, D., Bjorvatn, B., et al. (2004). Improvement in behavioural symptoms and advance of activity acrophase after short-term bright light treatment in severe dementia. *Psychiatry and Clinical Neurosciences,* 58, 343-347.

[86] Lockley, S.W., Evans, E.E., Scheer, F.A.J.L., Brainard, G.C., Czeisler, C.A. and Aeschbach, D. (2006). Short-wavelength sensitivity for the direct effects of light on alertness, vigilance, and the waking electroencephalogram in humans. *Sleep Physiology*, 29, 161-168.

[87] Vandewalle, G., Balteau, E., Phillips, C., Degueldre, C., Moreau, V., Sterpenich, V., Albouy, G., Darsaud, A., Deseilles, M., Dang-Vu, T.T., Peigneux, P., Luxen, A., Dijk, D.J. and Maquet, P. (2006). Daytime light exposure dynamically enhances brain responses. *Current Biology,* 16, 1616-1621.

[88] Arendt, J. (2000). In what circumstances is melatonin a useful sleep therapy? Consensus statement, WFSRS focus group, Dresden, November 1999. *Journal of Sleep Research,* 9, 397-398.

In: Progress in Circadian Rhythm Research
Editor: Anne-Laure Léglise, pp. 163-176

ISBN: 978-1-60021-876-7

Chapter VII

Circadian Rhythms and Physical Activity

Sebastien Racinais
ASPETAR, Qatar Orthopedic and Sports Medicine Hospital,
Exercise and Sports Science Department, Doha, Qatar

Abstract

In a neutral and controlled environment, physical performance during a short-duration exercise fluctuates during the day with an acrophase close to 18:00 h. The diurnal increase in muscle power is mainly attributed to a diurnal increase in muscle force. This diurnal increase in muscle force is mainly attributed to a diurnal increase in muscle contractile properties. Today, the physiological basis of this variation in muscle contractile function remains unclear and represents a challenging topic in the future because the circadian variation in muscle force is generally in phase with the circadian variation in body core temperature. It is hypothesized that the diurnal increase in core temperature could have a beneficial passive warm-up effect.

It is possible to shift the acrophase of the muscle power variation by a couple of hours following a few days by adapting sleep and meal schedules. Moreover, the use of another source of passive warm-up (e.g., a warm environment) seems to blunt the diurnal variation in muscle power. With long duration exercise, the thermal stress induced by physical exercise in a warm (and humid) environment should be higher in the evening, when the initial body temperatures are the highest.

Physical activities have to be considered for their effect on health. For example, cardiac events occur mainly between 06:00 - 12:00 h and a vigorous physical activity could be a trigger for them. Clinicians should advise at risk patients to avoid exercise at this time-of-day. Physical activity can also have some beneficial effects. For example, physical activity increases the amplitude of some circadian variations in elderly people allowing for the improvement both nocturnal sleep quality (i.e., less) and diurnal vigilance.

Introduction

The biological rhythms may influence the physical abilities, however participation in regular physical activity may influence the biological rhythms. The first part of this chapter will focus on the diurnal variation in physical attributes. Firstly, the effect of circadian rhythm upon different physical abilities will be discussed. However, because most physical activities are performed during day-time, and thus low nocturnal values have only a poor functional significance, the second part of this chapter will focus on the effect of time-of-day upon physical performance (i.e., morning versus afternoon). Thirdly, how to control the diurnal variation in physical ability will be discussed. The effect of body temperature and its diurnal variations on the exercising body will be emphased.

From a broader point of view, the influence of circadian rhythms constitutes a challenging issue in medicine. For example, for several decades it has been known that the time-of-day at which chemotherapy administrated influences relapse rates in children (e.g. Rivard *et al.* 1985). In parallel, physical activity is also generally considered to induce chronic adaptation that reduces risk factors for many medical conditions (long term). Consequently, the second part of this chapter aims to link the triptych circadian rhythms, health, and physical activity by relating the effects of physical activity on health in regard to the circadian rhythms. Three examples will be used to highlight this complex relationship. In the first example, the well-known circadian rhythm in body core temperature will be considered for its consequences on thermoregulation and health during exercise. The diurnal increase in core temperature generally observed at rest persists during physical exercise (Reilly *et al.* 1984, Reilly et Brooks 1990). The thermal stress induced by a physical exercise should be higher in the evening, when the initial body temperatures are the highest (Reilly et Garret 1998). This could be particularly important for exercising in warm and humid environments where sweating fails to compensate for the increase in body temperature increase or for a patient with spinal cord injuries, with a loss or reduction in the ability to dissipate heat from below the level of lesion. In the second example, physical exercise will be considered as a risk factor for a cardiac event (Siscovick *et al.* 1984, Giri *et al.* 1999). This will be discussed in relation to the diurnal variation in the occurrence of cardiac events (Muller *et al.* 1987, Willich *et al.* 1989). In the third example, physical activity will be considered as a zeitgeber helping to maintain the circadian sleep-wake cycle in the elderly as temperature variations of the circadian rhythm are reduced with aging (Atkinson *et al.* 1992). This decrease is associated with an alteration of the sleep-wake cycle (Flamer 1996, Vitiello *et al.* 1997). Regular physical activity, however, can induce chronic adaptations in the amplitude of circadian rhythms (Atkinson *et al.* 1993). An improvement in the sleep-wake cycle (i.e. with better daytime vigilance and less wake-up time overnight) has been seen in the elderly subjects after a physical training program (Mauvieux *et al.* 2003).

I. Diurnal Variations in Physical Ability

I.1. Physical Ability and Circadian Rhythms

The ability to perform in middle- and long-duration events is generally considered to be linked to the athlete's maximal oxygen consumption; however, the circadian rhythm of this parameter remains unclear. Some of the studies have concluded that maximal oxygen consumption does not significantly fluctuate during the day (Faria and Drummond 1982, Reilly and Brooks 1982 and 1990, Jones and Hill 1995), whereas others have observed diurnal variation (Hill *et al.* 1988 and 1989b, Williams and Hill 1995) with, in some instances, no signiticant diurnal variation in performance itself (e.g. time to exhaustion) (Hill *et al.* 1988).

In short duration performances, the data is unequivocal and displays a diurnal fluctuation. Recent research supports and explains the observation of Rodahl (1976) that high-level swimmer perform better in the evening than in the morning. A significant circadian variation in both back and leg force, with an acrophase of the circadian rhythm at 16:53 h and 18:20 h, respectively, (Coldwells *et al.* 1994) has been observed. Futhermore, the peak torque produced by elbow flexors confirms these observations by displaying a circadian rhythm peaking at 17:30 (Gauthier *et al.* 1996).

Additionally, this circadian variation in muscle force (i.e., a value obtained in a laboratory during an isolated isometric contraction) seems to lead to a circadian variation in performance (e.g., a time, a distance, a length…). Performances recorded during an isolated movement (maximal jump) have previously been recorded to display a circadian variation with an acrophase close to 17:45 h (Reilly and Down 1986 and 1992). A similar variation was also observed for an anaerobic test (test of Margaria) with a calculated acrophase at 17:26 h (Reilly and Down 1992).

In controlled conditions, most studies have observed an acrophase in performance during short-duration exercise close to 18:00 h. Furthermore, the circadian variation in this kind of performance is generally in phase with the circadian variation in body core temperature (Reilly and Down 1986, 1992). Several studies have suggested that the simultaneous increases in core temperature (from oral or rectal data) and muscular power are causally related because the diurnal increase in core temperature may have a beneficial passive warm-up effect (Bernard *et al.* 1998, Melhim 1993, Racinais *et al.* 2004 and 2005a). However, the fact that the circadian rhythms in both body temperature and short-duration performance, are in phase doesn't represent evidence of causal link. This question remains unclear. The next objective in these research topics is to investigate the existence of a circadian variation in muscle temperature.

I.2. Physical Ability and Time-of-Day

We saw in the preceding paragraph that most of the studies concluded a circadian variation of the muscle force, and thus of the performance, during a short-duration exercise. However, the statistical significance of this circadian rhythm could be linked to low values during night whereas most of the physical activities (sport events, leisure or work) are performed during the day. A functional approach of the diurnal variation in physical abilities could be to investigate the existence of fluctuation during day-time rather than nocturnal nadir. In the literature, studies on the whole solar cycle (i.e., 24 h) generally use the terminology of 'circadian rhythms' whereas studies focusing on the day-time use 'time-of-day'.

From this point of view, it is relevant to highlight that most of the studies observed an improvement in performance during short-duration exercise during the day. For example, better performance were observed in the afternoon than in the morning for: vertical jump height (Bernard *et al.* 1998, Racinais *et al.* 2004), sprint running velocity (Javiere *et al.* 1996), maximal cycling power output (Bernard *et al.* 1998, Racinais *et al.* 2005b), or total work during 30 seconds (Melhim 1993) and 60 seconds (Hill *et al.* 1992) cycle exercise.

This diurnal fluctuation in muscle power can be a consequence of a modification in neural drive and/or a modification in muscle contractile function. The later seems to be the dominant factor. Skeletal muscle contractility is significantly greater in the afternoon than in the morning (Gauthier *et al.* 1996, Martin *et al.* 1999, Racinais *et al.* 2005a), whereas this is not the case for the muscle electrical activity (Gauthier *et al.* 1996, Racinais *et al.* 2005a). Racinais *et al.* (2005a) observed a large (+12%) increase in isometric force of the knee extensors during the day despite a decrease in muscle electrical activity. Similar data were previously observed by Gauthier *et al.* (1996) leading them to conclude that the vulnerability of the organism could be reduced by having the acrophase of different physiological parameters at different times. From a general point of view, muscle electrical activity is higher in the morning suggesting that the body may well have a protective mechanism that partially compensates for the diurnal variation in the muscle contractile properties by an opposite variation in the central command (Gauthier *et al.* 1996, Racinais *et al.* 2005a).

Thus, in summary, the diurnal increase in muscle power is mainly attributed to a diurnal increase in muscle force (Racinais *et al.* 2005b). This diurnal increase in muscle force is mainly attributed to a diurnal increase in muscle contractile properties (Gauthier *et al.* 1996, Martin *et al.* 1999, Racinais *et al.* 2005a). However, the physiological basis of this variation in muscle contractile function remained unclear and will represent a challenging topic in the future.

One proposed theory is that the circadian rhythm in body temperature could act as a passive warm-up for increasing muscle contractility (Racinais *et al.* 2005a), suggests that the low morning temperatures do not match the optimal conditions for muscle efficiency. This theory is partly based on the previous observation of a greater Ca^{2+} retention by the sarcoplasmic reticulum with a decrease in muscle temperature (Segal *et al.* 1986, Stein *et al.* 1982). If we explain the variation in muscle contractility by a variation in calcium concentration in the muscle fiber , this would not influence the muscle contractility during sub-maximal contractions, when the needs are lower. In line with this hypothesis, it is relevant to note that muscle contractility is significantly higher in the afternoon than in the morning during high intensity contractions (i.e., from 75% to 100% of max), but not during contractions at 25% or 50% of max (Racinais *et al.* 2005a). However, Martin *et al.* (1999) advanced the hypothesis that these mechanisms were not only muscle temperature-dependent, but also controlled by the daytime hormonal fluctuations. Furthermore, it is generally considered that it is local temperature of the exercising limbs rather than core temperature which influences maximal muscle power (Falk *et al.* 1998), but the diurnal fluctuation in muscle temperature was never investigated yet. Thus, today, we can only conclude that there is currently insufficient data available to explain with certitude the cause of the variation in muscle contractile state as a function of time-of-day.

I.3. How to Control the Diurnal Variation in Physical Ability

The previous paragraph presented the diurnal variation in some physical abilities. But, if it is very useful to know and understand these fluctuations, athletes and coaches need also to know how to control them. The best and easiest way to obtain the best performance possible is to schedule the recording at the time-of-day corresponding to the acrophase of the athlete for this performance. However, this solution may be not possible, especially in official events. So, if the time of the event can not be adapted to the athlete, the athlete has to adapt himself to the time of the event. In this paragraph, we will see that this kind of adaptation is possible.

The first possibility existing for an athlete to have a small shift in his circadian rhythm is to have a shift in his zeitgebers. Generally, it is impossible to change the night-day cycle. However, if the alternation in light and dark are the main zeitgeber for most of the animals, the social activities are a major zeitgeber in humans. A very interesting study was performed by Javierre *et al.* (1996) with national level sprinters. In this experiment, they recorded the diurnal fluctuations in 80 m sprint performance after different modifications in sleep and/or meal schedules. Their data showed that to postpone both meal and sleep schedules by 2 hours shifts the initial fluctuation in performance by two hours later. Inversely, to advance both meal and sleep schedules by two hours lead a shift of the initial fluctuation in performance by two hours earlier. In summary, two points particularly relevant in their results: (i) it is possible to have a shift the acrophase of your performance in a task in which you specialise, and (ii) this shift was obtained within only two days (Javierre *et al.* 1996).

It is also possible to influence the natural diurnal fluctuation in performance by regularly practicing a physical activity at a given time-of-day. Indeed, muscle power seems to progress after training specifically with time-of-day of training (Souissi *et al.* 2002). Likewise, a greater improvement in aerobic performances seems also to occur at the time-of-day at which training was regularly performed (Hill *et al.* 1989a, 1998). A short-term adaptation to the time-of-day for exercising has also been emphasized from one day to the following day (Edwards *et al.* 2005).

While controlling sleep, meal and exercise schedules modifies or to shifts the diurnal variation in performance, some other conditions can completely blunt this diurnal fluctuation. This is especially the case of a warm environment. As detailed above, in temperate conditions, short-duration performance is generally increased by the end of the afternoon, at the peak of the core temperature curve (Coldwells *et al.* 1994, Melhim 1993, Reilly and Down 1986, 1992, Bernard *et al.* 1998, Deschenes *et al.* 1998). Even if the circadian rhythm in core temperature is not necessarily the cause of the rhythm in muscle performance, some studies have suggested that the simultaneous increases in both body temperature and muscle power are causally related (Coldwells *et al.* 1994, Melhim 1993, Bernard *et al.* 1998, Racinais *et al.* 2004a,b, 2005a). These studies based their hypothesis on previous work showing that an increase in muscle temperature could increase muscle power (i.e. Bergh and Ekblom 1979) and considered the diurnal increase in body temperature as a passive warm-up (Bernard *et al.* 1998, Melhim 1993, Racinais *et al.* 2004a,b) increasing muscle contractility (Racinais *et al.* 2005a).

In parallel with the diurnal variation in body temperature, a variation in environmental temperature can influence muscle power. A moderately hot environment has been shown to enhance jump performance (Hue *et al.* 2003) and cycle power (Falk *et al.* 1998). These observations are confirmed by a lot of studies showing that a local passive warm-up increased muscle performance (e.g., Bergh and Ekblom 1979, Bergh 1980, Davies and Young 1985, Sargeant 1987, Crowley *et al.* 1991). In summary, time-of-day (i.e., the diurnal increase in body temperature) and a warm environment seems to represent two factors increasing muscle performance by a passive warm-up effect. So, the question is to know whether an external passive warm-up controls the passive warm-up effect of the time-of-day. In line with this hypothesis, a study conducted in a tropical environment failed to show any circadian variation in muscle power, suggesting that the hot and humid environment (28.1°C and 63% of relative humidity) may have blunted the circadian rhythm in muscular power (Racinais *et al.* 2004a). This study raised intriguing questions about the cause of this stability. Two suggestions seemed reasonable: (i) a long-term adaptation to the environmental conditions or (ii) an instantaneous effect of the testing conditions. Some recent studies comparing diurnal variation (i.e., morning versus afternoon) in muscle power in the same subject in function of the environmental temperature observed an interaction effect within time-of-day and environmental conditions (Racinais *et al.* 2004b, 2005a), showing that it was the testing condition that blunted the diurnal variation in muscle power. Interestingly, a warm environment only facilitates improved muscle power (Racinais *et al.* 2004b) and contractility (2005a) in the morning, when body temperature is at its lowest. These findings suggest that these two passive warm-up effects have no additional effect on the improvement in muscle power and there is possibly a threshold for passive warm-up effects, above which additional sources of warm-up have no additional effect. (Racinais *et al.* 2006). In summary, athletes and coaches maybe able to blunt the diurnal variation in muscle power by using another source of passive warm-up (Racinais *et al.* 2004, 2005, and 2006).

II. Circadian Variation in Physical Ability: Effect on Health

II.1. Physical Activity and Circadian Variation in Temperature

One of the most known circadian rhythms is core temperature. Human core temperature is generally measured by rectally and displays an acrophase in the late afternoon, close to 18:00 h (Waterhouse *et al.* 1993). This circadian rhythm is mainly endogenous and persists in free-living conditions, without influence of the zeitgebers (Colin *et al.* 1968). Even if the absolute recorded values are dependent of the methodology (i.e., different depth of insertion within a probe and a clinical thermometer), it is well established that rectal temperature displays a diurnal increase whether measured by a rectal probe (Bernard *et al.* 1998, Racinais *et al.* 2005b) or clinical thermometer (Atkinson *et al.* 1993, Racinais *et al.* 2004, 2005a).

Circadian variations in body temperature are considered to lead many of other circadian variations but can have its own consequences on health, especially during exercise. During exercise, muscle contraction produces heat leading to a proportional relation within relative work intensity and core temperature (Saltin *et al.* 1968). For example, during moderate exercise, muscle temperature increases quickly from a basal value close to 35°C, rising above rectal temperature within 3 to 5 min before plateauing after 10 to 20 minutes. In a neutral environment, core temperature increases when muscle temperature becomes higher than rectal temperature but different thermoregulatory systems attempt to preserve homeostasis. During prolonged exercise the majority of heat dissipation is facilitated by evaporation, while convection, conduction, and radiation play are smaller role. A part of the heat is also stored in the body leading to an increase in core temperature. However, the storage possibilities are limited and more and more studies state core -and may be cerebral- temperature as the limiting factor of aerobic performance in warm environment (Caputa *et al.* 1986, Brück and Olschewski 1987, Gonzales-Alonzo *et al.* 1999). For trained subjects, cessation of activity in warm environment seems to occur at a given critical core temperature whatever the exercise intensity, the heat storage capacity, the acclimatizing state, the initial temperature or the skin temperature (Brück and Olshewski 1987, Gonzales-Alonso *et al.* 1999). Accordingly, rats stop exercising in hot environment at the same abdominal and cerebral temperature whatever the modification made to their initial temperature (Fuller *et al.* 1998, Walters *et al.* 2000). Furthermore, if the cerebral temperature of a goat is increased close to 42°C by a local warm-up, they reduced their velocity or refuse to move (Caputa *et al.* 1986).

In human body, from a basal temperature close to 37°C to a limit close to 40°C (Gonzales-Alonso *et al.* 1999), the heat storages possibilities seem inadequate. From this point of view, the natural circadian variation in core temperature seems to be an important point to consider. Indeed, an increase in body temperature before a long-duration exercise can decrease performance by altering the heat storage capacity (Nadel 1987, Gonzales-Alonso *et al.* 1999). Similarly, pre-cooling can increase the heat storage capacity and increase the running time to exhaustion (Lee and Haymes 1995, Gonzales-Alonso *et al.* 1999). Thus, we can hypothesis that the circadian variation in core temperature will modify the heat storage capacity and thus the ability to exercise. This hypothesis is valid only when heat dissipation fails to maintained homeostasis and thus the heat storage capacity becomes a limiting factor. This is notably the case in warm and humid environment, when required evaporative cooling exceeds the evaporative capacity of the environment (Shapiro *et al.* 1998). The diurnal variation in core temperature persisting during physical exercise (Reilly *et al.* 1984, Reilly and Brooks 1990) and the thermal stress induced by physical exercise in warm (and humid) environment should be higher in the evening, when the initial body temperatures are at its highest (Reilly and Garret 1998). However, this could auto regulated by the human body. Indeed, it seems that the increase in body temperature during prolonged exercise could be greater in the morning, leading to a reduction in the amplitude of the circadian variation as exercise progress (Aldemir *et al.* 2000).

Thus, this topic needs further research in warm environments is required to investigate the effect the circadian rhythm in core temperature on thermoregulation capacity during exercise. An additional source of thermoregulation limitation should also be investigated in regard to the circadian rhythm in core temperature in patients with spinal cord injuries can suffer from a reduction or loss in ability to dissipate heat from below the level of lesion.

II.2. Physical Activity and Circadian Variations in Cardiac Event

From a chronobiological point of view, the rate of occurrence in cardiac events seems to display circadian variations. Muller *et al.* (1987) observed a prominent circadian variation of sudden cardiac death, with a low incidence during the night and an increased incidence from 07:00 to 11:00. a.m. These data are confirmed by the observation of Willich *et al.* (1989) that the incidence of myocardial infarction was markedly increased between 06:00 and 12:00 a.m. compared with other times of day. The amplitude of variation is relatively important as myocardial infarction occurred 3.8 times more frequently between 08:00 and 09:00 (hour of maximum incidence) than between 00:00 and 01:00 a.m. (hour of minimum incidence) (Willich *et al.* 1989). However, these differences are also linked to the different levels of activity and to the stress involved with the wake-up itself (Atkinson *et al.* 2006).

Moreover, vigorous physical activity acutely increases the risk of sudden cardiac death (Siscovick *et al.* 1984)http://circ.ahajournals.org/cgi/content/full/107/24/3109 - R57-130157# R57-130157 and myocardial infarction (Giri *et al.* 1999)http://circ.ahajournals.org/cgi/content/full/107/24/3109 - R59-130157#R59-130157 among individuals with both diagnosed and occult heart disease. Thus, because acute cardiac events occur mainly between 06:00 - 12:00 a.m. and that physical activity is generally thought to be a trigger for such events, clinicians should advise patients already at risk not to exercise in the morning. However many people prefer this time-of-day for recreational activities and household chores, while the general preference of middle-aged and elderly people scheduling activity and sleep at earlier times than young adults (Atkinson *et al.* 2006).

It is important to know how to make morning recreational activities safe. At this point, it is particularly relevant to consider that supervised exercise programs seem to be safe for cardiac patients, irrespective of time-of-day (Atkinson *et al.* 2006), while the incidence of cardiac events is low in coronary artery disease patients during regular, sub-maximal exercise whether performed in the morning or afternoon (Murray *et al.* 1993). Moreover, from a general point of view, regular sub-maximal exercises are generally considered to be beneficial for health. Regular physical activity using large muscle groups, such as walking, running, or swimming, produces cardiovascular adaptations that increase exercise capacity, endurance, and skeletal muscle strength. Habitual physical activity also prevents the development of coronary artery disease and reduces symptoms in patients with established cardiovascular disease (Statement from the Council on Clinical Cardiology and Council on Nutrition, Physical Activity, and Metabolism, 2003). But, if exercise programs for cardiac patients have generally a 2- to 6-months supervised period of sub-maximal exercise, they are generally followed by unsupervised activity in the subsequent months (Atkinson *et al.* 2006).

Consequently, the question can be reformulated to know if the supervised exercise program can reduce the risk of a cardiac event if the patients exercise in the morning thereafter.

In this way, it is particularly relevant to consider than physical training can induce some adaptation specific to the time-of-day in which the training is performed. For example, as presented in the first section of this chapter, a greater improvement in aerobic performances seems to occur at the time-of-day at which training was regularly performed (Hill *et al.* 1989a, 1998). It may be useful for physicians to introduce supervised sub-maximal exercise in the morning during the supervised program. That could induce specific adaptations to reduce the risk of cardiac event in the subsequent physical activity, if, against medical advice, the patients perform strenuous activity in the morning.

II.3. Physical Activity and Circadian Rhythm in Elderly

On other important relation within biological rhythms, physical exercise and health concerns the beneficial effects of physical training. Chronic physical activity is considered to have beneficial effect in ederly on both physiological (e.g., the risk of cardiovascular event, Andrawes *et al.* 2005) and cognitive (Renaud and Bherer 2005) aging, but physical activity can also act as a zeitgeber. This point is particularly relevant in this population displaying a loss of the stability of their circadian rhythm.

Indeed, some circadian variations are blunted with aging (e.g., mood state, Atkinson *et al.* 1992) and others persist but with a reduction in amplitude (e.g., core temperature, Atkinson *et al.* 1992). In addition to the reduction in the amplitude of the circadian rhythms, an earlier acrophase is also observed in elderly (Atkinson *et al.* 1992). Moreover, aging leads an alteration of the sleep-wake rhythm with nap period during the day and a loss in quality/continuity of the nocturnal sleep (Flamer 1996, Vitiello *et al.* 1997).

However, regular physical activity induces chronic adaptations such as an increase in circadian variations amplitude in young subjects (Atkinson *et al.* 1993). The question is to know whether maintaining physical activity throughout the aging process can maintain the standard sleep-wake cycle. The beneficial effect of physical activity for better aging is well documented (for a review, see the position stand of the American College of Sports Medicine 1998). But the problem is that, in parallel of alterations in biological rhythm, physical activity generally decreases with aging in relation to a more sedentary lifestyle and/or a loss in mobility.

Concerning the biological rhythms, Mauvieux *et al.* (2003) observed a significant increase in the amplitude of the circadian rhythm in core temperature after four months of training. From a theoretical point of view, this observation confirms in an elderly population the observation of Atkinson *et al.* (1993) in young subjects. From a functional point of view, this increases in the amplitude of the circadian variation in core temperature was associated to an improvement in both nocturnal sleep quality (i.e., less wake-up) and diurnal vigilance (Mauvieux *et al.* 2003). It seems important to highlight that, despite some physiological adaptation to training persist for a long time, the zeitgeber action of the physical activity does not. Regular physical activity is needed to help maintain both the sleep-wake cycle and sleep quality. It is also interesting to consider that it is the increase in core temperature during exercise which could lead amelioration in sleep quality (Horne and Staff 1983). Indeed, a passive or active increase in core temperature allows for an increase in the slow-wave sleep in the subsequent night and thus improves the sleep-wake cycle (Horne et Staff 1983, Horne 1992).

Conclusion

In summary, the relationship between circadian rhythms and physical activity is complex and bidirectional. On one hand, biological rhythms influence physical abilities during short-duration exercise. This influence seems to be lead by a diurnal variation in muscle contractile properties, which may be related with the circadian rhythm in core temperature. On the other hand, physical activity can also influence the biological rhythms. This influence can allow modification in the amplitude of a rhythm (i.e., an active life generally leads to an increase in amplitude) or its synchronization in regard to the solar day (i.e., possibility to shift the acrophase by a few hours or to stabilize a rhythm). In this chapter, we tried to highlight some possible ways for future research. In view of the importance given to the circadian rhythm in core temperature and for the synchronization of the other rhythms, we think that the possibility of a diurnal variation in muscle temperature should be investigated. Indeed, it is generally considered that it is local temperature of the exercising muscle rather than core temperature that influences muscle power.

Future research should also continue to investigate the adaptation in patient, worker and athletes under the influence of perturbations of their endogenous circadian rhythms. The perturbations generally analyzed are notably shift work or jet lag but another source of perturbation should be investigated in athletes in the following years. During the holy month of Ramadan Muslims all over the world fast everyday for 4 weeks between the hours of sunrise to sunset. This special pattern of alimentation represents a perturbation of the habitual rhythm of life. Based on the 10 month lunar calendar Ramadan starts 10 days earlier each year, which could possibly impact on Muslim athletes yearly training and competition cycles (Grantham *et al.* 2006).

References

Aldemir H, Atkinson G, Cable T, Edwards B, Waterhouse J, Reilly T. (2000). A comparison of the immediate effects of moderate exercise in the early morning and late afternoon on core temperature and cutaneous thermoregulatory mechanisms. *Chronobiol. Int.* 17: 197-207.

American College of Sports Medicine. (1998). ACSM position stand on exercise and physical activity for older adults. *Med. Sci. Sports Exerc.* 30: 992-1008.

Andrawes WF, Bussy C, Belmin J. (2005). Prevention of cardiovascular events in elderly people. *Drugs Aging.* 22: 859-876.

Atkinson G, Coldwells A, Reilly T, Waterhouse J. (1992). Effects of age on circadian rhythms in physical performance and mood states. *J. Interdisciplin. Cycle Res.* 23: 186-188.

Atkinson G, Coldwells A, Reilly T, Waterhouse J. (1993). A comparison of circadian rhythms in work performance between physical active and inactive subjects. *Ergonomics.* 16: 273-281.

Atkinson G, Drust B, George K, Reilly T, Waterhouse J. (2006). Chronobiological considerations for exercise and heart disease. *Sports Med.* 36: 487-500.

Bergh U, Ekblom B. (1979). Influence of muscle temperature on maximal muscle strength and power output in human skeletal muscles. *Acta Physiol. Scand.* 107: 33-37.

Bergh U. (1980). Human power at submaximal body temperature. *Acta Physiol. Scand.* Suppl: 478.

Bernard T, Giacomoni M, Gavarry O, Seymat M, Falgairette G. (1998a). Time-of-day effects in maximal anaerobic leg exercise. *Eur. J. Appl. Physiol.* 77: 133-138.

Bruck K, Olschewski H. (1987). Body temperature related factors diminishing the drive to exercise. *Can. J. Physiol. Pharmacol.* 65: 1274-1280.

Caputa M, Feistkorn G, Jessen C. (1986). Effects of brain and trunk temperatures on exercise performance in goats. *Pflugers. Arch.* 406: 184–189.

Coldwells A, Atkinson G, Reilly T. (1994). Sources of variation in back and leg dynamometry. *Ergonomics.* 37: 79-86.

Colin J, Timbal J, Boutelier C, Houdas Y, Siffre M. (1968). Rhythm of the rectal temperature during a 6-month free-running experiment. *J. Appl. Physiol.* 25: 170-1 76.

Council on Clinical Cardiology (Subcommittee on Exercise, Rehabilitation, and Prevention) and Council on Nutrition, Physical Activity, and Metabolism (Subcommittee on Physical Activity): Thompson PD, Buchner D, Piña IL, Balady GJ, Williams MA, Marcus BH, Berra K, Blair SN, Costa F, Franklin B, Fletcher GF, Gordon NF, Pate RR, Rodriguez BL, Yancey AK, Wenger NK. (2003). Exercise and Physical Activity in the Prevention and Treatment of Atherosclerotic Cardiovascular Disease. *Circulation.* 107: 3109.

Crowley GC, Garg A, Lohn MS, Van Someren N, Wade AJ. (1991). Effects of cooling the legs on performance in a standard Wingate anaerobic test. *Br. J. Sports Med.* 25: 200-203.

Davies CT, Young K. (1985). Effect of heating on the contractile properties of triceps surae and maximal power output during jumping in elderly men. *Gerontology.* 31: 1-5.

Deschenes MR, Kraemer WJ, Bush JA, Doughty TA, Kim D, Mullen KM, Ramsey K. (1998). Biorhythmic influences on functional capacity of human muscle and physiological responses. *Med. Sci. Sports Exerc.* 30: 1399-1407.

Edwards BJ, Edwards W, Waterhouse J, Atkinson G, Reilly T. (2005). Can Cycling Performance in an Early Morning, Laboratory-Based Cycle Time-Trial be Improved by Morning Exercise the Day Before? *Int. J. Sports Med.* 26: 651-656.

Falk B, Radom-Isaac S, Hoffmann JR, Wang Y, Yarom Y, Magazanik A, Weinstein Y. (1998). The effect of heat exposure on performance of and recovery from high-intensity, intermittent exercise. *Int. J. Sports Med.* 19: 1-6.

Faria IE, Drummond BJ. (1982). Circadian changes in resting heart rate and body temperature, maximal oxygen consumption and perceived exertion. *Ergonomics.* 25: 381-386.

Flamer HE. (1996). Sleep disorders in the elderly. *Aust. N Z J. Med.* 26: 96-104.

Fuller A, Carter RN, Mitchell D. (1998). Brain and abdominal temperatures at fatigue in rats exercising in the heat. *J. Appl. Physiol.* 84: 877–883.

Gauthier A, Davenne D, Martin A, Cometti G, Van Hoecke J. (1996). Diurnal rhythm of the muscular performance of elbow flexors during isometric contractions. *Chronobiology International.* 13: 135-146.

Giri S, Thompson PD, Kiernan FJ, Clive J, Fram DB, Mitchel JF, Hirst JA, McKay RG, Waters DD. (1999). Clinical and angiographic characteristics of exertion-related acute myocardial infarction. *JAMA.* 282: 1731–1736.

Gonzalez-Alonso J, Teller C, Andersen SL, Jensen FB, Hyldig T, Nielsen B. (1999). Influence of body temperature on the development of fatigue during prolonged exercise in the heat. *J. Appl. Physiol.* 86: 1032–1039.

Grantham JR, Belhaj J, O'Brien B, Balasekaran G. (2006). The effect of 4 weeks fasting during Ramadan upon isokinetic knee flexor/extensor performance. *Med. Sci. Sports Exerc.* 38: S337.

Hill DW, Borden DO, Darnaby KM, Hendricks DN, Hill CM. (1992). Effect of time of day on aerobic and anaerobic responses to high-intensity exercise. *Can. J. Sport Sci.* 17: 316-319.

Hill DW, Cureton KJ, Collins MA, Grisham SC. (1988). Diurnal variations in responses to exercise of "morning types" and "evening types". *J. Sports Med.* 28: 213-219.

Hill DW, Cureton KJ, Collins MA. (1989a). Circadian specificity in exercise training. *Ergonomics.* 32: 79-92.

Hill DW, Cureton KJ, Collins MA. (1989b). Effect of time of day on perceived exertion at work rates above and below the ventilatory threshold. *R. Q. Exerc. Sports.* 60: 127-133.

Hill DW, Leiferman JA, Lynch NA, Dangelmaier BS, Burt SE. (1998). Temporal specificity in adaptations to high-intensity exercise training. *Med. Sci. Sports Exerc.* 30. 450-455.

Horne J, Staff L. (1983). Exercise and sleep: body-heating effects. *Sleep.* 1: 36-46.

Horne J. (1992). Human slow wave sleep: a review and appraisal of recent findings, with implications for sleep functions, and psychiatric illness. *Experientia.* 4: 941-954.

Hue O, Coman F, Blonc S, Hertogh C. (2003). Effect of tropical climate on performance during repeated jump-and-reach tests. *J. Sports Med. Phys. Fitness.* 43: 475-480.

Javierre C, Calvo M, Diez A, Garrido E, Segura R, Ventura JL. (1996). Influence of sleep and meal schedules on performance peaks in competitive sprinters. *Int. J. Sports Med.* 17: 404-408.

Jones HA, Hill DW. (1995). responses to incremental and constant power exercise tests at different times of day. *Med. Sci. Sports Exerc.* 27: S109.

Lee DT, Haymes EM. (1995). Exercise duration and thermoregulatory responses after whole body precooling. *J. Appl. Physiol.* 79: 1971-1976.

Martin A, Carpentier A, Guissard N, Van Hoecke J, Duchateau J. (1999). Effect of time of day on force variation in a human muscle. *Muscle Nerve.* 22: 1380-1387.

Mauvieux B, Davenne D, Gruau S, Sesboue B, Denise P. (2003). Physical training effect on the circadian rhythm of the temperature and the sleep-activity cycle in the elderly. *Science and Sports.* 18: 93-103.

Melhim AF. (1993). Investigation of circadian rhythms in peak power and mean power of female physical education students. *Int. J. Sports Med.* 14: 303-306.

Muller JE, Ludmer PL, Willich SN, Tofler GH, Aylmer G, Klangos I, Stone PH. (1987). Circadian variation in the frequency of sudden cardiac death. *Circulation.* 75:131-138.

Murray PM, Herrington DM, Pettus CW, Miller HS, Cantwell JD, Little WC. (1993). Should patients with heart disease exercise in the morning or afternoon? *Arch. Intern. Med.* 153: 833-836.

Nadel ER. (1987). Prolonged exercise and high and low ambient temperatures. *Can. J. Sport Sci.* 12: 140S-142S.

Racinais S, Blonc S, Hue O. (2005b). Effects of active warm-up and diurnal increase in temperature on muscular power. *Med. Sci. Sports Exerc.* 37: 2134-2139.

Racinais S, Blonc S, Jonville S, Hue O. (2005a). Time-of-day influences the environmental effects on muscle force and contractility. *Med. Sci. Sports Exerc.* 37: 256-261.

Racinais S, Chamari K, Hachana Y, Bartagi Z, Blonc S, Hue O. (2006). Effect of an acute hot and dry exposure in moderately warm and humid environment on muscle performance at different time-of-day. *Int. J. Sports Med.* 27: 49-54.

Racinais S, Hue O, Blonc S. (2004b). Time-of-day effects on anaerobic muscular power in a moderately warm environment. *Chronobiol. Int.* 21: 485-495.

Racinais S, Hue O, Hertogh C, Damiani M, Blonc S. (2004a). Time-of-day effects in maximal anaerobic leg exercise in tropical environment: a first approach. *Int. J. Sports Med.* 25: 186-190.

Reilly T, Brooks GA. (1982). Investigation of circadian rhythms in metabolic responses to exercise. *Ergonomics.* 25: 1093-1107.

Reilly T, Brooks GA. (1990). Selective persistence of circadian rhythms in physiological responses to exercise. *Chronobiol. Int.* 7: 59-67.

Reilly T, Down A. (1986). Circadian variation in the standing broad jump. *Perceptual and Motor Skills.* 62: 830.

Reilly T, Down A. (1992). Investigation of circadian rhythms in anaerobic power and capacity of the legs. *J. Sports Med. Phys. Fitness.* 32: 343-347.

Reilly T, Garret R (1998). Investigation of diurnal variation in sustained exercise performance. *Ergonomics.* 41: 1085-1094.

Reilly T, Robinson G, Minors DS. (1984). Some circulatory responses to exercise at different times of day. *Med. Sci. Sports Exerc.* 16: 477-482.

Renaud M, Bherer L. (2005). Impact on physical fitness on cognitive aging. *Psychol. Neuropsychiatr. Vieil.* 3: 199-206.

Rodahl A, O'Brien M, Firth RG. (1976). Diurnal variation in performance of competitive swimmers. *J. Sports Med. Phys. Fitness.* 16: 72-76.

Saltin B, Gagge AP, Stolwijk JA. (1968). Muscle temperature during submaximal exercise in man. *J. Appl. Physiol.* 25: 679-688.

Sargeant AJ. (1987). Effect of muscle temperature on leg extension force and short-term power output in humans. *Eur. J. Appl. Physiol. Occup. Physiol.* 56: 693-698.

Segal SS, Faulkner JA, White TP. (1986). Skeletal muscle fatigue in vitro is temperature dependent. *J. Appl. Physiol.* 61: 660-665.

Shapiro Y, Moran D, Epstein Y. (1998). Acclimatization strategies – preparing for exercise in the heat. *Int. J. Sports Med.* 19: S161-S163.

Siscovick DS, Weiss NS, Fletcher RH, Lasky T. (1984). The incidence of primary cardiac arrest during vigorous exercise. *N. Engl. J. Med.* 311: 874–877.

Souissi N, Gauthier A, Sesboüé B, Larue J, Davenne D. (2002). Effects of regular training at the same time of day on diurnal fluctuations in muscular performance. *J. Sports Sci.* 20: 929-937.

Stein RB, Gordon T, Shriver J. (1982). Temperature dependence of mammalian muscle contractions and ATPase activities. *Biophys. J.* 40: 97-107.

Vitiello MV, Schwartz RS, Davis MW. (1997). Sleep disorders and aging: understanding the causes. *J. Gerontol. A Biol. Sci. Med. Sci.* 5: M189-191.

Walters TJ, Ryan KL, Tate LM, Mason PA. (2000). Exercise in the heat is limited by a critical internal temperature. *J. Appl. Physiol.* 89: 799-806.

Waterhouse JM, Minors DS, Folkard S. (1993). Estimating the endogenous component of the circadian rhythm of the rectal temperature in humans undergoing normal sleep/activity schedules. *J. Interdisciplin Cycle Res.* 24: 197-211.

Williams CS, Hill DW. (1995). The effect of time of day on $\dot{V}O_2$ kinetics and $\dot{V}O_2$ max during high-intensity exercise. *Med. Sci. Sports Exerc.* 27: S109.

Willich SN, Linderer T, Wegscheider K, Leizorovicz A, Alamercery I, Schroder R. (1989). Increased morning incidence of myocardial infarction in the ISAM Study: absence with prior beta-adrenergic blockade. ISAM Study Group. *Circulation.* 80: 853-858.

In: Progress in Circadian Rhythm Research
Editor: Anne-Laure Léglise, pp. 177-193
ISBN: 978-1-60021-876-7

Chapter VIII

Intercellular and Intracellular Mechanisms of Circadian Coordination in the Mammalian Circadian Pacemaker

***Sara J. Aton*[1] *and Erik D. Herzog*[2].**
[1] Department of Neuroscience, University of Pennsylvania, Philadelphia, PA 19104.
[2] Department of Biology, Washington University, St. Louis, MO 63130.

Abstract

The suprachiasmatic nuclei (SCN) of the mammalian hypothalamus coordinate a wide range of circadian rhythms in behavior and physiology. Recent technical advances allow the recording of firing rate rhythms from individual SCN neurons and the measurement of clock gene expression rhythms from various cell types over several days. These techniques provide an opportunity to study the generation and synchronization of cellular rhythms *in vitro*. Studies of transgenic mice lacking vasoactive intestinal polypeptide (VIP) or its receptor $VPAC_2$ show that impaired behavioral rhythms in these mice stem from dysfunction at the level of individual SCN neurons. In both mutants, the proportion of SCN neurons with circadian rhythms in firing is reduced, while circadian synchrony between the remaining rhythmic neurons is absent. Studies mimicking or blocking GABAergic signaling between SCN neurons have shown that while daily GABA treatment can synchronize SCN neurons *in vitro*, endogenous GABA signaling is not required for SCN timekeeping or synchrony. Pharmacological manipulations of wild-type SCN neurons suggest that daily rhythms in G-protein mediated signaling pathways regulating adenylyl cyclase are required to maintain rhythms and synchrony within the SCN. It is hoped that such studies will clarify the signaling pathways involved in circadian rhythm coordination *in vivo*, leading to novel, targeted interventions for sleep disturbances and health problems stemming from circadian dysfunction.

The SCN - How the Brain Tells Time

In mammals, circadian (~24 h) rhythms in physiology and behavior are coordinated by a master "circadian pacemaker" in suprachiasmatic nuclei (SCN) of the hypothalamus. When the SCN are lesioned (Moore and Eichler, 1972;Stephan and Zucker, 1972;Ibuka et al., 1977;Eastman et al., 1984;Schwartz and Zimmerman, 1991) or electrically silenced *in vivo* (Schwartz et al., 1987;Schwartz, 1991), behavioral and physiological circadian rhythms (schematized in Figure 1A) disappear. SCN allografts are sufficient to restore daily behavioral rhythms (and a subset of physiological rhythms) - with the endogenous period of the donor animal - when transplanted to SCN-lesioned hosts (Ralph et al., 1990). Restoration of most circadian functions appears to be mediated by daily rhythms of firing and rhythmic release of diffusible synaptic signals from the SCN graft (Silver et al., 1996;Buijs et al., 2003;Lehman et al., 1987;Inouye and Kawamura, 1979). The nature of these output signals remains unknown, although candidates have been identified (Cheng et al., 2002;Kalsbeek et al., 1992;Teclemariam-Mesbah et al., 1997;Kramer et al., 2001;Kraves and Weitz, 2006).

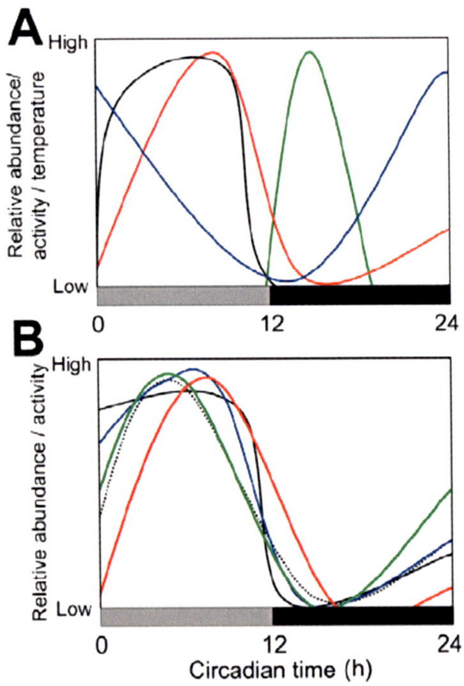

Figure 1. Circadian rhythms in physiology, behavior, and the SCN *in vivo*. (A) Daily rhythms in locomotor activity and feeding (black), body temperature (red), plasma growth hormone (green), and plasma cortisol (blue) are schematized for a diurnal mammal in constant conditions. Times are presented in circadian hours (normalized to a 24-h day), where CT 0 = beginning of the subjective day, and CT 12 = beginning of the subjective night. In nocturnal mammals, these rhythms maintain a similar phase relationship to each other, but are delayed by 12 h relative the prior environmental light cycle. Data are modified from Moore-Ede et al. (1982). (B) Schematic of daily rhythms in the SCN (based on *in vivo* and *in vitro* data) of 2-deoxyglucose utilization (black; (Schwartz and Gainer, 1977)), firing rate (blue; (Klein et al., 1991;Green and Gillette, 1982), cAMP content (red; (Murakami and Takahashi, 1983), CREB-mediated gene expression (green;(Obrietan et al., 1999), and VIP and AVP release (black dashed line; (Shinohara et al., 2000).

Early *in vivo* studies demonstrated that at the tissue level, the SCN of nocturnal rodents exhibit daily rhythms (schematized in Figure 1B) in 2-deoxyglucose uptake (Schwartz and Gainer, 1977), cAMP content (Murakami and Takahashi, 1983), CREB-mediated gene expression (Obrietan et al., 1999), electrical activity (Inouye and Kawamura, 1979), and neurotransmitter release (Schwartz and Reppert, 1985). The isolated SCN retain these rhythmic properties *in vitro* (Groos and Hendriks, 1982;Watanabe et al., 2000;Shinohara et al., 2000;Shibata and Moore, 1988). There is some disagreement in the literature regarding the kinetics of SCN cAMP content (Murakami and Takahashi, 1983;Prosser and Gillette, 1991;Yamazaki et al., 1994) and VIP release rhythms (Nakamura et al., 2001;Shinohara et al., 2000). However, all of the aforementioned rhythms appear to be approximately in phase with one another, with peak firing rate, VIP/AVP release, cAMP content and CREB-mediated transcription occurring at mid-to-late subjective day (in constant conditions, the point in the animal's free-running rhythm corresponding to mid-to-late day in a light:dark cycle; circadian time [CT] 6 to 9).

From Cells to Systems - The SCN as a Master Coordinator of Circadian Rhythms

With the advent of multielectrode array recording technology, it was possible to demonstrate that individual SCN neurons show circadian rhythms in firing, even after disruption of SCN cytoarchitecture and dissociation (Welsh et al., 1995). These data led many to hypothesize that all SCN neurons can function as cell-autonomous circadian clocks. In support of this hypothesis, spontaneous mutations and genetic screens have led to the discovery of several "clock genes" involved in circadian timekeeping, at the level of both SCN neurons and behavior (Dunlap, 1999;Emery and Reppert, 2004). Transgenic animals have been generated in which a bioluminescent (luciferase; *luc*) reporter is driven by clock gene promoters, primarily *Period1* (*Per1*) and *Per2*. Recent advances in imaging technology have made it possible to visualize clock reporter expression in individual cells. In support of the idea of cell-autonomous rhythm generation, *Per1::luc* and *PER2::luc* rhythms have been detected in individual mouse SCN neurons (Yamaguchi et al., 2003;Maywood et al., 2006b;Aton et al., 2006).

Studies of *Per1::luc* transgenic rats and *PER2::luc* fusion-protein expressing mice have revealed that similar rhythms in clock gene promoter activity occur throughout the body, in various organs and brain areas, both *in vitro* and *in vivo* (Abe et al., 2002;Saxena et al., 2007;Granados-Fuentes et al., 2004;Abraham et al., 2005;Yoo et al., 2005). These tissue-specific molecular rhythms are thought to mediate local circadian functions. Excellent examples of this are the circadian regulation of olfactory sensitivity mediated by a clock in the olfactory bulb (Granados-Fuentes et al., 2006), the regulation of body weight by the molecular clock in muscle tissue (McDearmon et al., 2006), and the regulation of various clock controlled genes within the liver by the molecular clock (Kornmann et al., 2007).

While many peripheral tissues show rhythms in clock gene expression that are not SCN-dependent, these tissues lose circadian synchrony with one another when the SCN are lesioned (Yoo et al., 2004). Thus the role of the SCN has been recently redefined, from the "master driver" of rhythms in the periphery to the "master synchronizer" of the body's various clocks. However, the signaling mechanism by which the SCN synchronizes these clocks is unknown. Candidates include cues from rhythmic feeding (Stokkan et al., 2001), glucocorticoid release rhythms (Balsalobre et al., 2000), and body temperature rhythms (Brown et al., 2002;Kornmann et al., 2007), all of which are regulated by the SCN.

SCN Structural and Functional Heterogeneity

By convention, the SCN are anatomically subdivided. The ventrolateral "core" contains vasoactive intestinal polypeptide (VIP)-expressing neurons, which project to the dorsomedial "shell" containing arginine vasopressin (AVP)-positive neuronal cell bodies. In addition to these and other neuropeptides, all SCN neurons appear to release GABA and express GABA receptors (Abrahamson and Moore, 2001;Gao et al., 1995;Moore et al., 2002). This basic structure and organization is highly conserved among mammalian species, including rodents, cats, nonhuman primates, and humans (Cassone et al., 1988;Moore, 1993;Hofman et al., 1988;Gao and Moore, 1996;Cayetanot et al., 2006). The SCN also shows functional heterogeneity, with subsets of neurons differing in the amplitude and timing of their rhythms (Schaap et al., 2003;Antle and Silver, 2005;Hastings and Herzog, 2004;Yamaguchi et al., 2003), their inputs (Abrahamson and Moore, 2001), and the outputs they regulate (LeSauter and Silver, 1998;Silver and Schwartz, 2005). The dorsal and ventral SCN appear to function as independent populations of oscillators under some conditions. APV and VIP can be released in antiphasic rhythms from the SCN *in vitro* (Shinohara et al., 1995), and certain light conditions, such as large shifts in the light:dark cycle, can cause rhythms of dorsal and ventral SCN to become antiphasic *in vivo* (De la Iglesia et al., 2004;Albus et al., 2005). This temporary condition is apparently due to a lag in synchronization of the dorsal SCN to signals from the retinorecipient ventral SCN.

The Critical Role for Intercellular Communication in SCN Function

Intercellular communication has long been thought to play a role in synchronization of rhythms across the SCN network. Individual SCN neurons express firing rate rhythms with widely varying circadian periods and phases when dispersed at low density (Welsh et al., 1995;Herzog et al., 1998). However, SCN neurons cultured in organotypic slices or dispersed at high density show relatively synchronized rhythms, suggesting a role for diffusible neuronal signals in SCN synchrony (Aton et al., 2006;Aton et al., 2005;Nakamura et al., 2001). Disruption of action potentials with tetrodotoxin (TTX) also leads to desynchrony across the SCN (Yamaguchi et al., 2003), and depolarization can resynchronize SCN neurons (Maywood et al., 2006b). When dorsal and ventral SCN are separated by a knife cut, synchrony is lost among individual neurons in the dorsal half of the nuclei, but not the ventral half (Yamaguchi et al., 2003). Altogether, these data suggest that neurotransmitter release from ventral SCN neurons is required for circadian synchrony.

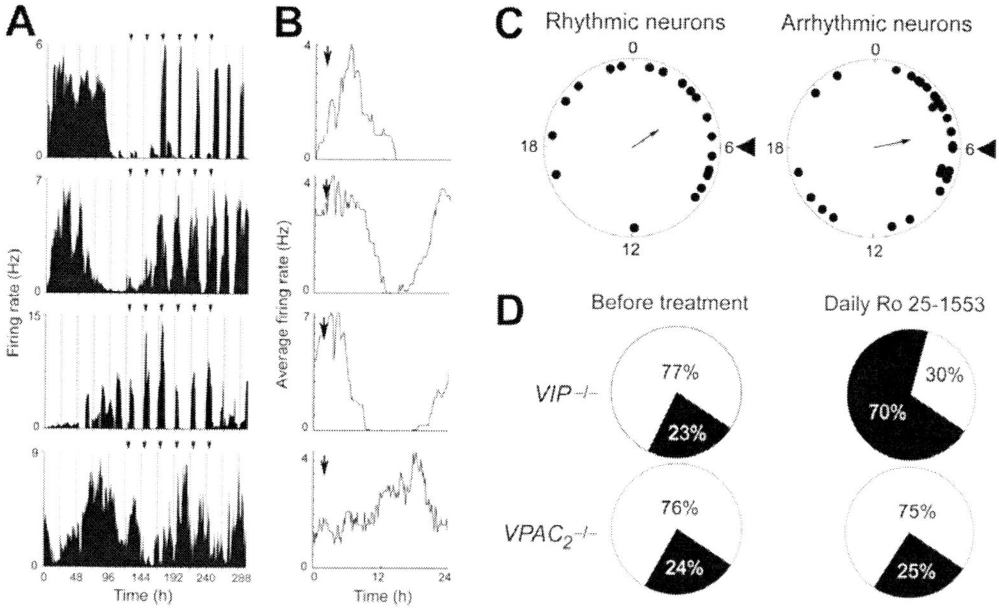

Figure 2. Daily application of $VPAC_2$ agonist Ro 25-1553 restores rhythmicity to *VIP-/-* SCN neurons. (A) Firing rate records of three initially arrhythmic *VIP-/-* neurons and a rhythmic neuron treated with 150 nM Ro 25-1553 every 24 hours for 6 days (arrows). The top two traces show examples from representative neurons in which rhythmicity was restored and the bottom trace shows a neuron that remained arrhythmic. The third trace shows a representative rhythmic neuron entrained by agonist treatment. Rhythmicity persisted for at least 48 h following the last treatment, indicating that cyclic firing was entrained by drug application. (B) Firing patterns of rhythmic neurons, averaged over the last four days of treatment, increased in anticipation of the drug treatment, indicating that the daily change in firing rate was not simply an acute response to the agonist. (C) Rayleigh plots showing the relative peak phases of initially rhythmic ($n = 20$, left) and arrhythmic neurons with restored rhythms ($n = 28$, right) from a representative *VIP-/-* culture on the sixth day of Ro 25-1553 treatment. Arrowheads at hour 6 denote the time of agonist application. Both subsets of neurons showed statistically significant synchronization (Rayleigh test, $p < 0.05$ and $p < 0.005$ for rhythmic and arrhythmic neurons, respectively) with the mean peak firing (arrows) 1-2 hours prior to final agonist application. (D) The relative proportion of rhythmic (black) to arrhythmic (white) neurons from *VIP-/-* cultures was restored by daily $VPAC_2$ agonist treatment to wild-type levels, while there was little change in the proportion of rhythmic *Vipr2-/-* neurons with treatment. Modified from Aton *et al.*, 2005.

Several candidate synchronizing signals have been identified in the SCN (Aton and Herzog, 2005). Recent work from our lab (Aton et al., 2005), and others (Maywood et al., 2006b;Brown et al., 2005), has identified VIP as a critical synchronizing signal in the SCN. A role for VIP in circadian timekeeping was demonstrated in mice lacking the VIP peptide (Colwell et al., 2003) or its cognate receptor $VPAC_2$ (encoded by the *Vipr2* gene; (Harmar et al., 2002)), which show very similar defects in circadian behavior (Aton et al., 2005). Under constant conditions, both mutants show reduced-amplitude behavioral rhythms, with approximately 70% showing multiple free-running circadian periods simultaneously, and 30% free-running with an abnormally short circadian period. Consistent with the behavior of these mutants, individual SCN neurons from both *VIP-/-* and *Vipr2-/-* mice show reduced circadian synchrony *in vitro*, measured by both *Per1::luc* bioluminescence (Maywood et al., 2006b) and firing rate recordings (Aton et al., 2005). Individual neurons from these mice also have reduced rhythm amplitudes, and only about 30% of all recorded neurons exhibit detectable rhythmicity (Aton et al., 2005;Maywood et al., 2006b). This is consistent with findings from an earlier study, showing that TTX treatment of SCN slices - which would block activity-dependent release of VIP - not only desynchronized rhythms in SCN neurons, but also reduced their amplitude (Yamaguchi et al., 2003). Thus rhythmicity itself, once thought to be a cell-autonomous property of all SCN neurons, is dependent on VIPergic intercellular communication. Importantly, daily treatment with a $VPAC_2$ receptor agonist restores rhythms and synchrony to *VIP-/-* (but not *Vipr2-/-*) neurons *in vitro* (Figure 2), indicating that VIP-$VPAC_2$ signaling is not only necessary, but sufficient, for these circadian functions.

GABA is a ubiquitous neurotransmitter in the SCN which has long been implicated in SCN rhythmicity and synchrony (Aton and Herzog, 2005). When applied daily to SCN neurons dispersed at low density, GABA is sufficient to synchronize firing rhythms *in vitro* (Liu and Reppert, 2000). GABA signaling between dorsal and ventral SCN also appears to be necessary for coordinating entrainment to large shifts in the environmental light:dark cycle (Albus et al., 2005). However, a recent study from our lab shows that long-term treatments with GABA antagonists do not impair synchrony or rhythmicity in mammalian clock neurons (Aton et al., 2006). In fact, SCN neuronal rhythms show greater amplitude and cycle-to-cycle precision under conditions of GABA antagonism. Thus GABA signaling may play a role in entrainment to environmental cycles, but it is not required for entrainment of SCN neurons to one another.

Intracellular Pathways Involved in Circadian Rhythm Generation and Synchrony in the SCN

An important unanswered question is how signals between SCN neurons are converted into intracellular cues which can interact with the molecular clock. While much is known about pathways mediating SCN phase shifts in response to light input (Meijer and Schwartz, 2003), relatively little is known about pathways mediating mutual entrainment between SCN neurons (Aton and Herzog, 2005). CREB-mediated induction of *Period* genes is a likely target of both entrainment pathways (Tischkau et al., 2003;Travnickova-Bendova et al., 2002), and activity of this pathway appears to be clock-regulated in the SCN in vivo (see Figure 1B). VIP is known to induce *Period* gene transcription in a circadian-phase-dependent manner, through $VPAC_2$- and G_s-mediated activation of adenylyl cyclase, PKA, and CREB-dependent signaling (Travnickova-Bendova et al., 2002;Yamazaki et al., 1994;Itri and Colwell, 2003). We and others have suggested that this pathway is critical for synchronization and amplification of rhythms in the SCN by VIP signaling (Figure 3)(Aton and Herzog, 2005;Hao et al., 2006).

In addition to pathways which activate adenylyl cyclase, we have recently demonstrated a role for inhibitory G-protein signaling (mediated by $G_{i/o}$) in SCN timekeeping (Aton et al., 2006). When SCN neurons are treated with $G_{i/o}$ inhibitor pertussis toxin (PTX) over several days *in vitro*, both their *PER2::luc* bioluminescence rhythms and firing rate rhythms decrease in amplitude, with the majority losing rhythmicity entirely (Figure 4A and B). The proportion of neurons that remains rhythmic under PTX treatment – about 30% - is strikingly similar to

VIP-/- and Vipr2-/- SCN neurons. Furthermore, like VIP-/- and Vipr2-/- SCN cultures, the neurons that maintain rhythmicity in PTX-treated cultures lose the ability to synchronize to one another (Figure 4A and B).

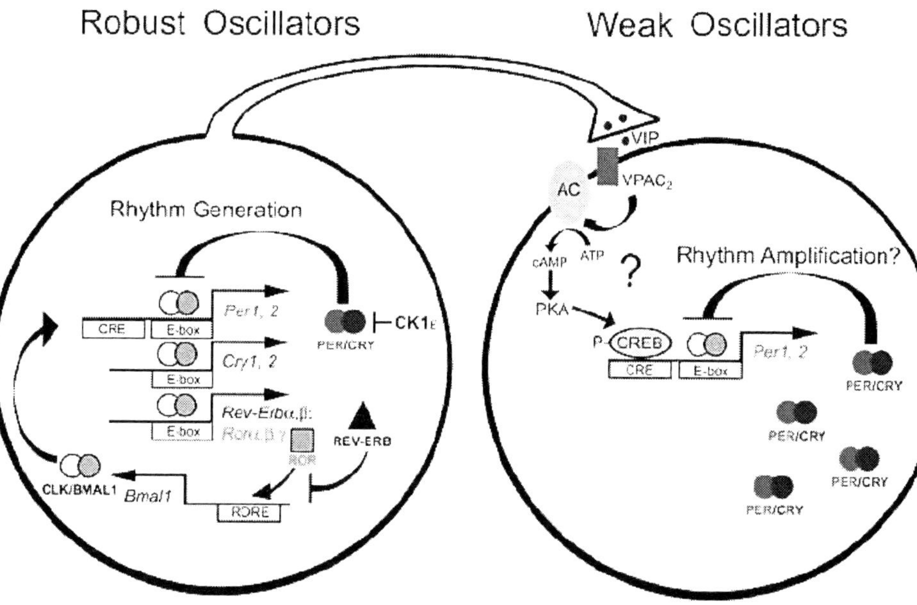

Figure 3. Amplification and synchronization of rhythms by neuropeptidergic signaling within the SCN. While some neurons within the SCN ("robust oscillators") maintain high-amplitude rhythms of firing rate and gene expression in the absence of VIP signaling, more than half of all SCN neurons require VIP/VPAC$_2$ signaling to maintain these rhythms. VPAC$_2$ is known to activate the cAMP/PKA/CREB pathway in SCN neurons (Itri and Colwell, 2003), and this pathway is known to induce *Per* expression (Travnickova-Bendova et al., 2002), thus providing a link between intercellular communication and the core clock mechanism. From Aton and Herzog, 2006.

The similarity of phenotype between VIP-/- or Vipr2-/- mutations and PTX treatment suggests that they impinge on common intracellular pathways mediating SCN rhythms and synchrony. Because VPAC2 is thought to activate Gs, not Gi/o, in SCN neurons (Itri and Colwell, 2003), we tested whether daily VPAC2 activation alone could restore rhythms and synchrony to VIP-/- neurons in the absence of Gi/o function (Aton et al., 2006). We applied VPAC2 agonist Ro 25-1553 daily to VIP-/- SCN cultures in the continuous presence of PTX over 6 days (Figure 4C). The proportion of neurons with rhythmic firing was similar before (32%) and during (23%) daily Ro 25-1553 application in the presence of PTX. Daily agonist application also failed to restore synchrony among rhythmic neurons treated with PTX. We conclude that in the absence of $G_{i/o}$ signaling, daily VPAC$_2$ activation is insufficient to restore function to *VIP-/-* SCN neurons.

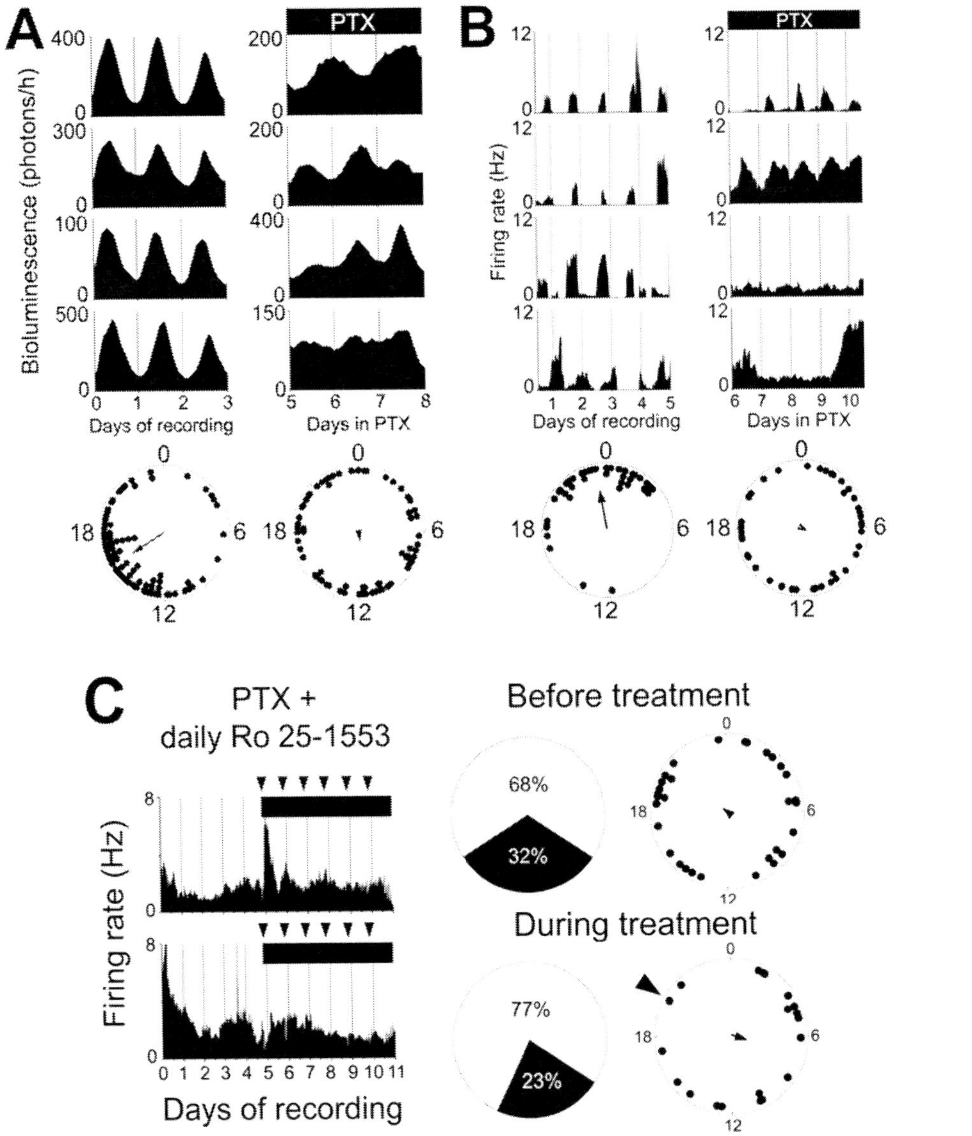

Figure 4. PTX treatment impairs rhythmicity and synchrony of mouse SCN neurons. (A) Representative *PER2::luc* bioluminescence rhythms for individual SCN neurons in SCN slices and (B) representative firing rate rhythms for individual SCN neurons in high-density dispersed cultures, under control conditions (left), over the last 5 days of a 10-day PTX treatment (right). Rayleigh plots show the distribution of peak bioluminescence or firing for each neuron recorded in a single culture, on the last day of control treatment (left), and following several days of PTX treatment (right). Note that circadian neurons reliably peaked at similar times during control treatment, but were less likely to maintain rhythmicity, showed lower-amplitude rhythms, and had unstable phase relationships during PTX treatment. (C) Firing rate traces for two arrhythmic *VIP-/-* SCN neurons treated with Ro 25-1553 (arrowheads) every 24 hours for 6 days in the continuous presence of PTX (left). PTX prevented the restoration of rhythmicity to SCN neurons by daily Ro 25-1553 treatment (right). Distributions of peak firing rate phases for rhythmic *VIP-/-* neurons within representative cultures were random both before and during daily Ro 25-1553 + PTX treatment. Modified from Aton *et al.*, 2006.

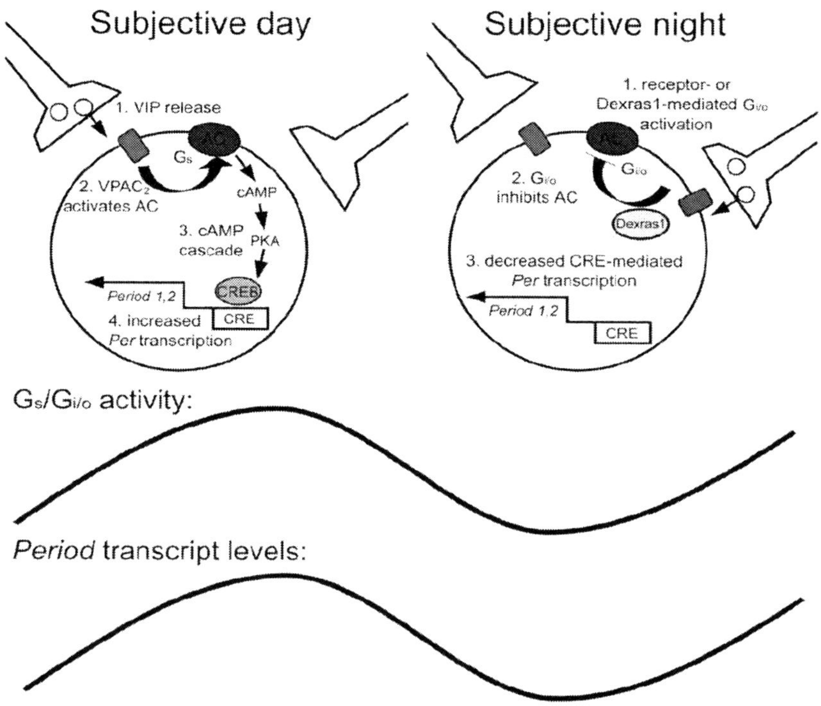

Figure 5. Model of circadian regulation of G-protein signaling, which maintains rhythmicity and synchrony of SCN neurons. During the subjective day, VIP release activates $VPAC_2$ receptors (red), which activate adenylyl cyclase (AC; blue) via G_s, increasing intracellular cAMP levels. Subsequent activation of protein kinase A (PKA) and phosphorylation of cAMP response element binding protein (CREB; green) induces *Period1* and *Period2* transcription via CREB response elements (CREs) on their respective promoters. During the subjective night, VIP release and $G_{\alpha s}$-mediated signaling diminish. $G_{i/o}$ becomes activated by Dexras1 (yellow), or by neurotransmitter-mediated activation of an unknown $G_{i/o}$-coupled receptor (violet). $G_{i/o}$ inhibits AC, down regulating the cAMP cascade and, ultimately, decreases *Period* transcription. In this model, the daily rhythm in the ratio of G_s to $G_{i/o}$ activity synchronizes pacemaking neurons and sustains rhythms in non-pacemaking neurons. Thus, inhibition of $VPAC_2$-mediated G_s signaling or blocking $G_{i/o}$ signaling with PTX disrupts rhythms in SCN neurons and prevents their synchronization. Modified from Aton *et al.*, 2006.

Taken together, our data suggest that G_s and $G_{i/o}$ share common targets which are necessary for maintaining and synchronizing rhythms in SCN neurons. We propose a model in which G_s and $G_{i/o}$ act on adenylyl cyclase at different times of day to maintain rhythms in intracellular cAMP signaling (Figure 5). In this model, VIP released during the day activates G_s coupled to $VPAC_2$, which in turn activates adenylyl cyclase, increasing intracellular cAMP. At night, when VIP release is low, $G_{i/o}$ activity predominates, and cAMP levels drop. This second messenger rhythm leads to rhythmic CREB-mediated *Period* expression, amplifying and synchronizing rhythms in target neurons.

One unresolved question is how $G_{i/o}$ is activated in SCN neurons. One likely candidate is the non-receptor-associated $G_{i/o}$ activator Dexras1, which is expressed rhythmically in the SCN with peak expression during the subjective night (*i.e.*, in antiphase to the *Period* genes)(Ueda et al., 2002). Dexras1 negatively regulates cAMP levels and CREB activity in the SCN *in vivo* (Cheng et al., 2006), and is necessary for behavioral synchrony under constant light conditions (Cheng et al., 2004). Some neurotransmitter receptors expressed by SCN neurons also couple to $G_{i/o}$ – for example, nociceptin/orphanin FQ (Meis, 2003;Allen et al., 1999;Reinscheid et al., 1995), neuropeptide Y (Obrietan and van den Pol, 1996), melatonin (Hastings et al., 1997), and vasopressin receptors (Abel et al., 2000). Although these pathways may activate $G_{i/o}$ *in vivo*, the ligands for at least some of these receptors are unnecessary for behavioral rhythms in constant darkness (Boer et al., 1999;Cassone et al., 1993). On the other hand, as is the case for Dexras1, rodents lacking melatonin show behavioral desynchrony under constant light (Cassone, 1992;Aguilar-Roblero and Vega-Gonzalez, 1993). Melatonin has also been shown to inhibit CREB activity in the SCN *in vivo* (von Gall et al., 2000;von Gall and Weaver, 2006). It is also possible that multiple, redundant upstream signaling pathways regulate $G_{i/o}$ activity. It will be important to assess which of these pathways are required to maintain and synchronize rhythms in SCN neurons.

Toward a Cohesive Cellular Model of Circadian Clock Function

Recent data have implicated other inter- and intracellular pathways in the maintenance and synchronization of rhythms in SCN neurons. For example, gastrin-releasing peptide (GRP) has been demonstrated to restore circadian rhythmicity and rhythmicity to SCN neurons lacking the $VPAC_2$ receptor when exogenously applied (Brown et al., 2005;Maywood et al., 2006b). However, endogenous GRP signaling does not appear to be sufficient to maintain normal circadian functions in the absence of VIP/$VPAC_2$ signaling (Maywood et al., 2006b;Aton et al., 2005;Harmar et al., 2002). Nonetheless GRP and GRP receptors are heavily expressed in the SCN (Karatsoreos et al., 2006;Abrahamson and Moore, 2001), and intra-SCN GRP signaling has been hypothesized to maintain clock function by communicating phase information from "gate" cells in the ventral SCN to oscillators in the dorsal SCN (Antle et al., 2007;Antle et al., 2005). While emphasis has been placed on depolarization as an effecter mechanism for GRP in the SCN (Brown et al., 2005), it is possible – even likely – that GRP compensates for lack of VIP signaling by activating AC activity (*i.e.*, GRP receptor-mediated stimulation of cAMP production fills in for a lack of $VPAC_2$-mediated cAMP production (Garcia et al., 1997)). If it is indeed the case that cAMP – not membrane potential – is the critical mediator of rhythmicity and synchrony in the SCN, it seems likely that cAMP signaling could mediate rhythms and synchrony in other, non-neuronal cell types throughout the body (Welsh et al., 2004;Yoo et al., 2004;Nagoshi et al., 2004).

Implications for Human Health and Healthcare

While our understanding of the circadian system is improving, less is known about how dysfunction in central and peripheral clocks leads to human disease. However, circadian clock dysfunction has been implicated in cardiovascular disease, cancer, and metabolic disorders, as well as some neurological diseases (Maywood et al., 2006a). In particular, desynchrony within the SCN, or between the SCN and peripheral clocks, can adversely affect human sleep, physiology, and cognitive functioning; this is clearly seen in aging populations, and during jet lag and shift work. While light and melatonin are now commonly used "chronotherapy" for such internal desynchrony (Revell and Eastman, 2005), as our understanding of SCN function improves, so to will our ability to maintain clock function under such adverse circumstances.

Acknowledgements

This work was supported by a National Science Foundation Graduate Research Fellowship (to S.J.A.), a Pickwick Postdoctoral Fellowship from the National Sleep Foundation, and National Institutes of Health Grants MH63104 (to E.D.H.) and MH073302 (to S.J.A.).

References

Abe M, Herzog ED, Yamazaki S, Straume M, Tei H, Sakaki Y, Menaker M, Block GD (2002) Circadian rhythms in isolated brain regions. *J. Neurosci.* 22: 350-356.

Abel A, Wittau N, Wieland T, Schultz G, Kalkbrenner F (2000) Cell cycle-dependent coupling of the vasopressin V1a receptor to different G proteins. *J. Biol. Chem.* 275: 32543-32551.

Abraham U, Prior JL, Granados-Fuentes D, Piwnica-Worms DR, Herzog ED (2005) Independent circadian oscillations of Period1 in specific brain areas in vivo and in vitro. *J. Neurosci.* 25: 8620-8626.

Abrahamson EE, Moore RY (2001) Suprachiasmatic nucleus in the mouse: retinal innervation, intrinsic organization and efferent projections. *Brain Res.* 916: 172-191.

Aguilar-Roblero R, Vega-Gonzalez A (1993) Splitting of locomotor circadian rhythmicity in hamsters is facilitated by pinealectomy. *Brain Res.* 605: 229-236.

Albus H, Vansteensel MJ, Michel S, Block GD, Meijer JH (2005) A GABAergic Mechanism Is Necessary for Coupling Dissociable Ventral and Dorsal Regional Oscillators within the Circadian Clock. *Curr. Biol.* 15: 886-893.

Allen CN, Jiang ZG, Teshima K, Darland T, Ikeda M, Nelson CS, Quigley DI, Yoshioka T, Allen RG, Rea MA, Grandy DK (1999) Orphanin-FQ/nociceptin (OFQ/N) modulates the activity of suprachiasmatic nucleus neurons. *J. Neurosci.* 19: 2152-2160.

Antle MC, Foley NC, Foley DK, Silver R (2007) Gates and oscillators II: zeitgebers and the network model of the brain clock. *J. Biol. Rhythms.* 22: 14-25.

Antle MC, Kriegsfeld LJ, Silver R (2005) Signaling within the master clock of the brain: localized activation of mitogen-activated protein kinase by gastrin-releasing peptide. *J. Neurosci.* 25: 2447-2454.

Antle MC, Silver R (2005) Orchestrating time: arrangements of the brain circadian clock. *Trends Neurosci.* 28: 145-151.

Aton SJ, Colwell CS, Harmar AJ, Waschek J, Herzog ED (2005) Vasoactive intestinal polypeptide mediates circadian rhythmicity and synchrony in mammalian clock neurons. *Nat. Neurosci.* 8: 476-483.

Aton SJ, Herzog ED (2005) Come together, right...now: synchronization of rhythms in a mammalian circadian clock. *Neuron.* 48: 531-534.

Aton SJ, Huettner JE, Straume M, Herzog ED (2006) GABA and Gi/o differentially control circadian rhythms and synchrony in clock neurons. *Proc. Natl. Acad. Sci. USA.* 103: 19188-19193.

Balsalobre A, Brown SA, Marcacci L, Tronche F, Kellendonk C, Reichardt HM, Schutz G, Schibler U (2000) Resetting of circadian time in peripheral tissues by glucocorticoid signaling. *Science.* 289: 2344-2347.

Boer GJ, van Esseveldt KE, van der Geest BA, Duindam H, Rietveld WJ (1999) Vasopressin-deficient suprachiasmatic nucleus grafts re-instate circadian rhythmicity in suprachiasmatic nucleus-lesioned arrhythmic rats. *Neuroscience.* 89: 375-385.

Brown S, Zumbrunn G, Fleury-Olela F, Preitner N, Schibler U (2002) Rhythms of mammalian body temperature can sustain peripheral circadian clocks. *Curr. Biol.* 12: 1574.

Brown TM, Hughes AT, Piggins HD (2005) Gastrin-releasing peptide promotes suprachiasmatic nuclei cellular rhythmicity in the absence of vasoactive intestinal polypeptide-VPAC2 receptor signaling. *J. Neurosci.* 25: 11155-11164.

Buijs RM, Van Eden CG, Goncharuk VD, Kalsbeek A (2003) The biological clock tunes the organs of the body: timing by hormones and the autonomic nervous system. *J. Endocrinol.* 177: 17-26.

Cassone VM (1992) The pineal gland influences rat circadian activity rhythms in constant light. *J. Biol. Rhythms.* 7: 27-40.

Cassone VM, Speh JC, Card JP, Moore RY (1988) Comparative anatomy of the mammalian hypothalamic suprachiasmatic nucleus. *J. Biol. Rhythms.* 3: 71-91.

Cassone VM, Warren WS, Brooks DS, Lu J (1993) Melatonin, the pineal gland, and circadian rhythms. *J. Biol. Rhythms.* 8 Suppl: S73-S81.

Cayetanot F, Bentivoglio M, Aujard F (2006) Arginine-vasopressin and vasointestinal polypeptide rhythms in the suprachiasmatic nucleus of the mouse lemur reveal aging-related alterations of circadian pacemaker neurons in a non-human primate. *Eur. J. Neurosci.* 22: 902-910.

Cheng HY, Dziema H, Papp J, Mathur DP, Koletar M, Ralph MR, Penninger JM, Obrietan K (2006) The molecular gatekeeper Dexras1 sculpts the photic responsiveness of the mammalian circadian clock. *J. Neurosci.* 26: 12984-12995.

Cheng HY, Obrietan K, Cain SW, Lee BY, Agostino PV, Joza NA, Harrington ME, Ralph MR, Penninger JM (2004) Dexras1 potentiates photic and suppresses nonphotic responses of the circadian clock. *Neuron.* 43: 715-728.

Cheng MY, Bullock CM, Li C, Lee AG, Bermak JC, Belluzzi J, Weaver DR, Leslie FM, Zhou QY (2002) Prokineticin 2 transmits the behavioural circadian rhythm of the suprachiasmatic nucleus. *Nature.* 417: 405-410.

Colwell CS, Michel S, Itri J, Rodriguez W, Tam J, Lelievre V, Hu Z, Liu X, Waschek JA (2003) Disrupted circadian rhythms in VIP and PHI deficient mice. *Am. J. Physiol. Regul. Integr. Comp. Physiol.* 285: R939-R949.

De la Iglesia HO, Cambras T, Schwartz WJ, Diez-Noguera A (2004) Forced Desynchronization of Dual Circadian Oscillators within the Rat Suprachiasmatic Nucleus. *Curr. Biol.* 14: 796-800.

Dunlap JC (1999) Molecular bases for circadian clocks. *Cell.* 96: 271-290.

Eastman CI, Mistlberger RE, Rechtschaffen A (1984) Suprachiasmatic nuclei lesions eliminate circadian temperature and sleep rhythms in the rat. *Physiol. Behav.* 32: 357-368.

Emery P, Reppert SM (2004) A rhythmic ror. *Neuron.* 43: 443-446.

Gao B, Fritschy JM, Moore RY (1995) GABA $_A$-receptor subunit composition in the circadian timing system. *Brain Res.* 700: 142-156.

Gao B, Moore RY (1996) Glutamic Acid Decarboxylase Message Isoforms In Human Suprachiasmatic Nucleus. *J. Biol. Rhythms.* 11: 172-179.

Garcia LJ, Pradhan TK, Weber HC, Moody TW, Jensen RT (1997) The gastrin-releasing peptide receptor is differentially coupled to adenylate cyclase and phospholipase C in different tissues. *Biochim. Biophys. Acta.* 1356: 343-354.

Granados-Fuentes D, Prolo LM, Abraham U, Herzog ED (2004) The suprachiasmatic nucleus entrains, but does not sustain, circadian rhythmicity in the olfactory bulb. *J. Neurosci.* 24: 615-619.

Granados-Fuentes D, Tseng A, Herzog ED (2006) A circadian clock in the olfactory bulb controls olfactory responsivity. *J. Neurosci.* 26: 12219-12225.

Green DJ, Gillette R (1982) Circadian rhythm of firing rate from single cells in the rat suprachismatic brain slice. *Brain Res.* 245: 198-200.

Groos G, Hendriks J (1982) Circadian rhythms in electrical discharge of rat suprachiasmatic neurones recorded in vitro. *Neurosci. Lett.* 34: 283-288.

Hao H, Zak DE, Sauter T, Schwaber J, Ogunnaike BA (2006) Modeling the VPAC2 activated cAMP/PKA signaling pathway: From receptor to circadian clock gene induction. *Biophys J.*

Harmar AJ, Marston HM, Shen S, Spratt C, West KM, Sheward WJ, Morrison CF, Dorin JR, Piggins HD, Reubi JC, Kelly JS, Maywood ES, Hastings MH (2002) The VPAC(2) Receptor Is Essential for Circadian Function in the Mouse Suprachiasmatic Nuclei. *Cell.* 109: 497-508.

Hastings MH, Duffield GE, Ebling FJ, Kidd A, Maywood ES, Schurov I (1997) Non-photic signalling in the suprachiasmatic nucleus. *Biol. Cell.* 89: 495-503.

Hastings MH, Herzog ED (2004) Clock genes, oscillators and cellular networks in the suprachiasmatic nuclei. *J. Biol. Rhythms.* 19: 400-413.

Herzog ED, Takahashi JS, Block GD (1998) Clock controls circadian period in isolated suprachiasmatic nucleus neurons. *Nat. Neurosci.* 1: 708-713.

Hofman MA, Fliers E, Goudsmit E, Swaab DF (1988) Morphometric analysis of the suprachiasmatic and paraventricular nuclei in the human brain: sex differences and age-dependent changes. *Journal of Anatomy.* 160: 127-143.

Ibuka N, Inouye SI, Kawamura H (1977) Analysis of sleep-wakefulness rhythms in male rats after suprachiasmatic nucleus lesions and ocular enucleation. *Brain Research.* 122: 33-47.

Inouye ST, Kawamura H (1979) Persistence of circadian rhythmicity in a mammalian hypothalamic "island" containing the suprachiasmatic nucleus. *Proc. Natl. Acad. Sci. USA.* 76: 5962-5966.

Itri J, Colwell CS (2003) Regulation of Inhibitory Synaptic Transmission by Vasoactive Intestinal Peptide (VIP) in the Mouse Suprachiasmatic Nucleus. *J. Neurophysiol.* 90: 1589-1597.

Kalsbeek A, Buijs RM, van Heerikhuize JJ, Arts M, van der Woude TP (1992) Vasopressin-containing neurons of the suprachiasmatic nuclei inhibit corticosterone release. *Brain Res.* 580: 62-67.

Karatsoreos IN, Romeo RD, McEwen BS, Silver R (2006) Diurnal regulation of the gastrin-releasing peptide receptor in the mouse circadian clock. *Eur. J. Neurosci.* 23: 1047-1053.

Klein DC, Moore RY, Reppert SM (1991) *Suprachiasmatic nucleus: The mind's clock.* New York: Oxford University Press.

Kornmann B, Schaad O, Bujard H, Takahashi JS, Schibler U (2007) System-driven and oscillator-dependent circadian transcription in mice with a conditionally active liver clock. *PLoS Biol.* 5: e34.

Kramer A, Yang FC, Snodgrass P, Li X, Scammell TE, Davis FC, Weitz CJ (2001) Regulation of daily locomotor activity and sleep by hypothalamic EGF receptor signaling. *Science.* 294: 2511-2515.

Kraves S, Weitz CJ (2006) A role for cardiotrophin-like cytokine in the circadian control of mammalian locomotor activity. *Nat. Neurosci.* 9: 212-219.

Lehman MN, Silver R, Gladstone WR, Kahn RM, Gibson M, Bittman EL (1987) Circadian rhythmicity restored by neural transplant. Immunocytochemical characterization of the graft and its integration with the host brain. *J. Neurosci.* 7: 1626-1638.

LeSauter J, Silver R (1998) Output signals of the SCN. *Chronobiology International.* 15: 535-550.

Liu C, Reppert SM (2000) GABA synchronizes clock cells within the suprachiasmatic circadian clock. *Neuron.* 25: 123-128.

Maywood ES, O'Neill J, Wong GK, Reddy AB, Hastings MH (2006a) Circadian timing in health and disease. *Prog. Brain Res.* 153: 253-269.

Maywood ES, Reddy A, Wong GK, O'Neill JS, O'Brien B, McMahon D, Harmar AJ, Okamura H, Hastings M (2006b) Synchronization and maintenance of timekeeping in suprachiasmatic circadian clock cells by neuropeptidergic signaling. *Curr. Biol.* 16: 599-605.

McDearmon E, Patel K, Ko CH, Walisser JA, Schook AC, Chong JL, Wilsbacher LD, Song EJ, Hong H-K, Bradfield CA, Takahashi JS (2006) Dissecting the functions of the mammalian clock protein BMAL1 by tissue-specific rescue in mice. *Science.* 314: 1304-1308.

Meijer JH, Schwartz WJ (2003) In search of the pathways for light-induced pacemaker resetting in the suprachiasmatic nucleus. *J. Biol. Rhythms.* 18: 235-249.

Meis S (2003) Nociceptin/orphanin FQ: actions within the brain. *The Neurscientist.* 9: 158-168.

Moore RY (1993) Organization of the primate circadian system. [Review]. *J. Biol. Rhythms.* 8: Suppl:S3-9.

Moore RY, Eichler VB (1972) Loss of a circadian adrenal corticosterone rhythm following suprachiamatic lesions in rat. *Brain Res.* 42: 201-206.

Moore RY, Speh JC, Leak RK (2002) Suprachiasmatic nucleus organization. *Cell Tissue Res.* 309: 89-98.

Murakami N, Takahashi K (1983) Circadian rhythm of adenosine-3',5'-monophosphate content in suprachiasmatic nucleus (SCN) and ventromedial hypothalamus (VMH) in the rat. *Brain Res.* 276: 297-304.

Nagoshi E, Saini C, Bauer C, Laroche T, Naef F, Schibler U (2004) Circadian Gene Expression in Individual Fibroblasts; Cell-Autonomous and Self-Sustained Oscillators Pass Time to Daughter Cells. *Cell.* 119: 693-705.

Nakamura W, Honma S, Shirakawa T, Honma K-I (2001) Regional pacemakers composed of multiple oscillator neurons in the rat suprachiasmatic nucleus. *Eur. J. Neurosci.* 14: 1-10.

Obrietan K, Impey S, Smith D, Athos J, Storm DR (1999) Circadian regulation of cAMP response element-mediated gene expression in the suprachiasmatic nuclei. *J. Biol. Chem.* 274: 17748-17756.

Obrietan K, van den Pol AN (1996) Neuropeptide Y depresses GABA-mediated calcium transients in developing suprachiasmatic nucleus neurons: a novel form of calcium long-term depression. *J. Neurosci.* 16: 3521-3533.

Prosser RA, Gillette MU (1991) Cyclic changes in cAMP concentration and phosphodiesterase activity in a mammalian circadian clock studied in vitro. *Brain Res.* 568: 185-192.

Ralph MR, Foster RG, Davis FC, Menaker M (1990) Transplanted suprachiasmatic nucleus determines circadian period. *Science.* 247: 975-978.

Reinscheid RK, Nothacker HP, Bourson A, Ardati A, Henningsen RA, Bunzow JR, Grandy DK, Langen H, Monsma FK, Jr., Civelli O (1995) Orphanin FQ: a neuropeptide that activates an opioidlike G protein-coupled receptor. *Science.* 270: 792-794.

Revell VL, Eastman CI (2005) How to trick mother nature into letting you fly around or stay up all night. *J. Biol. Rhythms.* 20: 353-365.

Saxena MT, Aton SJ, Hildebolt C, Prior JL, Abraham U, Piwnica-Worms D, Herzog ED (2007) Bioluminescence imaging of period1 gene expression in utero. *Mol. Imaging.* 6: 68-72.

Schaap J, Albus H, Tjebbe VH, Eilers PH, Detari L, Meijer JH (2003) Heterogeneity of rhythmic suprachiasmatic nucleus neurons: Implications for circadian waveform and photoperiodic encoding. *Proc. Natl. Acad. Sci. U S A.* 100: 15994-15999.

Schwartz WJ (1991) Further evaluation of the tetrodotoxin-resistant circadian pacemaker in the suprachiasmatic nuclei. *J. Biol. Rhythms.* 6: 149-158.

Schwartz WJ, Gainer H (1977) Suprachiasmatic nucleus: use of 14C-labeled deoxyglucose uptake as a functional marker. *Science.* 197: 1089-1091.

Schwartz WJ, Gross RA, Morton MT (1987) The suprachiasmatic nuclei contain a tetrodotoxin-resistant circadian pacemaker. *Proc. Natl. Acad. Sci. USA.* 84: 1694-1698.

Schwartz WJ, Reppert SM (1985) Neural regulation of the circadian vasopressin rhythm in cerebrospinal fluid: a pre-eminent role for the suprachiasmatic nuclei. *J. Neurosci.* 5: 2771-2778.

Schwartz WJ, Zimmerman P (1991) Lesions of the suprachiasmatic nucleus disrupt circadian locomotor rhythms in the mouse. *Physiol. Behav.* 49: 1283-1287.

Shibata S, Moore RY (1988) Electrical and metabolic activity of suprachiasmatic nucleus neurons in hamster hypothalamic slices. *Brain Res.* 438: 374-378.

Shinohara K, Funabashi T, Mitushima D, Kimura F (2000) Effects of gap junction blocker on vasopressin and vasoactive intestinal polypeptide rhythms in the rat suprachiasmatic nucleus in vitro. *Neurosci. Res.* 38: 43-47.

Shinohara K, Honma S, Katsuno Y, Abe H, Honma K-I (1995) Two distinct oscillators in the rat suprachiasmatic nucleus in vitro. *Proc. Natl. Acad. Sci. USA.* 92: 7396-7400.

Silver R, LeSauter J, Tresco PA, Lehman MN (1996) A diffusible coupling signal from the transplanted suprachiasmatic nucleus controlling circadian locomotor rhythms. *Nature.* 382: 810-813.

Silver R, Schwartz WJ (2005) The Suprachiasmatic Nucleus is a Functionally Heterogeneous Timekeeping Organ. *Methods Enzymol.* 393: 451-465.

Stephan FK, Zucker I (1972) Circadian rhythms in drinking behavior and locomotor activity of rats are eliminated by hypothalamic lesions. *Proc. Natl. Acad. Sci. USA.* 69: 1583-1586.

Stokkan KA, Yamazaki S, Tei H, Sakaki Y, Menaker M (2001) Entrainment of the circadian clock in the liver by feeding. *Science.* 291: 490-493.

Teclemariam-Mesbah R, Kalsbeek A, Pevet P, Buijs RM (1997) Direct vasoactive intestinal polypeptide-containing projection from the suprachiasmatic nucleus to spinal projecting hypothalamic paraventricular neurons. *Brain Res.* 748: 71-76.

Tischkau SA, Mitchell JW, Tyan SH, Buchanan GF, Gillette MU (2003) Ca2+/cAMP response element-binding protein (CREB)-dependent activation of Per1 is required for light-induced signaling in the suprachiasmatic nucleus circadian clock. *J. Biol. Chem.* 278: 718-723.

Travnickova-Bendova Z, Cermakian N, Reppert SM, Sassone-Corsi P (2002) Bimodal regulation of mPeriod promoters by CREB-dependent signaling and CLOCK/BMAL1 activity. *Proc. Natl. Acad. Sci. USA.* 99: 7728-7733.

Ueda HR, Chen W, Adachi A, Wakamatsu H, Hayashi S, Takasugi T, Nagano M, Nakahama KI, Suzuki Y, Sugano S, Iino M, Shigeyoshi Y, Hashimoto S (2002) A transcription factor response element for gene expression during circadian night. *Nature.* 418: 534-539.

von Gall C, Weaver DR (2006) Loss of responsiveness to melatonin in the aging mouse suprachiasmatic nucleus. Neurobiol. *Aging Epub.* ahead of print.

von Gall C, Weaver DR, Kock M, Korf HW, Stehle JH (2000) Melatonin limits transcriptional impact of phosphoCREB in the mouse SCN via the Mel1a receptor. *Neuroreport.* 11: 1803-1807.

Watanabe K, Vanecek J, Yamaoka S (2000) In vitro entrainment of the circadian rhythm of vasopressin-releasing cells in suprachiasmatic nucleus by vasoactive intestinal polypeptide. *Brain Res.* 877: 361-366.

Welsh DK, Logothetis DE, Meister M, Reppert SM (1995) Individual neurons dissociated from rat suprachiasmatic nucleus express independently phased circadian firing rhythms. *Neuron.* 14: 697-706.

Welsh DK, Yoo SH, Liu AC, Takahashi JS, Kay SA (2004) Bioluminescence imaging of individual fibroblasts reveals persistent, independently phased circadian rhythms of clock gene expression. *Curr. Biol.* 14: 2289-2295.

Yamaguchi S, Isejima H, Matsuo T, Okura R, Yagita K, Kobayashi M, Okamura H (2003) Synchronization of cellular clocks in the suprachiasmatic nucleus. *Science.* 302: 1408-1412.

Yamazaki S, Maruyama M, Cagampang FR, Inouye ST (1994) Circadian fluctuations of cAMP content in the suprachiasmatic nucleus and the anterior hypothalamus of the rat. *Brain Research.* 651: 329-331.

Yoo S-H, Yamazaki S, Lowrey PL, Shimomura K, Ko CH, Buhr ED, Siepka SM, Hong H-K, Oh WJ, Yoo OJ, Menaker M, Takahashi JS (2004) Period2::luciferase real-time reporting of circadian dynamics reveals persistent circadian oscillations in mouse peripheral tissues. *Proc. Natl. Acad. Sci. U S A.* Feb. 12: 1-8.

Yoo SH, Ko CH, Lowrey PL, Buhr ED, Song EJ, Chang S, Yoo OJ, Yamazaki S, Lee C, Takahashi JS (2005) A noncanonical E-box enhancer drives mouse Period2 circadian oscillations in vivo. *Proc. Natl. Acad. Sci. U S A.*

In: Progress in Circadian Rhythm Research
Editor: Anne-Laure Léglise, pp. 195-208
ISBN: 978-1-60021-876-7

Chapter IX

Neuroimmune Regulation of the Circadian Clock

*Andrew N. Coogan**
Institute of Life Science and School of Medicine,
Swansea University, Swansea, SA2 8PP, United Kingdom

Abstract

Circadian rhythmicity in mammals is chiefly governed by a master pacemaker located in the suprachiasmatic nucleus of the hypothalamus. The functioning of this pacemaker is known to be sensitive to a number of centrally and peripherally derived signals, such as neuropeptides and circulating hormones. Recently attention has been drawn as to how the immune system may impinge on circadian rhythms and alter the endogenous timekeeping system under basal conditions and immune challenge.

The brain is now realised to be far from the immune privileged site it was once thought to be, and as such the roles of immune modulators such as cytokines have been much studied in the central nervous system, with regard to both neuropathological processes and normal neurophysiology. Recent evidence has implied that circadian rhythms may be modulated by peripheral immune challenge, by infections such as trypanosmiasis, as well as by treatment with specific cytokines (e.g. interferon-γ, tumour necrosis factor-α). There is also the suggestion that age-related changes in brain cytokine levels may contribute to alterations in clock function with senescence. Furthermore, actions on the circadian clock of neuro-immune modulators may contribute to the syndrome of "sickness behaviour" by producing alterations in sleep, feeding, attention and memory, and the roles of cytokines in the psychopathology of affective illness may also be partly mediated via circadian rhythm disturbance. This chapter will further examine these interactions and look forward to exciting avenues of research that are about to be opened in the field.

* Andrew N. Coogan: Ph: 00-44-1792-602173; e-mail:a.coogan@swan.ac.uk

Introduction

Current understanding is that the central nervous system is by no means the "immune-privileged" site it was once considered to be (Lucas et al, 2006). Indeed the brain and spinal cord show signs of inflammation in response to injury/infection, and this immune-response is co-ordinated by a number of mediators, including cytokines, prostaglandins, which may be produced by, and act on, both glial and neuronal cells (Rivest, 2003). Systemically produced inflammatory mediators may also cross the blood-brain barrier and act on specific receptors expressed on neuronal and glial cells (Pollmacher et al, 2002; Perry et al, 2004). Activation of these systems is a hallmark of a number of neurodegenerative conditions, such as Alzheimer's disease, and appears to contribute to the pathological processes of these conditions (Weisman et al, 2006). Aside from promoting cell loss, pro-inflammatory mediators have been described to elicit a host of pathophysiological effects on neurons. For example, synaptic transmission and plasticity is impaired in vitro by the pro-inflammatory cytokines interleukin-1β (Il-1β; Coogan et al, 1999), tumour necrosis factor-α (TNF-α; Cunningham et al, 1996), interleukin-6 (Bellinger et al, 1995) and interferon-γ (IFN-γ; D'arcangelo et al, 1991). Indeed, upregulation of hippocampal Il-1β with age is proposed to contribute to age-related deficits in long-term potentiation, and by corollary, memory (Lynch, 2002).

Such are the wide-spectrums of actions of such neuroimmune signalers in the mammalian CNS that a syndrome of behavioural alterations that occur during infection, termed "sickness behaviour" has been ascribed to the central action of immune signalers (Dantzer, 2006). The physiological alterations seen during sickness behaviour during the acute phase of the immune response (lethargy, weakness, listlessness, decreased food intake, altered sleep/wake patterns and a tendency towards depression) have previously been described to have a circadian component (e.g.Wirz Justice 2006). Circadian rhythms are recurring patterns in behavioural, endocrine and physiological parameters that exhibit periodicities of approximately 24 hours (Levi and Schibler, 2007). In mammals the master circadian pacemaker is localized to the suprachiasmatic nucleus (SCN) of the anterior, ventral hypothalamus (Ralph et al, 1990), although several other central and peripheral sites are now understood to express quasi-autonomous circadian clocks (Gachon et al, 2004). Output from the SCN to target sites is via both humeral and neuronal connections, and clock output is known to have significant influence on the architecture of the sleep/wake cycle, food intake, attention, learning and memory and other cognitive functions (Kalsbeek et al, 2006).

Thus, the question arises, to what extent the behavioural effects of centrally acting cytokines and other immune mediators are mediated via the circadian clock. There is a reasonable literature on the circadian modulation of immune function (for recent review see Carillo-Vico et al, 2005), but there has been much less work on the converse relationship. There is evidence that substantial immune challenge, such as sepsis, systemic LPS treatment or trypanosomiasis produce significant defects in circadian timekeeping (Bauhofer et al, 2001; Bentivoglio et al, 1994; Marpegan et al, 2005). There is also evidence that experimental models of inflammatory processes, such as arthritis, display altered circadian rhythms (Cardinali and Esquifino, 2003). I will examine the current evidence for neuroimmune modulation of the circadian clock, and look towards the future in what promises to be an important and exciting topic in neuroimmunology and chronobiology.

Interleukin-1

Interleukin-1 (Il-1) was the first cytokine to be characterized, initially as endogenous pyrogen (Dinarello, 1979). Over the past three decades Il-1 has been characterized as a highly pluripotent proinflammatory mediator, with a host of central actions (Lucas et al, 2006). In brain, the beta form of Il-1 is predominantly expressed, and will be discussed further here. Il-1β, one of the two subforms of IL-1 (the other being Il-1α) has been found to be expressed by neurons, glia and endothelia in the brain (Rivest, 2003). Circulating peripheral IL-1 can induce the expression of central IL-1 (Parnet et al, 2002), and brain expression of this cytokine is significantly upregulated at numerous sites in neurodegerative conditions (Weismann et al, 2006). Il-1β signals by binding to the type 1 interleukin-1 receptor (IL-1R1), stimulating recruitment of the IL-1R accessory protein to the receptor and stimulation of a number of singalling cascades, including mitogen-activated protein kinase pathways and NFκB (O'Neill, 2006). There is also a truncated form of the receptor, the IL-1RII which lacks a significant portion of the cytoplasmic domain and ligand binding to this receptor does not elicit signal transduction. Therefore, this receptor appears to act as a decoy. Further control of the IL-1 system is provided by the interleukin-1 receptor antagonist (IL-1RA), and endogenously produced antagonist for the Il-1R1.

Expression and distribution studies of the IL-1R1 suggests that it is strongly expreesed in the cerebral vasculature, circumventricular organs and meninges (Konsman et al, 2004). Neuronal expression of IL-1R1 mRNA is described in the hippocampus, amygdala, basal forebrain, hypothalamus and cerebellum (Ericson et al, 1995). Currently, there is only limited data available on the expression of the IL-1R1 protein in mouse and rat brain. One study has shown expression of the IL-R1 by immunohistochemistry confined mostly to the hippocampus and the cerebellum, although it is worth noting that this study is based on just four mouse brains (French et al, 1999). Interestingly, expression of the interleukin-1 receptor accessory protein appears to be much more widespread than that of the Il-1R1 (Parnet et al, 2002), calling into questions its sole role as an accessory. Indeed, novel brain receptors for Il-1 have been postulated to mediate some of its central actions (Andre et al, 2005).

IL-1β has been reported to be widely expressed by glial and neuronal cells in the rodent brain (e.g. Lechan et al, 1990). The hypothalamus is reported as having amongst the highest basal level of Il-1β expression (Cearley et al, 2003). Peripheral treatment with IL-1, TNF or Lipopolysaccharide (LPS) is found to significantly upregulate the amount of IL-1 expressed across the brain, including the hypothalamus (Churchill et al, 2006). IL-1β has been described to elicit neurophysiological effects in a number of hypothalamic regions (e.g. preoptic and anterior hypothalamus, Tabaraen et al, 2006; Paraventricular nucleus, Ferri and Fergusson, 2003) and has been described to play a role in numerous hypothalamic-mediated phenomena, such as REM and non-REM sleep (Opp, 2005), hypothalamus-pituitary-adrenal axis activation (Dunn, 2000) to defensive rage (Hassanain et al, 2005).

With direct reference to the circadian clock of the SCN, there are a number of threads of evidence that suggest that IL-1 may impinge on rhythm generation by the master clock. Il-1β immunoreactivity has been observed in the SCN (Lechan et al, 1990), although at present the physiological significance of this is not known. Intriguingly, brain levels of IL-1β have been reported to display diurnal variation, with hypothalamic levels being higher during the lights on period (Taishi et al, 1997). Il-1β treatment of SCN derived astroglia resulted in the expression of NF-kB mediated transcription (Leone et al, 2006). Since expression of glial-fibrillary acidic protein in the SCN shows distinct diurnal modulation (Lavialle and Serviere, 1993), it is possible that Il-1β mediated changes in astroglial function may impact on the functional organization of the SCN clock. For example, SCN astrocytes are postulated to function by modulating photic input to the SCN via the retinohypothalamic tract, possibly by means of glutamate uptake from RHT-SCN synapses (Lavialle et al, 2001; Spanagel et al, 2005). Another point of interest is that NFκB siganlling has been postulated to play a role in photic resetting of the clock, with inhibition of NFκB signaling attenuating the magnitude of photic-phase shifts (Marpegan et al, 2004).

The effect of Il-1β on SCN astrocytes could also be elicited by TNF-α or LPS treatment (Leone et al, 2006). Indeed, systemic LPS challenge has been found to induce phase-delays in free-running hamsters (Marpgan et al, 2005) in a manner that was not additive with photic-induced phase-shifts. This systemic LPS challenge also induced c-Fos expression in a population of dorsomedial SCN cells, an area known to produce arginive vasopressin (AVP; Antle et al, 2005). Treatment of SCN slice cultures with LPS leads to an increase in AVP production, suggesting a role of SCN derived signals in decreased water intake during immune challenge (Nava et al, 2000). The mode of action of LPS treatment in these paradigms in unascertained. It is possible that the SCN express toll like receptors (TLR4) and CD14 to which LPS binds to in order to alter circadian parameters, although there is no data on this presently (Rivest, 2003). Part of the effect of LPS treatment of SCN slices may involve the induction of pro-inflammatory cytokine expression. Thus, Il-1β may be involved in mediating LPS effects on the SCN. Precedent for this type of mechanism can be seen in the hippocampus, where the LPS induced inhibition of LTP in vitro is attenuated by treatment with the Il-1RA (Cunningham et al, 1996). Likewise, the behavioural suppression invoked by systemic LPS treatment is attenuated by central Il-1RA (Laye et al, 2000). For systemic treatment with LPS, there is a well-characterised upregulation of circulating proinflammatory cytokines. These peripheral cytokines can then stimulate the expression of central cytokines, perhaps via actions at the circumventricular organs, or via activation of vagal afferents (Parnet et al, 2002).

Tumour Necrosis Factor

Tumour necrosis factor-α (TNF-α) is another pleiotropic pro-inflammatory cytokine, and similar to IL-1, acts as a key component of the inflammatory mechanism (Aggarwal, 2003). TNF-α is expressed in the brain under basal conditions, with astrocytes, microglia and neurons shown to express it (Breder et al, 1993). TNF-α expression is markedly upregulated in the brain during a number of pathological conditions, and is believed to be an important mediator of acute-phase reactions in the brain and spinal cord (Lucas et al, 2006).

Immunohistochemical analysis of TNF-α expression in the normal brain shows immunoreactivity in neurons of the hypothalamus, basal forebrain and the brainstem (Breder et al, 1993). TNF-α innervation was observed in a number of CNS sites, including a dense innervation of the dorsal border of the SCN (Breder et al, 1993). TNF-α has been postulated to play roles in modulating glutamatergic synaptic transmission and plasticity (reviewed in Pickering et al, 2005). For example, TNF-α has been shown to inhibit hippocampal long-term potentiation (Cunningham et al, 1996; Butler et al, 2004). Also, glial derived TNF-α has recently been shown to be involved in "synaptic scaling", the process by which long-term changes in neuronal activity lead to universal changes in all synapses on a given neurone (Stellwagen and Malenka, 2006). Thus, TNF-α appears to play critical roles in neurophysiological processes.

There are two receptors for TNF-α, the p55 and p75 receptors. The p55 receptor is constitutively expressed in a number of CNS regions, with highest expression present in the brainstem, as well as in circumventricular organs, chorid plexus, leptomeninges and ependymal cells (Bette et al, 2003; Nadeau and Rivest, 1999). The p75 receptor is barely detectable under basal conditions by in situ hybridization, but is upregulated by mitogen treatment (Bette et al, 2003). During CNS insult, the p55 and p75 are proposed to exert opposing effects, with p55 activation being deleterious and p75 activation neuroprotective (Fontaine et al, 2002). Signalling by these receptors is via formation of a TRADD/TRAF2 complex that enables activation of the MAP kinase and NFκB siganlling cascades (Wajant et al, 2003).

Similar to Il-1β, there is a mature literature on the roles of TNF-α in sleep regulation (Opp, 2005). TNF-α application to the preoptic hypothalamic area increases non-REM sleep in rodents (Kubota et al, 2002), and TNF-α application to the somatosensory cortex during non-REM sleep increases slow cortical waves on the EEG (Yoshida et al, 2003). Expression of TNF-α mRNA in the brain shows a diurnal variation that parallels the amount of non-REM sleep (Bredow et al, 1997), and levels across the forebrain, diencephalons and brainstem are higher at Zeitgeber time (ZT) 2 than at ZT11 (Cearley et al, 2003). TNF-α may also exert a circadian influence on the spinal cord, as spinal, but not brain, uptake of TNF-α shows circadian variation (Pan et al, 2002).

Thus, there is evidence that TNF may play roles in behavioural state-setting, and there may be a chronobiological component to theses effects. Indeed, there is some direct evidence for TNF-actions in the master circadian clock. Trypanosomiasis, the cause of sleeping sickness, alters the induction of c-Fos in the SCN in response to photic stimulation (Peng et al, 1994), as well as spontaneous c-Fos expression in the SCN (Bentivoglio et al, 1994). Trypanosomiasis is known to be accompanied by induction of TNF-α and interferon (IFN) expression in the brains of infected rats (Quan et al, 1999). Given this, Lundkvist et al (2002) examined the effects of trypanosomiasis on the neurophysioloigcal properties of the SCN, and found that infection lead to a decrease in the frequency, but not amplitude, of excitatory post-synaptic currents, indicating a decrease in transmitter release probability. The effects of trypanosome infection was mimicked by application of a cocktail of TNF-α, LPS and IFN-γ, which caused a decrease in the frequency of both excitatory and inhibitory currents

(Lundkvist et al, 2002). Similarly, this cocktail of inflammatory mediators also ablated the circadian rhythm in spontaneous neuronal firing frequency of the SCN, reducing firing frequency to nadir levels across the cycle. Interestingly, this flattening of the SCN firing rate rhythm by these cytokines is similar to that seen in aged mice and hamsters (Nygard et al, 2005; Watanabe et al, 1995), indicating that cytokine action in the SCN may play a role in age-related circadian dysfunction.

Expression of receptors for TNF-α has been demonstrated in the SCN by means of RT-PCR, with expression levels showing a diurnal rhythm in both young and aged mice (Sadki et al, 2007). Interestingly, the SCN also exhibits rhythmic expression of suppressor of cytokine signaling (SOCS) 1 and 3, molecules that constitute an intracellular inhibitory feedback mechanism on cytokine signaling (Sadki et al, 2007). The same study then examined the effect of TNF-α/IFN-γ i.c.v microinjection on SCN expressed c-Fos. Treatment of young animals with these cytokines during the subjective night was found to induce c-Fos expression in the "shell" region of the SCN, a region known to rhythmically express clock genes and produce humeralclock output signals such as AVPand cardiotrophin-like cytokine (Antle et al, 2004; Kraves and Weitz, 2006). In young animals treated with TNF-α and IFN during the early subjective night, c-Fos was induced, but in the "core" region of the SCN, an area known to directly responsive to photic stimulation. This temporal regional variation in c-Fos induction by TNF-α and IFN-γ was not seen in the SCN of aged mice, where induction of c-Fos occurred in the shell region during both the lights on and off periods (Sadki et al, 2007).

Further evidence for the actions of TNF-α in the SCN comes from the finding that treatment of SCN-derived astrocytes with TNF-α induces NFκB activity (Leone et al, 2006). Also, i.c.v injection of TNF-α/IFN-γ induced marked astroglial activation in both young and old mice, although the aged group showed a more exaggerated response to the cytokine treatment (Bentivoglio et al, 2006). The same study also showed activation of SCN micoglia in reponse to treatment with TNF-α/IFN-γ. Uniocular inoculation of the anterior chamber with herpes simplex virus induces the expression of TNF-α in the SCN of mice, and this upregulation of TNF-α in the SCN was sensitive to thalidomide treatment, indicating that a source of it was macrophages (Fields et al, 2006). The timing of TNF expression in the SCN in response to herpes simplex viral infection is co-ordinated with the time of T-cell infiltration to the SCN (Archin and Atherton, 2002).

Aside from its direct role in the SCN, TNF-α has also been shown to effect the production of the pineal nocturnal hormone melatonin. Melatonin is considered to be a key endocrine signaler of photoperiod, and in mammals melatonin production by the pineal gland is under the control of a noradrenergic sympathetic innervation emerging from the superior cervical ganglion. This projection to the pineal is in turn regulated by the SCN, via projections to the paraventricular nucleus of the hypothalamus. Thus, the SCN circadian clock gates the production of melatonin by the pineal. Melatonin has been described to have myriad effects on immunity, forming an important link between the circadian and immune systems (Carillo Vico et al, 2005). In humans, it has been reported that the nocturnal surge in melatonin is inhibited during an inflammatory rise in TNF-α (Pontes et al, 2006). Treatment of rat pinealocytes with TNF-α decreased the transcription of *aa-nat*, the gene encoding for *a*-arkylamine-N-acetyltransferase, a rate limiting enzyme in the melatonin biosynthesis pathway (Fernandes et al, 2006). In the same study, TNF-α also decreased the production of N-acetylserotnin, a precursor of melatonin. These findings tie in well with previous reports that severe sepsis in critically ill patients is accompanied by loss of the diurnal rhythm in melatonin (Mundigler et al, 2002).

Interferons

The interferons are a highly conserved groups of molecules that play roles in the immune response as well as physiological processes such as development and metabolism (Campbell et al, 1999). The interferons are broadly classed into type I interferons (IFN-α and IFN-β), molecules with immunomodulatory properties, and type II interferons (IFN-γ), which are not structurally related to the type I interferons, but also have immunoregulatory and immunoactivating properties (Takaoka and Yanai, 2006). Interferons have been reported to be present in brain, and are produced by both neurones and glia (Dafny and Yang, 2005). Intriguingly, interferon treatment for a number of clinical conditions may be accompanied by CNS mediated side-effects such as insomnia, depression, anorexia and cognitive dysfunction, indicating that circulating systemic IFNs cross the blood-brain barrier and mediate a number of CNS effects (Capuron and Miller, 2004).

IFNs signal through either class I or class II receptors, and these receptors are found on both glia and neurons in the CNS (Dafny and Young, 2005). These receptors signal via a classical tyrosine kinase JAK/STAT pathway impinging on gene expression and cellular activation. The SCN has been shown to express class II receptors (responsible for binding IFN-γ), and this expression in the SCN is regulated by development stage and exposure to the light/dark cycle (Lundkvist et al, 1998, 1999). The diurnal pattern of the receptor in the SCN was mirrored by the levels of JAK1 and JAK2, as well as STAT1 (Lundkvist e al, 1998). The presence of receptors for IFN-γ is in agreement with the above discussed effects of IFN-γ, when applied with TNF-α and/or LPS, in inducing changes in gene expression and electrophysiological parameters (Lundkvist and Kristensson, 2002; Sadki et al, 2007). Further to these experiments, it has been shown that i.c.v. microinjection of IFN-γ in hamsters leads to phase-advances when applied during the subjective day, but not during the subjective night (Boggio et al, 2003). This type of phase-dependency suggests that IFN-γ may reset the SCN pacemaker in a non-photic manner, although a more comprehensive phase-curve is required prior to making any firm deductions. As mentioned above, the IFN-γ receptor mRNA ceases to show significant rhythmicity in its SCN expression in aged mice, suggesting a modification of the system with increasing age that may contribute to altered circadian timekeeping in senescence (Sadki et al, 2007).

IFN-α has also been reported to have a number of effects on the circadian system. There are diurnal variations in the efficacy and tolerance of IFN-α immunotherapy (Abrams et al, 1985; Koren and Fleischman, 1993), and these variations may be due to temporal variations in the number of receptors expressed on IFN-α responsive lymphocytes and the metabolism of IFN-α (Ohdo et al, 2000). Similar diurnal variation in receptor expression may also account for the variability of IFN-β's antiviral properties (Takane et al, 2002). The role of interferon treatment on the SCN clock was then examined, as alterations in the central pacemaker could, via endocrine signalers, lead to alterations in peripheral immunity. Dosing of mice with a single treatment of IFN-α at "dusk" (ZT12) led to significant blunting of the SCN rhythm in *Per1, 2, 3 and Bmal1* expression, whilst dosing at "dawn" (ZT0) had no such effect (Ohdo et al, 2001). Dosing of mice with IFN-γ for 6 days also significantly depressed SCN *Per* expression, when the time of treatment was ZT12 but not ZT0. Repeated dosing with IFN-α at ZT12 was also found to attenuate the induction of *Per* by a nocturnal light-pulse (Ohdo et al, 2001). These changes in gene expression profiles following IFN-α treatment at ZT12 were mirrored in the blunting of locomotor and body temperature rhythms. Thus, these authors have proposed a chronological schedule for dosing with IFN that would minimize the effects on the circadian pacemaker, which in turn might contribute to the adverse CNS side-effects that such treatment often entails.

Further, treatment with IFN-α was then described as down-regulating the expression of *Clock* and *Bmal1* in a human hepatocyte cell line, in a STAT1 dependent mechanism (Koynagi and Ohdo, 2002). Continuous infusion of IFN-α via an osmotic minipump also downregulated the Clock and Bmal-1 immunoreactivity in the SCN (Koynagi and Ohdo, 2002). This continuous application of IFN-α also blunts the SCN rhythms in the core clock genes *Per1, Per2, Cry1,* and the outputs *avp* (SCN) and *dbp* (liver). The locomotor rhythm of mice receiving this infusion of IFN-α became fragmented under a light/dark schedule, with decreased nocturnal activity accompanied by an increase in activity during the lights on period (Koyngai and Ohdo, 2002).

Other Neuroimmune Signallers and Future Directions

At present there is little work published directly assessing the actions of cytokines other than the abovementioned on the SCN circadian clock. Cytokines, such as IL-2 and IL-6, have been shown to display diurnal variations in their serum levels (Lissoni et al, 1998) possibly gated by melatonin rhythms. Thus, there exists the possibility that these circulating cytokines can act on the SCN, providing an indirect feedback from the SCN-pineal pathway. Indeed, IL-6 has been shown to induce the expression of *Per1* in human hepatoma cells (Motzkus et al, 2002). Much work needs to be done to define the responsiveness of the SCN clock to such immune regulators.

Recently, a non-classical cytokine, cardiotrophin-like cytokine (CLC) has been postulated to act as SCN locomotor inhibitory output (Kraves and Weitz, 2006). This cytokine signals through the GP130 complex, which has been described in the periventricular region of the hypothalamus, an area known to be important in regulating locomotor rhythms. CLC was then found to be expressed in a rhythmic manner by a subset of SCN neurons, and when infused i.c.v. into the third ventricle it was found to depress locomotor activity. Ciliary neurotrophic factor, another cytokine that signals through the same GP130 complex as CLC also caused this profound attenuation of motor output, although IL-6 and IL-11, which signal through different GP130 containing receptor complexes, did not (Kraves and Weitz, 2006). These effects of CLC on locomotor rhythms are the first brain-actions of this cytokine described, and appear to be separate from any immune/inflammatory role CLC may play, either in the periphery or the brain. This finding suggests that other cytokines may be found to have physiological effects on the circadian system that are independent of their immune functions.

Conclusion

As the literature currently stands, there is good reason to imagine that future studies will reveal exciting details. However, there is much work still to do. By and large there is little anatomical data regarding the expression of cytokines and their receptors and signalers in the SCN, under both basal and immune-challenged conditions. Likewise, there needs to be further, comprehensive studies of the impact of various cytokines on SCN neuronal function and ultimately on the integrity of the SCN circadian oscillator. There also needs to be mechanistic studies by which we can understand how cytokine systems can effect timekeeping processes at the molecular and neurophysiological levels. And finally there needs to be more study of circadian rhythms in disease, both in patients and animal models, as circadian dysfunction can have very significant negative impact on quality of life indices.

References

Abrams PG, McClamrock E, Foon KA (1985) Evening administration of alpha interferon. *N. Engl. J. Med.* 312:443-444.

Aggarwal BB (2003) Signalling pathways of the TNF superfamily: a double-edged sword. *Nat. Rev. Immunol.* 3:745-756.

Andre R, Lerouet D, Kimber I, Pinteaux E, Rothwell NJ (2005) Regulation of expression of the novel IL-1 receptor family members in the mouse brain. *J. Neurochem.* 95:324-330.

Antle MC, Silver R (2005) Orchestrating time: arrangements of the brain circadian clock. *Trends Neurosci.* 28:145-151.

Archin NM, Atherton SS (2002) Infiltration of T-lymphocytes in the brain after anterior chamber inoculation of a neurovirulent and neuroinvasive strain of HSV-1. *J. Neuroimmunol.* 130:117-127.

Bellinger FP, Madamba SG, Campbell IL, Siggins GR (1995) Reduced long-term potentiation in the dentate gyrus of transgenic mice with cerebral overexpression of interleukin-6. *Neurosci. Lett.* 198:95-98.

Bentivoglio M, Grassi-Zucconi G, Peng ZC, Kristensson K (1994) Sleep and timekeeping changes, and dysregulation of the biological clock in experimental trypanosomiasis. *Bull. Soc. Pathol. Exot.* 87:372-375.

Bentivoglio M, Deng XH, Nygard M, Sadki A, Kristensson K (2006) The aging suprachiasmatic nucleus and cytokines: functional, molecular, and cellular changes in rodents. *Chronobiol. Int.* 23:437-449.

Bette M, Kaut O, Schafer MK, Weihe E (2003) Constitutive expression of p55TNFR mRNA and mitogen-specific up-regulation of TNF alpha and p75TNFR mRNA in mouse brain. *J. Comp. Neurol.* 465:417-430.

Boggio VI, Castrillon PO, Perez Lloret S, Riccio P, Esquifino AI, Cardinali DP, Cutrera RA (2003) Cerebroventricular administration of interferon-gamma modifies locomotor activity in the golden hamster. *Neurosignals.* 12:89-94.

Breder CD, Tsujimoto M, Terano Y, Scott DW, Saper CB (1993) Distribution and characterization of tumor necrosis factor-alpha-like immunoreactivity in the murine central nervous system. *J. Comp. Neurol.* 337:543-567.

Bredow S, Guha-Thakurta N, Taishi P, Obal F, Jr., Krueger JM (1997) Diurnal variations of tumor necrosis factor alpha mRNA and alpha-tubulin mRNA in rat brain. *Neuroimmunomodulation.* 4:84-90.

Butler MP, O'Connor JJ, Moynagh PN (2004) Dissection of tumor-necrosis factor-alpha inhibition of long-term potentiation (LTP) reveals a p38 mitogen-activated protein kinase-dependent mechanism which maps to early-but not late-phase LTP. *Neuroscience.* 124:319-326.

Campbell IL, Krucker T, Steffensen S, Akwa Y, Powell HC, Lane T, Carr DJ, Gold LH, Henriksen SJ, Siggins GR (1999) Structural and functional neuropathology in transgenic mice with CNS expression of IFN-alpha. *Brain Res.* 835:46-61.

Cardinali DP, Esquifino AI (2003) Circadian disorganization in experimental arthritis. *Neurosignals.* 12:267-282.

Carrillo-Vico A, Guerrero JM, Lardone PJ, Reiter RJ (2005) A review of the multiple actions of melatonin on the immune system. *Endocrine.* 27:189-200.

Churchill L, Taishi P, Wang M, Brandt J, Cearley C, Rehman A, Krueger JM (2006) Brain distribution of cytokine mRNA induced by systemic administration of interleukin-1beta or tumor necrosis factor alpha. *Brain Res.* 1120:64-73.

Coogan AN, O'Neill LA, O'Connor JJ (1999) The P38 mitogen-activated protein kinase inhibitor SB203580 antagonizes the inhibitory effects of interleukin-1beta on long-term potentiation in the rat dentate gyrus in vitro. *Neuroscience.* 93:57-69.

Cunningham AJ, Murray CA, O'Neill LA, Lynch MA, O'Connor JJ (1996) Interleukin-1 beta (IL-1 beta) and tumour necrosis factor (TNF) inhibit long-term potentiation in the rat dentate gyrus in vitro. *Neurosci Lett.* 203:17-20.

Dafny N, Yang PB (2005) Interferon and the central nervous system. *Eur. J. Pharmacol.* 523:1-15.

Dantzer R (2006) Cytokine, sickness behavior, and depression. *Neurol. Clin.* 24:441-460.

D'Arcangelo G, Grassi F, Ragozzino D, Santoni A, Tancredi V, Eusebi F (1991) Interferon inhibits synaptic potentiation in rat hippocampus. *Brain Res.* 564:245-248.

Dinarello CA (1979) Production of endogenous pyrogen. *Fed. Proc.* 38:52-56.

Dunn AJ (2000) Effects of the IL-1 receptor antagonist on the IL-1- and endotoxin-induced activation of the HPA axis and cerebral biogenic amines in mice. *Neuroimmunomodulation.* 7:36-45.

Ericsson A, Liu C, Hart RP, Sawchenko PE (1995) Type 1 interleukin-1 receptor in the rat brain: distribution, regulation, and relationship to sites of IL-1-induced cellular activation. *J. Comp. Neurol.* 361:681-698.

Fernandes PA, Cecon E, Markus RP, Ferreira ZS (2006) Effect of TNF-alpha on the melatonin synthetic pathway in the rat pineal gland: basis for a 'feedback' of the immune response on circadian timing. *J. Pineal Res.* 41:344-350.

Ferri CC, Ferguson AV (2003) Interleukin-1 beta depolarizes paraventricular nucleus parvocellular neurones. *J. Neuroendocrinol.* 15:126-133.

Fields M, Zheng M, Zhang M, Atherton SS (2006) Tumor necrosis factor alpha and macrophages in the brain of herpes simplex virus type 1-infected BALB/c mice. *J. Neurovirol.* 12:443-455.

Fontaine V, Mohand-Said S, Hanoteau N, Fuchs C, Pfizenmaier K, Eisel U (2002) Neurodegenerative and neuroprotective effects of tumor Necrosis factor (TNF) in retinal ischemia: opposite roles of TNF receptor 1 and TNF receptor 2. *J. Neurosci.* 22:RC216.

French RA, VanHoy RW, Chizzonite R, Zachary JF, Dantzer R, Parnet P, Bluthe RM, Kelley KW (1999) Expression and localization of p80 and p68 interleukin-1 receptor proteins in the brain of adult mice. *J. Neuroimmunol.* 93:194-202.

Gachon F, Nagoshi E, Brown SA, Ripperger J, Schibler U (2004) The mammalian circadian timing system: from gene expression to physiology. *Chromosoma.* 113:103-112.

Hassanain M, Bhatt S, Zalcman S, Siegel A (2005) Potentiating role of interleukin-1beta (IL-1beta) and IL-1beta type 1 receptors in the medial hypothalamus in defensive rage behavior in the cat. *Brain Res.* 1048:1-11.

Kalsbeek A, Palm IF, La Fleur SE, Scheer FA, Perreau-Lenz S, Ruiter M, Kreier F, Cailotto C, Buijs RM (2006) SCN outputs and the hypothalamic balance of life. *J. Biol. Rhythms.* 21:458-469.

Konsman JP, Vigues S, Mackerlova L, Bristow A, Blomqvist A (2004) Rat brain vascular distribution of interleukin-1 type-1 receptor immunoreactivity: relationship to patterns of inducible cyclooxygenase expression by peripheral inflammatory stimuli. *J. Comp. Neurol.* 472:113-129.

Koren S, Fleischmann WR, Jr. (1993) Circadian variations in myelosuppressive activity of interferon-alpha in mice: identification of an optimal treatment time associated with reduced myelosuppressive activity. *Exp. Hematol.* 21:552-559.

Koyanagi S, Ohdo S (2002) Alteration of intrinsic biological rhythms during interferon treatment and its possible mechanism. *Mol. Pharmacol.* 62:1393-1399.

Kraves S, Weitz CJ (2006) A role for cardiotrophin-like cytokine in the circadian control of mammalian locomotor activity. *Nat. Neurosci.* 9:212-219.

Kubota T, Li N, Guan Z, Brown RA, Krueger JM (2002) Intrapreoptic microinjection of TNF-alpha enhances non-REM sleep in rats. *Brain Res.* 932:37-44.

Lavialle M, Serviere J (1993) Circadian fluctuations in GFAP distribution in the Syrian hamster suprachiasmatic nucleus. *Neuroreport.* 4:1243-1246.

Lavialle M, Begue A, Papillon C, Vilaplana J (2001) Modifications of retinal afferent activity induce changes in astroglial plasticity in the hamster circadian clock. *Glia.* 34:88-100.

Laye S, Gheusi G, Cremona S, Combe C, Kelley K, Dantzer R, Parnet P (2000) Endogenous brain IL-1 mediates LPS-induced anorexia and hypothalamic cytokine expression. *Am. J. Physiol. Regul. Integr. Comp. Physiol.* 279:R93-98.

Lechan RM, Toni R, Clark BD, Cannon JG, Shaw AR, Dinarello CA, Reichlin S (1990) Immunoreactive interleukin-1 beta localization in the rat forebrain. *Brain Res.* 514:135-140.

Leone MJ, Marpegan L, Bekinschtein TA, Costas MA, Golombek DA (2006) Suprachiasmatic astrocytes as an interface for immune-circadian signalling. *J. Neurosci. Res.* 84:1521-1527.

Levi F, Schibler U (2007) Circadian Rhythms: Mechanisms and Therapeutic Implications. *Annu. Rev. Pharmacol. Toxicol.* 47:593-628.

Lissoni P, Rovelli F, Brivio F, Brivio O, Fumagalli L (1998) Circadian secretions of IL-2, IL-12, IL-6 and IL-10 in relation to the light/dark rhythm of the pineal hormone melatonin in healthy humans. *Nat. Immun.* 16:1-5.

Lucas SM, Rothwell NJ, Gibson RM (2006) The role of inflammation in CNS injury and disease. *Br. J. Pharmacol.* 147 Suppl 1:S232-240.

Lundkvist GB, Robertson B, Mhlanga JD, Rottenberg ME, Kristensson K (1998) Expression of an oscillating interferon-gamma receptor in the suprachiasmatic nuclei. *Neuroreport.* 9:1059-1063.

Lundkvist GB, Andersson A, Robertson B, Rottenberg ME, Kristensson K (1999) Light-dependent regulation and postnatal development of the interferon-gamma receptor in the rat suprachiasmatic nuclei. *Brain Res.* 849:231-234.

Lundkvist GB, Hill RH, Kristensson K (2002) Disruption of circadian rhythms in synaptic activity of the suprachiasmatic nuclei by African trypanosomes and cytokines. *Neurobiol. Dis.* 11:20-27.

Lynch MA (2002) Interleukin-1 beta exerts a myriad of effects in the brain and in particular in the hippocampus: analysis of some of these actions. *Vitam. Horm.* 64:185-219.

Marpegan L, Bekinschtein TA, Costas MA, Golombek DA (2005) Circadian responses to endotoxin treatment in mice. *J. Neuroimmunol.* 160:102-109.

Motzkus D, Albrecht U, Maronde E (2002) The human PER1 gene is inducible by interleukin-6. *J. Mol. Neurosci.* 18:105-109.

Mundigler G, Delle-Karth G, Koreny M, Zehetgruber M, Steindl-Munda P, Marktl W, Ferti L, Siostrzonek P (2002) Impaired circadian rhythm of melatonin secretion in sedated critically ill patients with severe sepsis. *Crit. Care Med.* 30:536-540.

Mundigler G, Delle-Karth G, Koreny M, Zehetgruber M, Steindl-Munda P, Marktl W, Ferti L, Siostrzonek P (2002) Impaired circadian rhythm of melatonin secretion in sedated critically ill patients with severe sepsis. *Crit. Care Med.* 30:536-540.

Nadeau S, Rivest S (1999) Regulation of the gene encoding tumor necrosis factor alpha (TNF-alpha) in the rat brain and pituitary in response in different models of systemic immune challenge. *J. Neuropathol. Exp. Neurol.* 58:61-77.

Nava F, Carta G, Haynes LW (2000) Lipopolysaccharide increases arginine-vasopressin release from rat suprachiasmatic nucleus slice cultures. *Neurosci. Lett.* 288:228-230.

Nygard M, Hill RH, Wikstrom MA, Kristensson K (2005) Age-related changes in electrophysiological properties of the mouse suprachiasmatic nucleus in vitro. *Brain Res. Bull.* 65:149-154.

Ohdo S, Wang DS, Koyanagi S, Takane H, Inoue K, Aramaki H, Yukawa E, Higuchi S (2000) Basis for dosing time-dependent changes in the antiviral activity of interferon-alpha in mice. *J. Pharmacol. Exp. Ther.* 294:488-493.

Ohdo S, Koyanagi S, Suyama H, Higuchi S, Aramaki H (2001) Changing the dosing schedule minimizes the disruptive effects of interferon on clock function. *Nat. Med.* 7:356-360.

Ohdo S (2007) Chronopharmacology focused on biological clock. *Drug Metab. Pharmacokinet.* 22:3-14.

O'Neill LA (2006) How Toll-like receptors signal: what we know and what we don't know. *Curr. Opin. Immunol.* 18:3-9.

Opp MR (2005) Cytokines and Sleep. *Sleep Med. Rev.* 6: 355-364.

Pan W, Cornelissen G, Halberg F, Kastin AJ (2002) Selected contribution: circadian rhythm of tumor necrosis factor-alpha uptake into mouse spinal cord. *J. Appl. Physiol.* 92:1357-1362; discussion 1356.

Parnet P, Kelley KW, Bluthe RM, Dantzer R (2002) Expression and regulation of interleukin-1 receptors in the brain. Role in cytokines-induced sickness behavior. *J. Neuroimmunol.* 125:5-14.

Peng ZC, Kristensson K, Bentivoglio M (1994) Dysregulation of photic induction of Fos-related protein in the biological clock during experimental trypanosomiasis. *Neurosci. Lett.* 182:104-106.

Perry VH (2004) The influence of systemic inflammation on inflammation in the brain: implications for chronic neurodegenerative disease. *Brain Behav. Immun.* 18:407-413.

Pickering M, Cumiskey D, O'Connor JJ (2005) Actions of TNF-alpha on glutamatergic synaptic transmission in the central nervous system. *Exp. Physiol.* 90:663-670.

Pollmacher T, Haack M, Schuld A, Reichenberg A, Yirmiya R (2002) Low levels of circulating inflammatory cytokines--do they affect human brain functions? *Brain Behav. Immun.* 16:525-532.

Pontes GN, Cardoso EC, Carneiro-Sampaio MM, Markus RP (2006) Injury switches melatonin production source from endocrine (pineal) to paracrine (phagocytes) - melatonin in human colostrum and colostrum phagocytes. *J. Pineal Res.* 41:136-141.

Quan N, Mhlanga JD, Whiteside MB, McCoy AN, Kristensson K, Herkenham M (1999) Chronic overexpression of proinflammatory cytokines and histopathology in the brains of rats infected with Trypanosoma brucei. *J. Comp. Neurol.* 414:114-130.

Ralph MR, Foster RG, Davis FC, Menaker M (1990) Transplanted suprachiasmatic nucleus determines circadian period. *Science.* 247:975-978.

Rivest S (2003) Molecular insights on the cerebral innate immune system. *Brain Behav. Immun.* 17:13-19.

Sadki A, Bentivoglio M, Kristensson K, Nygard M (2007) Suppressors, receptors and effects of cytokines on the aging mouse biological clock. *Neurobiol. Aging.* 28:296-305.

Spanagel R, Pendyala G, Abarca C, Zghoul T, Sanchis-Segura C, Magnone MC, Lascorz J, Depner M, Holzberg D, Soyka M, Schreiber S, Matsuda F, Lathrop M, Schumann G, Albrecht U (2005) The clock gene Per2 influences the glutamatergic system and modulates alcohol consumption. *Nat. Med.* 11:35-42.

Stellwagen D, Malenka RC (2006) Synaptic scaling mediated by glial TNF-alpha. *Nature.* 440:1054-1059.

Tabarean IV, Korn H, Bartfai T (2006) Interleukin-1beta induces hyperpolarization and modulates synaptic inhibition in preoptic and anterior hypothalamic neurons. *Neuroscience.* 141:1685-1695.

Taishi P, Bredow S, Guha-Thakurta N, Obal F, Jr., Krueger JM (1997) Diurnal variations of interleukin-1 beta mRNA and beta-actin mRNA in rat brain. *J. Neuroimmunol.* 75:69-74.

Takane H, Ohdo S, Baba R, Koyanagi S, Yukawa E, Higuchi S (2002) Relationship between 24-hour rhythm in antiviral effect of interferon-beta and interferon-alpha/beta receptor expression in mice. *Jpn. J. Pharmacol.* 90:304-312.

Takaoka A, Yanai H (2006) Interferon signalling network in innate defence. *Cell Microbiol.* 8:907-922.

Wajant H, Pfizenmaier K, Scheurich P (2003) Tumor necrosis factor signaling. *Cell Death Differ.* 10:45-65.

Watanabe A, Shibata S, Watanabe S (1995) Circadian rhythm of spontaneous neuronal activity in the suprachiasmatic nucleus of old hamster in vitro. *Brain Res.* 695:237-239.

Weisman D, Hakimian E, Ho GJ (2006) Interleukins, inflammation, and mechanisms of Alzheimer's disease. *Vitam. Horm.* 74:505-530.

Wirz-Justice A (2006) Biological rhythm disturbances in mood disorders. *Int. Clin. Psychopharmacol.* 21 Suppl 1:S11-15.

Yoshida H, Kubota T, Krueger JM (2003) A cyclooxygenase-2 inhibitor attenuates spontaneous and TNF-alpha-induced non-rapid eye movement sleep in rabbits. *Am. J. Physiol. Regul. Integr. Comp. Physiol.* 285:R99-109.

In: Progress in Circadian Rhythm Research
Editor: Anne-Laure Léglise, pp. 209-222
ISBN: 978-1-60021-876-7

Chapter X

Do Scales for Measuring Morningness-Eveningness Exist All Over the World?

Hervé Caci[1*], Vincenzo Natale[2] and Ana Adan[3]
[1] CHU de Nice, Hôpital Archet 2, Service de Pédiatrie. 06202 Nice Cedex 3, France.
[2] Department of Psychology, University of Bologna. Bologna, Italy.
[3] Department of Psychiatry and Clinical Psychobiology,
School of Psychology. Barcelona, Spain.

Abstract

Morningness-Eveningness has been theorized as a continuum along which subjects can be arranged or classified into coarsely determined categories: owls (or Evening-types), intermediates and larks (or Morning-types). This trait is thought to be universal chiefly because it is biologically and genetically determined. The number of categories changes to 5 after adding the "Definitively Evening" and "Definitively Morning" extreme types. This procedure is justified for epidemiological, clinical and research purposes, and it is common in psychology, but it also raises a number of concerns we shall discuss in this chapter.

To proceed, one first needs some valid instrument: the Morningness-Eveningness Questionnaire is the most widely used scale, though others exist that deserve attention. Reliability is a necessary but insufficient property on the road toward validity of a scale's score. Next, beyond the fact that the commonly used methods for categorisation are not theoretically sound or equivocal (i.e. for what reasons should we use extreme deciles or extreme quartiles? how are the cut-off scores determined?), the most important criticism is that demographical variables such as age and gender are not taken into consideration. Some recent data emphasize that evening-types are not rare in middle-aged, an erroneous conclusion derived from the rigid categorization. To say it simpler, norms for Morningness scores have not been determined. Moreover, we, like others, have claimed that cross-cultural validity is a desirable property for scales intended to measure

* Hervé Caci: Email : caci.h@chu-nice.fr

personality traits including Morningness-Eveningness. This property might be extended to the factors extracted in the studied cultures leading to the problem of factorial invariance.

In this chapter, we review the existing scales and their general properties in different cultures before focusing on the problem of categorization with examples showing its limits (e.g. lack of coherence between types determined on different scales). We will eventually propose some remedies. Finally, we will address the issue of cross-gender and cross-cultural factorial invariance of these scales. We conclude that the way is long that will lead to valid, between-group comparisons in terms of Morningness-Eveningness scores and/or frequencies of the corresponding types.

Introduction

The idea of measuring the mind was first mentioned by Leonardo da Vinci. In 1764, the Swiss philosopher Charles Bonet forged the word "psychometer" which was later defined in 1879 by Sir Francis Galton. In 1901, the True Score theory was grounded on the coefficient of correlation invented by Pearson, as were the related mathematical tools: Principal Component Analysis and Exploratory Factor Analysis, which appeared within the subsequent thirty years. At the same time, methods for analyzing covariance matrices were developed, such as path analysis. Powerful analytical methods were later developed in replacement of geometrical ones, and they were implemented in statistical packages for widespread use. Hence, while psychometric tools are now available to almost everyone, but it is doubtful that end users master the entire theory. This discussion is beyond the scope of this chapter, and the interested reader should refer to criticisms published here and there [9].

In the 1970s, the pioneering works of Öquist recalled the old characteristic morningness and eventually made it a measurable individual characteristic [39]. Once his questionnaire was revised and made available in English, it became very popular and was widely used in different languages without much control. *De facto,* it attained the role of gold-standard. Since then, other instruments have been devised as concurrent measures (and therefore their properties compared to the gold standard) or as tools to provide a more comprehensive approach of circadian rhythms. Furthermore, it is expected that morningness is not a simple and unitary construct. Nowadays, studies are conducted on the Internet using these morningness scales or other personality questionnaires, and data is collected on very large samples. At this point it is necessary to know the factorial structure of these instruments and whether their cross-gender and cross-cultural validities have been ascertained. Before providing an attempt to answer these questions, we briefly present the existing scales.

Existing Scales

Morningness-Eveningness Questionnaire (MEQ)

Öquist constructed a Swedish language Morningness-Eveningness Questionnaire to distinguish morning-type and evening-type individuals in patterns of food intake and oral temperature [39]. Later he re-designed this questionnaire to use in shift work studies [40], and eventually Horne and Östberg produced an English version adding new questions and deleting others, and they revised the scoring using either a four-choice format or a time scale [25]. One hundred and fifty subjects with ages ranging between 18 and 32 were enrolled in the first validation study with sleep agenda and body temperature as a external criteria for a subsample of 48 randomly selected subjects over a three-week period. With regard to the total score, but without any rationale, a five-category classication was proposed: Definitively evening-type (16-30), Moderately evening-type (31-41), Neiher type (42-58), Moderately morning-type (59-69) and Definitively morning-type (70-86).

The MEQ total score, and therefore its empirical transformation into a five-category circadian typology, received many validation elements, the enumeration of which is outside the scope of this chapter. Hence, the MEQ appeared as the undisputed gold standard measure of morningness, explicating the very small number of studies questioning its psychometric properties. The cutoff scores remained mostly unchallenged, though Horne and Östberg mentioned that these values might not be valid in older samples which was confirmed later in a sample of 566 good sleepers aged between 44 and 58 years [52]. Later in this chapter, we will turn to the cross-cultural validity of the MEQ and see that the picture is even more problematic.

The internal properties of the MEQ have been questioned from time to time. The 3-month temporal stability is good (r = +0,88) on a sample of 74 students [29]. The reliability, as measured by Cronbach's α, is regularly above 0.80. The mean inter-item correlation ranges from +0.20 [49] to +0.40 [29] suggesting that the scale is not unidimensional. After discarding six items with weak measurement properties, a three-factor structure has been proposed: Morning/Effort (items 1, 9, 11, 13, 15, 17 and 18), Evening (items 2, 10 and 12) and Morning Alertness (items 4, 5 and 7) [50]. Using the entire scale, we recently reported a similar four-factor structure [12].

The length of the MEQ and the fact that item 19 is merely a "validity check" item (since it asks about the subject's self-perceived typology in four categories) has been discussed [49]. Also, the longer the scale the higher the probability that the measure will not be unidimensional. Some authors proposed shortened versions of the MEQ and of other morningness scales, as we will see later in this chapter, therefore positing that morningness should be a unitary construct. The properties of the reduced MEQ (rMEQ) proposed by Adan and Almirall in a sample of Spanish workers with various schedules [1,2] have been studied in various samples of students in the USA, Italy and France [12,17,36]. The rMEQ counts five items, whose numbers in the MEQ are 1, 7, 10, 18 and 19, and cutoff scores were defined as follows: Definitively Evening-type (4 to 7), Moderately Evening-type (8 to 11), Neither Type (12 to 17), Moderately Morning-type (18 to 21) and Definitively Morning-type (22 to 25) . The MEQ and rMEQ total scores correlate above 0.88 in all samples with a satisfactory level of sensivity around 0.80 which represents the proportion of subjects classified in the same category with both scales [17]. In the Spanish and French sample, the rMEQ score is independent of sex when age is taken into consideration and varies in the same sense of age; that is, older subjects score higher on the rMEQ [2,12]. In the Italian sample, the rMEQ shows a fair external validity using actigraphy: evening-types present a significant phase delay versus intermediates and morning-types [36].

Circadian Type Questionnaire (CTQ) and Circadian Type Inventory (CTI)

The original 19-item CTQ was devised as a "predictive test of adjustment to shift work" [21] yielding three factors: Rs (rigidity of sleeping habits; items 1 to 8), V (vigour; items 9 to 13) and M (morningness; items 14 to 19) related to the stability, amplitude and phasing of the respondent's circadian rhythms, respectively. Unfortunately, the instrument had weak psychometric properties [23,49]: the M scale had to be replaced by the MEQ, while the V and Rs scales had to be revised although they showed an encouraging predictive value for adaptation.

The CTQ was revised and became the 30-item Circadian Type Inventory (CTI) [6,18,51]. It measured Flexibility/Rigidity of sleep habits (FR; Rigid-types prefer to eat and sleep at regular times) and Languid/Vigourous (LV; Languid-types find it difficult to overcome drowsiness and feel lethargic following reduced sleep). It was again shortened to 18 items and included in the Standard Shiftwork Index (SSI) [6]. An even shorted version was suggested with five items measuring FR and six measuring LV [18,19]. In light of the validity arguments collected for the CTQ/CTI in longitudinal studies, these authors commented that "it may be time for researchers to abandon the notion that a single circadian rhythm characteristic is the best predictor of adjustment to shiftwork and consider Smith et al.'s (1989) recommendation that 'a weighted combination' (p. 738) may better serve this goal given the variety of both individual differences and shift schedules in operation." (p. 1303).

Diurnal Type Scale (DTS)

The DTS was devised as a brief and homogeneous instrument capable of being completed regardless of the work schedules of the individuals [53]. There are seven 4-point items (items 1, 3, 6 and 7 are reverse coded). The psychometric properties of the DTS have been studied in samples of male, industrial workers by its authors, which might seem a clear limitation, and in two samples of undergraduated students [22,49]. The last two studies report that the DTS has marginal psychometric properties. The scale is not unidimensional; there are probably two factors: "Morningness" (items 1, 4, 6 and 7) and "Eveningness" (items 2 and 5). The DTS total score correlated only +0.50 with the MEQ and was found much more stable over a three-month period (+0,78; n=40) than over a nine-month period (+0.43; n=40). But the author commented that "the cause of the longer term instability might be related to exposure to night or shift work and not simply be due to transient effects of current lifestyles" [22]. These results tend to support Kerkhof's recommendation that "instruments which attempt to assess morningness-eveningness need to be validated on each population for which their use is considered" [27]. Finally, we have been unable to find any translation of the DTS into other languages.

Composite Scale of Morningness (CSM) and Basic Language Morningness (BALM)

Acknowledging the psychometric flaws in the measurement of morningness-eveningness, Smith et al. (1989) devised the CSM by extracting the 13 'best' items of the MEQ (items 1, 2, 4, 5, 7, 9, 10, 11 and 19) and the DTS (items 1, 4, 6 and 7) [49]. The MEQ items were slightly modified to allow all the items to be rated on a 1 to 4- ou 5-point scale. The total score ranges from 13 to 55, and the authors proposed a three-category classification taking the percentile 10 and percentile 90 as cutoff scores, i.e. 22 and 44 respectively. The total score was predictive of patterns of American students behavior and performance [24]; the reader interested into the relationships between morningness and cognition is referred to the chapter by Adan, Natale and Caci.

Many studies replicated the sound psychometric properties of the CSM (e.g. reliability in the 0.80-0.90 range) and a correlation with the MEQ greater than 0.90, but the cutoff scores logically appeared to be sample-dependent. The instrument yielded three factors: Morning Activities (items 1, 6, 8 to 11, and 13), Morning Affect (items 3, 4, 5 and 12) and Time of Retiring (items 2 and 7). The CSM has been translated into many languages including Czech, Dutch, French [13], German [45], Italian [5], Korean [56], Polish, Portuguese, Spanish [3,20], and Thai [42]. We analyzed five national samples and found a three-factor structure that was highly congruent between genders in all cultures [11]. The Morning Affect factor proposed by Smith et al. (1989) was retrieved and reported to be the most congruent between cultures. Very recently, a similar three-factor structure has been reported partially invariant by Confirmatory Factor Analysis between German and Spanish samples [45]. This is a very exciting result that will be commented later.

Additionally, Pornpitakpan proposed the short-CSM, a reduced version, by discarding items that seemed redundant (leaving items 2, 5 and 8 to 12) [42] and Brown proposed a version with simplified wording at 7th grade (12-13 years) reading level, he called the BAsic Language scale of Morningness (BALM) [10,43]. Interestingly, the structure of the BALM was close to that reported by Smith et al. (1989) but in two samples each of 150 shiftworkers and fixed schedule dayworkers. This was an interesting attempt to put on some consistency in the measurement of morningness within a large range of age regardless of the level of education. As for the MEQ and the CSM, a shortened version of the BALM was proposed yielding two factors: Morningness/Effort (items 1, 8 to 11, and 13) and Morning Alertness (items 4, 5 and 12). To add to the confusion, Pornpitakpan (2000) shortened the BALM to seven items yielding two components: "Self-Reported Morningness-Eveningness/Voluntary Schedules" (items 6, 8, 9, 10 and 11) and "Alertness After Waking up" (items 5 and 12). Hence the short-CSM and the short-BALM are totally different instruments.

Morningness Scales for Youngsters

Some attempts have been made to derive measures of morningness for children and adolescents from the MEQ (Junior-MEQ) [26] or from the CSM (Morningness-Eveningness Scale for Children, MESC) [15].

Table 1. Pattern matrices after oblique PROMAX rotation on three independent samples (ninety-nine 9th graders and 176 inpatients with both mean age of 15.2, and one hundred seventeen 6th graders with mean age of 12.7). Loadings higher than 0.400 are in bold. Taken from [14] with permission

	Factor 1 Morningness	Factor 2 Planification
Item 1	0.382 0.529 0.378	0.243 0.143 0.082
Item 2	0.771 0.788 0.687	-0.110 -0.061 0.138
Item 3	0.502 0.497 0.493	0.009 0.006 -0.177
Item 4	-0.007 0.251 0.185	0.205 0.414 0.238
Item 5	0.250 0.204 0.240	0.469 0.420 0.313
Item 6	-0.088 -0.124 -0.010	0.669 0.804 0.630
Item 7	0.654 0.725 0.542	-0.083 -0.101 0.005
Item 8	-0.152 -0.043 -0.018	0.509 0.832 0.667
Item 9	0.658 0.739 0.676	0.107 -0.000 -0.024
Item 10	0.631 0.679 0.615	0.005 0.003 -0.112
Correlation	—	0.330 0.531 0.347

The MESC has been translated into several languages including French [14] and Portugese [37]. We have reported a congruent factorial structure in three independent samples of 6th, 7th and 9th graders [14]: congruence coefficients after Procrustes rotations were greater the 0.96 between homologuous factors (table 1). Factor 1 was called *Morningness* (items 1-3, 7, 9 and 10) and factor 2 was called *Planification* (items 5, 6 and 8). Additionally, we reported in two of these samples an effect of the month of birth similar to that reported by others using the MEQ on student samples [33-35]. Unlike others, our work didn't rely on the chronotype but rather regressed the month of birth on the MESC total score using a cosinual function and controlling for age and gender. Evening-oriented adolescents tend to be born in months characterized by a long photoperiod (i.e. 26,43±4,94 for march-april) while morning-oriented adolescents tend to be born in months characterized by a short photoperiod (i.e. 29,82±4,78 for september-october). In our sample of inpatients this relationship was significant only for those hospitalized in the suit of a suicide attempt (e.g. medicine overdose or wrist-cutting) and characterized by their higher level of impulsivity.

Another interesting point is the variation of morningness during puberty. Although only cross-sectional designs were used, many authors reported a shift toward eveningness around 12 years in girls and 13 years in boys regardless of culture [15,28,37]. Some of them even suggest that the period of puberty would be bounded by a shift toward eveningness at its beginning and a shift toward morningness at its end [46].

Cross-Gender Results

Kerkhof advocated that "well-controled experiments under temporal isolation are of crucial relevance" to answer the question of sex differences in circadian rhythms [27]; he cites Wever who reported that "the overall period of the rhythms of sleep-wake and rectal temperature is 17-28 minutes shorter for females than for males" [55]. Therefore, it was reasonable to hypothesize that women would be exhibit a higher level of morningness than females. The phase advance of their biological clock is estimated to less than one hour, which is probably a too small value to be reflected by self-assessment questionnaires. One solution is to increase the sample size hence the statistical power of the test, but one could object that a statistically significant tiny difference might not be clinically relevant. Age is another problem with group comparisons as the covariable is often omitted. Indeed, most studies present simple Student *t*-tests whereas ANCOVAs (with age as covariable and gender as cofactor) are indicated. Our rationale goes through with the question of the age range. Since morningness is somewhat age-dependent, between-gender comparisons should compare groups of the same age range. It is likely that all these are the reasons for the discrepancies in cross-gender comparisons. Some authors report a difference [4], some don't [16,31,38,41,44]. Interestingly, a higher level of morningness in females was reported in twin study [54].

A more serious criticism, although not directed against morningness scales but rather against all rating scales, is that any between-group comparison takes for granted the measurement invariance between groups. In a provocative way, we would ask what relevant difference between the two fruit sellers would reflect the comparison of the mean diameters of apples on one stall and oranges on another one? This is a long known but still under-recognized point in psychometry [32].

Cross-Cultural Results

Biological and genetical mechanisms underly personality traits such as morningness. Therefore we expect a between-group consistency in the psychometric properties of the morningness construct. When comparing the structure of a morningness scale between groups (e.g. males vs. females, or national groups), the same items should load on the same factors not on others (i.e. factorial form) and with a similar magnitude (i.e. equal loadings). Given that the morningness construct would function similarly across groups, true differences in mean scores could be measured [8]. This is a far more comprehensive approach than the simple comparison of unweighted sums of item scores.

It has been speculated that socio-cultural differences could explain the observed differences in morningness scores across countries because some questions explicitly refer to specific hours of the day [27]. For example, an unexpected large proportion of morning-types was reported with in Italy and Brazil [7,30]. Obviously, this is a direct consequence of the use of rigid cutoff scores, but other explications are plausible. The reader has noticed that while the most translated instruments are the MEQ and the CSM, few studies assessed the translation quality with careful back-translations and contrasted translations from different countries or cultures, either for the MEQ [48,50] or the CSM [11,45]. This is true for adults and adolescents too [47].

Smith et al. (1991) reported a thorough study of the equivalence between the English and Japanese versions of the MEQ. Their analyses were restricted to thirteen items because of weak mesurement characteristics of the six discarded items. The factorial structure was far more complex in the Japanese sample but a *Morningness alertness* factor (items 4, 5 and 7) appeared as equivalent across samples with Japanese respondent reporting a higher mean value. As a consequence, the Japanese version had to be revised to clear content discrepancies casting doubt on results of studies that used the former version.

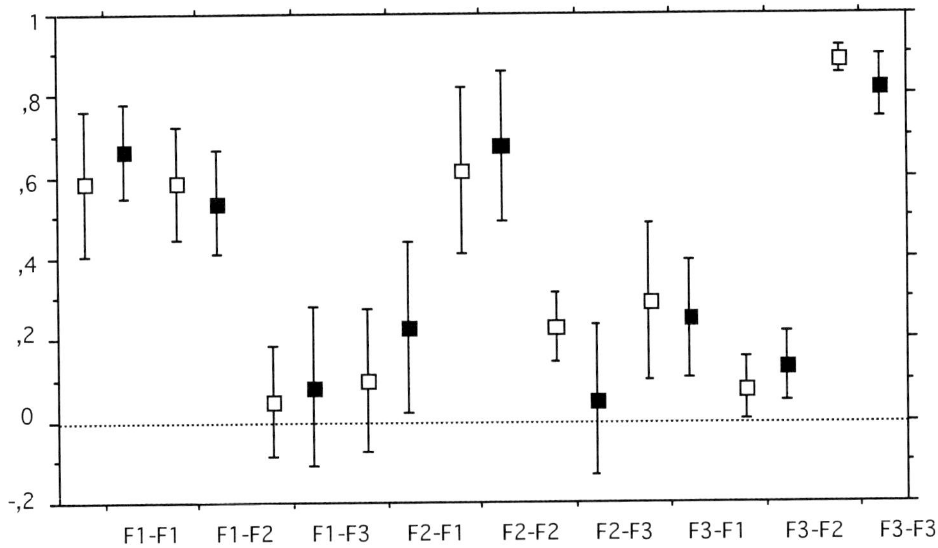

Figure 1. Tucker's congruence coefficients for homolog and heterolog factors with 95% confidence interval, split by sex (white boxes for females and black boxes for males). Taken from [11] with permission.

Table 2. Pattern matrix of the CSM after oblique PROMAX rotation (N=436). Loadings higher than 0.400 are in bold. Taken from [12] with permission

Variable	Factor 1	Factor 2	Factor 3	Uniqueness
Item 1	0.545	0.008	0.294	0.369
Item 2	-0.087	0.043	0.794	0.397
Item 3	0.350	0.475	-0.025	0.446
Item 4	-0.089	0.896	0.059	0.254
Item 5	0.080	0.821	-0.041	0.256
Item 6	0.617	0.142	-0.174	0.580
Item 7	-0.137	-0.075	0.742	0.549
Item 8	0.405	0.029	0.209	0.612
Item 9	0.378	0.051	0.533	0.261
Item 10	0.752	-0.154	0.099	0.466
Item 11	0.785	0.015	-0.086	0.428
Item 12	-0.153	0.762	0.061	0.523
Item 13	0.375	0.076	0.488	0.290
Factor 2	0.662	1		
Factor 3	0.554	0.371	1	

Table 3. Pattern matrix of the MEQ after oblique PROMAX rotation (N=436). Loadings higher than 0.400 are in bold. Taken from [12] with permission

Variable	Factor 1	Factor 2	Factor 3	Factor 4	Uniqueness
Item 1	0.357	-0.025	0.238	0.398	0.330
Item 2	0.006	-0.025	0.760	0.178	0.288
Item 3	0.049	0.013	-0.109	0.496	0.673
Item 4	0.162	0.643	-0.012	0.099	0.297
Item 5	-0.056	0.969	0.016	-0.055	0.141
Item 6	-0.048	0.234	0.134	-0.021	0.839
Item 7	-0.048	0.972	-0.012	-0.070	0.156
Item 8	-0.321	0.012	0.468	0.418	0.639
Item 9	0.395	0.211	-0.261	0.306	0.472
Item 10	0.156	-0.031	0.592	-0.061	0.518
Item 11	0.751	0.005	0.103	-0.202	0.446
Item 12	0.040	0.074	0.593	-0.141	0.580
Item 13	-0.086	-0.086	0.101	0.608	0.639
Item 14	0.047	0.091	0.201	0.013	0.826
Item 15	0.520	-0.124	-0.018	0.146	0.654
Item 16	0.156	-0.061	0.612	-0.165	0.525
Item 17	0.673	-0.014	0.121	0.010	0.429
Item 18	0.519	-0.030	0.300	0.019	0.435
Item 19	0.327	0.147	0.450	0.159	0.273
Factor 2	0.547	1			
Factor 3	0.472	0.279	1		
Factor 4	0.518	0.577	0.204	1	

In our study we used exploratory factor analysis and reach a conclusion similar to that of Smith et al. (1991), though we compared five versions of the CSM: English, French, Italian, Spanish and Thai [11]. The *Morning Affect* factor (items 3, 4, 5 and 12) encompasses the items of their *Morning Alertness* factor, and was found to be the most congruent across countries (figure 1). To bring support to our model, we recently found in the same sample an identical structure for the CSM embedded within the MEQ [12] (tables 2 and 3). Interestingly, a Multigroup Confirmatory Factor Analysis of the CSM very recently showed partial factorial invariance between German (N=432) and Spanish (N=439) students with a 3-factor model similar to the one we had proposed [11]: *Morningness* (items 1, 3, 6, 8, 10 and 11), *Morning alertness* (items 4, 5 and 12) and *Activity planning* (items 2, 7, 9 and 13). The first two factors were found to be invariant, and German scored higher on both; that is, they are more morning-oriented than Spanish students.

Conclusion

The obvious conclusion is that the studied instruments may not, as a whole, be equivalent between cultures. As Smith et al. (1991) pointed out, the guidelines for scale translation may not have been followed in some cases, "which precludes clearly separating linguistic or semantic and population differences in item responses" [50]. The positive issue is that researchers may indifferently use the *Morning Alertness* factor from the MEQ and the *Morning Affect* factor from the CSM as a tool to explore sleep inertia regardless of culture. The negative issue is that although everyone agrees on the multifactorial nature of the construct of morningness, researchers are forced to rely on a total score which is the unweighted sum of the item scores. More disquieting, this continuous variable is coarsely transformed into a three- or five-categorical typology following a rigid rule. Indeed, proposed cutoff scores are sample-dependent (i.e. 10-90 percentile rule for the DTS or the CSM) or arbitrary in the case of the MEQ. Of course, this is less harmful when comparing extreme chronotypes of similar age but this leaves unexplored the larger range of the continuum (i.e. individuals belonging to the intermediate chronotype). For these reasons, we propose that future translations of morningness scales should be placed under some authority to assure the respect of the guidelines (e.g. direct- and back-translations by independent translators). Next, cross-cultural studies should be fostered in different groups with the same methodology and with regard to age, gender, work schedule and maybe some other socio-demographic variables so that normative data could be derived for different groups. Finally, the categorization of the total score or of the factor scores in sex- and/or age-subgroups could rely on the percentile method or on T-scores, for example. Of course, to achieve a comprehensive assessment of circadian rhythms it is advisable to include measures of stability and amplitude such as those proposed in the CTI.

In this chapter, we reviewed the psychometric properties of existing morningness scales. At the cross-cultural level, we showed that most of the work remains undone. Additionally, we emphasized the problem of the scoring and of the coarse transformation into a chronotype allocation. Unless progress are made in these directions, we feel that the generalizability of results obtained in any culture will remain weak or moderate.

References

[1] Adan A. and Almirall H. (1990). Standardization of a reduced scale of morningness in Spanish population: individual differences. *Psicothema,* 2, 137-149.

[2] Adan A. and Almirall H. (1991). Horne and Östberg morningness-eveningness questionnaire: a reduced scale. *Personality and Individual Differences,* 12, 241-253.

[3] Adan A., Caci H., and Prat G. (2005). Reliability of the Spanish version of the Composite Scale of Morningness. *European Psychiatry,* 20, 503-509.

[4] Adan A. and Natale V. (2002). Gender differences in morningness-eveningness preference. *Chronobiology International,* 19, 709-720.

[5] Alzani A. and Natale V. (1998). Uno strumento per la valutazione delle differenze individuali nei ritmi circadiani: une versione italiana della Composite Scale of Morningness. *Testing Psicom. Metodol,* 5, 19-31.

[6] Barton J., Spelten E., Totterdell P., Smith L., Folkard S., and Costa G. (1995). The standard shiftwork index: a battery of questionnaires for assessing shiftwork-related problems. *Work Stress,* 9, 4-30.

[7] Benedito-Silva A. A., Menna-Barreto I. S., Cipolla-Neto J., Marques N., and Tenreiro S. (1989). A self-evaluation questionnaire for the determination of morningness-eveningness types in Brazil. *Chronobiologia,* 16, 311.

[8] Bollen KA: (1989). Structural Equations with Latent Variables. New York, Wiley.

[9] Borsboom D. (2006). The attack of the psychometricians. *Psychometrika,* 71, 425-440.

[10] Brown F. M. (1993). Psychometric equivalence of an improved Basic Language Morningness (BALM) Scale using industrial population within comparisons. *Ergonomics*, 36, 191-197.

[11] Caci H., Adan A., Bohlé P., Natale V., Pornpitakpan C., and Tilley A. (2005). Transcultural properties of the Composite Scale of Morningness: the relevance of the 'Morning Affect' factor. *Chronobiology International,* 22, 523-540.

[12] Caci H., Deschaux O., Adan A., and Natale V. (Submitted). Psychometric comparison between three morningness scales in French-speaking students.

[13] Caci H., Nadalet L., Staccini P., Myquel M., and Boyer P. (1999). Psychometric properties of the French version of the Composite Scale of Morningness in adults. *European Psychiatry,* 14, 284-290.

[14] Caci H., Robert Ph., Dossios C., and Boyer P. (2005). L'échelle de Matinalité pour enfants et adolescents: propriétés psychométriques et effet du mois de naissance. [Morningness-Eveningness for Children Scale: psychometric properties and month of birth effect]. *L'Encéphale,* 31, 56-64.

[15] Carskadon M. A., Vieira C., and Acebo C. (1993). Association between puberty and delayed phase preference. *Sleep,* 16, 258-262.

[16] Chelminski I., Ferraro F. R., Petros T., and Plaud J. J. (1997). Horne and Östberg questionnaire: a score distribution in a large sample of young adults. *Personality and Individual Differences,* 23, 647-652.

[17] Chelminski I., Petros T. V., Plaud J. J., and Ferraro F. R. (2000). Psychometric properties of the reduced Horne and Östberg questionnaire. *Personality and Individual Differences,* 29, 469-478.

[18] Di Milia L., Smith P. A., and Folkard S. (2004). Refining the psychometric properties of the Circadian Type Inventory. *Personality and Individual Differences,* 36, 1953-1964.

[19] Di Millia L., Smith P. A., and Folkard S. (2005). A validation of the revised circadian type inventory in a working sample. *Personality and Individual Differences.*

[20] Diaz-Morales J. F. and Sanchez-Lopez M. P. (2005). Composite scales of morningness and preferences: preliminary validity data in Peruvian undergraduates. *Ergonomics,* 48, 354-363.

[21] Folkard S., Monk T., and Lobban M. (1979). Towards a predictive test of adjustment to shift work. *Ergonomics,* 22, 79-91.

[22] Greenwood K. M. (1991). Psychometric properties of the Diurnal Type Scale of Torsvall and Åkerstedt (1980). *Ergonomics,* 34, 435-443.

[23] Greenwood K. M. (1995). An evaluation of the Circadian Type Questionnaire. *Ergonomics,* 38, 347-360.

[24] Guthrie J. P., Ash R. A., and Bendapudi V. (1995). Additional validity evidence for a measure of Morningness. *Journal of Applied Psychology,* 80, 186-190.

[25] Horne J. A. and Östberg O. (1976). A self-assessment questionnaire to determine morningness-eveningness in human circadian rhythms. *International Journal of Chronobiology,* 4, 97-110.

[26] Ishihara K., Honma Y., and Miyake S. (1990). Investigation of the children's version of the Morningness-Eveningness Questionnaire with primary and junior high school pupils in Japan. *Perceptual and Motor Skills,* 71, 1353-1354.

[27] Kerkhof G. A. (1985). Inter-individual differences in the human circadian system: A review. *Biological Psychology,* 20, 83-112.

[28] Kim S., Dueker G. L., Hasher L., and Goldstein D. (2002). Children's time of day preference: age, gender and ethnic differences. *Personality and Individual Differences,* 33, 1083-1090.

[29] Larsen R. J. (1985). Individual differences in circadian activity rhythm and personality. *Personality and Individual Differences,* 6, 305-311.

[30] Mecacci L. and Zani A. (1983). Morningness-eveningness preferences and sleep-waking diary data of morning and evening types in student and worker samples. *Ergonomics,* 26, 1147-1153.

[31] Mecacci L., Zani A., Rocchetti G., and Lucioli R. (1986). The relationships between morningness-eveningness, ageing and personality. *Personality and Individual Differences,* 7, 911-913.

[32] Meredith W. (1964). Notes on factorial invariance. *Psychometrika,* 29, 177-185.

[33] Mongrain V., Paquet J., and Dumont M. (2006). Contribution of the photoperiod at birth to the association between season of birth and diurnal preference. *Neuroscience Letters,* 406, 113-116.

[34] Natale V. and Adan A. (1999). Season of birth modulates morningness-eveningness preference in humans. *Neuroscience Letters,* 274, 139-141.

[35] Natale V., Adan A., and Chotai J. (2002). Further results on the association between morningness-eveningness preference and the season of birth in human adults. *Neuropsychobiology,* 46, 209-214.

[36] Natale V., Esposito M. J., Martoni M., and Fabbri M. (2006). Validity of the reduced version of the Morningness Eveningness Questionnaire. *Sleep of Biological Rhythms,* 4, 72-74.

[37] Natale VB, Pellarin L, Alves RC, et al: Morningness tendency among 10 to 15 years-old normal population. *Sleep.* 1999;22:S184(Abstract)

[38] Neubauer A. C. (1992). Psychometric comparison of two circadian rhythm questionaires and their relationship with personality. *Personality and Individual Differences,* 13, 125-131.

[39] Öquist O: Kartlaggning av individuella dygnsrytmer. Thèse du département de Psychologie de l'Université de Göteborg, *Suède.* 1970.(UnPub)

[40] Östberg O. (1973). Interindividual differences in circadian fatigue patterns of shift workers. *Br. J. Indust. Med,* 30, 341-351.

[41] Paine S.-J., Gander P. H., and Travier N. (2006). The epidemiology of morningness/eveningness: influence of age, gender, ethnicity and socioeconomic factors in adults (30-49 years). *Journal of Biological Rhythms,* 21, 68-76.

[42] Pornpitakpan C. (1998). Psychometric properties of the composite scale of morningness: a shortened version. *Personality and Individual Differences,* 25, 699-709.

[43] Pornpitakpan C. (2000). Additional validity of the Basic Language Morningness (BALM) scale. *Personality and Individual Differences,* 28, 59-72.

[44] Posey T. B. and Ford J. A. (1981). The morningness-eveningness preference of college students as measured by the Horne and Östberg questionnaire. *International Journal of Chronobiology,* 7, 141-144.

[45] Randler C. and Diaz-Morales J. F. (2007). Morningness in German and Spanish students: a comparative study. *European Journal of Personality, 21, 419-427*

[46] Roenneberg T., Kuehnle T., Pramstaller P. P., Ricken J., Havel M., Guth A., and Merrow M. (2004). A marker for the end of adolescence. *Current Biology,* 14, R1038.

[47] Shinomiya H., Takeuchi H., Martoni M., Natale V., and Harada T. (2004). Comparative study on circadian typology of Japanese and Italian students aged 12-18 years. *Sleep and Biological Rhythms,* 2, 93-95.

[48] Smith C. S., Folkard S., Schmieder R. A., Parra L. F., Spelten E., Almiral H., Sen R. N., Sahu S., Perez L. M., and Tisak J. (2002). Investigation of morning-evening orientation in six countries using the preferences scale. *Personality and Individual Differences,* 32, 949-968.

[49] Smith C. S., Reilly C., and Midkiff K. (1989). Evaluation of three circadian rhythm questionnaires with suggestions for an improved measure of morningness. *Journal of Applied Psychology,* 74 , 728-738.

[50] Smith C. S., Tisak J., Bauman T., and Green E. (1991). Psychometric equivalence of a translated circadian rhythm questionnaire: implications for between- and within-population assessments. *Journal of Applied Psychologyl,* 76, 628-636.

[51] Smith P. A., Brown D. F., Di Milia L., and Wragg C. (1993). The use of the circadian type inventory as a measure of the constructs of vigour and rigidity. *Ergonomics,* 36, 169-175.

[52] Taillard J., Philip P., Chastang J.-F., and Bioulac B. (2004). Validation of Horne and Östberg Morningness-Eveningness questionnaire in a middle-aged population of French workers. *Journal of Biological Rhythms,* 19, 76-86.

[53] Torsvall L. and Åkerstedt T. (1980). A diurnal type scale. Construction, consistency and validation in shift work. *Scandinavian Journal of Work, Environment and Health.,* 6, 283-290.

[54] Vink J. M., Groot A. S., Kerkhof G. A., and Boomsma D. I. (2001). Genetic analysis of morningness and eveningness. *Chronobiology International,* 18, 809-822.

[55] Wever R. A. (1984). Sex differences in human circadian rhythms: Intrinsic periods and sleep fractions. *Experientia,* 40, 1226-34.

[56] Yoon J.-S., Shin S.-M., Kook S.-H., and Lee H.-Y. (1997). A preliminary study on the Korean translation of Composite Scale (KtCS) to measure morningness-eveningness. *Journal of Korean Neuropsychiatric Association,* 36, 122-131.

In: Progress in Circadian Rhythm Research
Editor: Anne-Laure Léglise, pp. 223-234

ISBN: 978-1-60021-876-7

Chapter XI

The Cortisol Awakening Response (CAR): A Useful Tool for Ambulant Assessment of Hypothalamus-Pituitary-Adrenal (HPA) Axis Activity

Brigitte M. Kudielka*[*] *and Stefan Wüst
Graduate School of Psychobiology,
Department of Theoretical and Clinical Psychobiology
University of Trier, Johanniterufer 15, 54290 Trier, Germany

Abstract

In this brief review on the cortisol awakening response (CAR), we provide a description of this useful and easy to measure marker of hypothalamic-pituitary-adrenal (HPA) axis activity. The CAR is a discrete entity superimposing the circadian cycle. To date, the measurement of the CAR by collecting saliva samples in the first hour after awakening proofed to be useful in psychosomatic medicine and psychoneuroendocrinological research. In particular, the CAR technique allows the assessment of basal HPA axis functioning in ambulatory settings. Beside technical issues, known sources of inter- and intraindividual variability are discussed. Finally, the existing literature on associations with psychological states, chronic stress, and psychopathology is presented.

[*] Corresponding author: Brigitte M. Kudielka, Ph.D. Graduate School of Psychobiology; Department of Theoretical and Clinical Psychobiology; University of Trier; Johanniterufer 15; D-54290 Trier, Germany; Phone: +49-651-201-2981; Fax: +49-651-201-3690; e-mail: kudielka@uni-trier.de

Cortisol and the Hypothalamic-Pituitary-Adrenal (HPA) Axis

In the last decade, the so-called cortisol awakening response (CAR) has been established as a useful and easy to measure marker of hypothalamic-pituitary-adrenal (HPA) axis activity. The HPA axis is a central regulatory and control system of the organism that connects the central nervous system (CNS) with the endocrine system. This hierarchical hormone system encompasses the hypothalamus, the pituitary gland and the adrenal cortex with their respective hormones. Beside its role in stress regulation, the HPA axis is vital for supporting normal physiological functioning.

Under basal conditions, the adrenal hormone cortisol shows a marked circadian rhythm with typically lowest secretion during the first half of night time sleep (= quiescent period), an abrupt elevation during the second half of sleep, peak levels shortly after awakening (= awakening response) and continuously decreasing levels over the remainder of the day, except for stress-related cortisol surges that superimpose on the normal circadian rhythm (Born et al 1999; Czeisler and Klerman 1999; Späth-Schwalbe et al 1992; Van Cauter 1990). The two major processes, which control the overall diurnal cortisol variation are the circadian signal generated by the circadian pacemaker located in the suprachiasmatic nucleus (SCN) of the hypothalamus, and the alteration of wakefulness and sleep. In face of an internal or external challenge, nerval stimulation of the hypothalamus initiates the secretion of corticotrophin-releasing hormone (CRH), which provokes the release of adrenocorticotrophic hormone (ACTH) from the pituitary while ACTH triggers cortisol secretion from the adrenal cortex. In humans, the glucocorticoid cortisol has a wide range of physiological effects since virtually every nucleated cell in the body expresses glucocorticoid receptors and significant amounts of glucocorticoids penetrate the blood-brain barrier. Cortisol has tremendously diverse regulatory roles in development, metabolism, the immune system, cognitive as well as emotional processes and in control of the HPA axis itself (Belanoff et al 2001a, 2001b; Chrousos 1998; Het et al 2005; McEwen 2003; Sapolsky et al 2000).

In blood, cortisol is predominantly (90 to 95%) bound to binding proteins, only 5 to 10% of the total plasma cortisol circulates biologically active as unbound, "free" cortisol (for a recent overview see Kudielka and Kirschbaum 2007). Since alterations in basal HPA axis activity or HPA axis responses to stress appear to be a close correlate or even a determining factor of the onset of different diseases or disease progression (Chrousos and Gold 1992; Heim et al 2000; Holsboer 1989; Raison and Miller 2003; Stratakis and Chrousos 1995; Tsigos and Chrousos 1994, 2002), the assessment of the integrity and functioning of an individual's HPA axis appears to be of major interest in the field of psychosomatic medicine and psychoneuroendocrinology. However, since the HPA axis is a highly adaptive system which is characterized by marked inter- and intraindividual variability (Mason 1968), the development of such markers has been, and still is, quite a challenge.

The Cortisol Awakening Response (CAR)

As outlined above, the cortisol awakening response (CAR) is a response to morning awakening that is distinct from the circadian rise in hypothalamo-pituitary-adrenal (HPA) activity in the morning hours. Although the cortisol awakening response is modulated by circadian influences, it primarily reflects phasic psychophysiological processes specific to the sleep-wake transition (Edwards et al 2001a; Wilhelm et al 2007).

Pruessner and colleagues (1997) were the first who proposed that the assessment of the cortisol increase after awakening might represent a useful and easy to measure index of adrenocorticol activity. With strict reference to morning awakening time, free (salivary) as well as total (plasma) cortisol levels show a marked increase during the first hour after wake-up. On average, peak values are mostly observed during the first 30-45 minutes (Wilhelm et al 2007; Wüst et al 2000b). Across several studies in healthy adults it was reported that salivary cortisol levels increase about 50 up to over 100% (Clow et al 2004). In a highly controlled study under sleep laboratory conditions, we could recently show that the CAR is a genuine response to awakening, and that the transition from sleep to wake is essential for the cortisol morning rise to occur (Wilhelm et al 2007). While all subjects in this sleep lab study (N= 15) showed a distinct cortisol rise, we observed a significant morning cortisol rise in about 75% of a large sample of healthy participants in an ambulatory setting (Wüst et al 2000b; see also Dockray et al in press). To-date, it is still unclear under which circumstances no morning rise occurs and if this lack of an increase reflects a true flat morning cortisol profile after awakening. With the help of electronic monitoring devices, we demonstrated that the absence of a CAR in a remarkable part of our samples simply did not reflect absent morning increases but measurement error based on participant's non-compliance with the saliva sampling procedure (Broderick et al 2004; Kudielka et al 2003; Kudielka et al 2007c). If a subject does not start sampling right after awakening (maybe because of dozing or inability to follow instructions) and/or if subsequent samples are delayed (or collected to early), an existing cortisol increase might be missed with the collected samples.

The assessment of the CAR by saliva samples has many advantages. The sample collection is easy and non-invasive. Subjects either salivate directly in small pre-labelled plastic tubes or use devices such as the Salivette. Meanwhile, a standard procedure has been established encompassing four sampling time points (directly after awakening, 30, 45, and 60 minutes after awakening), while older studies included up to five time points for saliva collection (directly after awakening, 15, 30, 45, and 60 minutes after awakening). Storage and mailing of samples is also uncomplicated since cortisol in saliva is rather stable over time. Analysis in our own lab revealed that cortisol in saliva is usually stable at room temperature over a one month period. It should be noted, however, that we occasionally identified single samples with a lower stability. Garde and Hansen (2005) did not observe effects on cortisol concentrations after storage of saliva at 5°C for up to three months or at -20°C and -80°C for up to one year. Also, repeated freezing and thawing of samples up to four times before analysis does not affect the measured concentrations of cortisol. In sum, samples can easily be stored in the refrigerator over shorter time periods and express-mailed to a laboratory without cooling. Though, to avoid molding and to preserve samples over longer time periods, samples should be stored at -20°C or lower.

A major advantage of the CAR assessment by saliva samples is the possibility to obtain data in ambulatory settings, like the subject's home or the work place. For example, there is increasing research activity in the area of stress at work. In the last years, the CAR has been assessed in relation to chronic work stress in terms of job strain, time pressure, work load, effort-reward-imbalance, overcommitment, or financial strain (Eller et al 2006; Lundberg and Hellström 2002; Steptoe et al 2005; Steptoe et al 2000; Steptoe et al 2004b). Also, specific populations can be examined, like children (Gunnar and Vazquez 2001; Rosmalen et al 2005), pregnant women (de Weerth and Buitelaar 2005; Harville et al 2006), diverse ethnic groups (Bennett et al 2004), school teachers (Bellingrath et al in press; Pruessner et al 1999), victims of mobbing/bullying at the work place (Hansen et al 2006; Kudielka and Kern 2004), or shift workers across different shifts and across different shift rotas (Kudielka et al 2007b). The CAR technique also allows performing large cross-sectional or longitudinal studies that can only be performed by mailing (for reviews see Kirschbaum and Hellhammer 1989, 1994, 2000). Though, researchers should be aware of the issue of potential non-compliance in ambulatory saliva collections. Several techniques are available to improve subjects' adherence to the sampling procedure (e.g., electronic monitoring devices, pre-programmed wristwatches, pocket-palm-computers, reminders via SMS on cell phones, etc.).

Sources of Intra- and Intraindividual Variability

Although it is a reasonable and understandable wish to define normal values for cortisol concentrations after awakening and/or the net cortisol rise, research in the last decade revealed how difficult this endeavour is. The HPA axis is a highly adaptive system that is characterized by marked intra- and interindividual variability. The CAR shows medium to high stability over time ($r = .63$ for the areas under the curve across two consecutive days; Wüst et al 2000b), but repeated daily measurements appear to be necessary to reliably assess the cortisol rise after awakening since the CAR of a single day is determined to a great extent by situational factors (Hellhammer et al 2007). In line with this, recent analysis have begun to focus in more detail on day-to-day dynamics of the CAR (Adam et al 2006). Although its magnitude and time course is, in part, determined by genetic factors (Bartels et al 2003; Kupper et al 2005; Wüst et al 2000a), it varies, for example, with gender and age (Kudielka and Kirschbaum 2003; Pruessner et al 1997; Schlotz et al 2004; Wüst et al 2000b), socioeconomic status (Kunz-Ebrecht et al 2004b; Steptoe et al 2003; Wright and Steptoe 2005) and is also dependent of day of week (Kunz-Ebrecht et al 2004a; Schlotz et al 2004) and time of awakening (Edwards et al 2001b; Federenko et al 2004; Kudielka and Kirschbaum 2003). Although men and women show identical peak levels after awakening, females display a delayed decrease compared to males. Investigating the association between age and the CAR yielded mixed results. While age was positively associated with cortisol concentrations immediately after awakening and negatively with the following increase in one study (Kudielka and Kirschbaum 2003), other studies did not observe significant age effects (Edwards et al 2001a; Pruessner et al 1997; Wüst et al 2000b). Steptoe and coworkers demonstrated that low social status was associated with a larger CAR though there might be differential effects in men and women (Kunz-Ebrecht et al 2004b; Steptoe et al 2003; Wright and Steptoe 2005). In respect to day of week and awakening time, several studies report that the CAR is greater on work than weekend days (Kunz-Ebrecht et al 2004a; Schlotz et al 2004) and that early morning awakeners show a significantly higher CAR compared to late awakeners (Edwards et al 2001b; Federenko et al 2004; Kudielka and Kirschbaum 2003; Williams et al 2005). Light exposure, but not the postural shift from supine to standing (Hucklebridge et al 2002), can modulate the CAR, providing evidence for a possible mechanistic pathway. Although the CAR is apparent when subjects stay in complete darkness (Wilhelm et al 2007), the CAR is significantly larger if subjects are exposed to bright light (800 Lux; Scheer and Buijs 1999). Also, simulated dawn before awakening heightened the CAR (Thorn et al 2004).

The impact of smoking, coffee, the female menstrual cycle phase, and sleep length appears to be minor (Badrick et al 2007; Clow et al 2004; Harris et al 2007; Kudielka and Kirschbaum 2003; Steptoe and Ussher 2006; Wüst et al 2000b). Significant group differences in the CAR in smokers versus non-smokers only emerged in very large study samples showing relatively low explained variances and results are contradictory in the direction of the effect (Badrick et al 2007; Steptoe and Ussher 2006; Wüst et al 2000b). Systematic sleep restriction appears to alter the peak cortisol secretion after awakening, resulting in a lower peak cortisol secretion after sleep restriction compared to unrestricted sleep (Vgontzas et al 2004). Furthermore, the timing of the cortisol rise may be influenced by sudden changes in the sleep-wake schedule (Davidson et al 1991). In women, parity does not seem to alter the CAR (Federenko et al 2006). We also tested associations between selected personality traits and the cortisol awakening response. Our data supports the idea that morning relative to evening chronotypes ("larks" relative to "owls") show higher cortisol levels in the first hour after awakening (Kudielka et al 2007a; Kudielka et al 2006b). In another study, the CAR was reported to be positively associated with neuroticism (Portella et al 2005) though we could not find evidence for a close relationship between Eysenck's personality traits extraversion, neuroticism or psychoticism and the circadian cortisol rhythm (Schommer et al 1999).

Associations with Acute Psychological States, Chronic Stress and Psychopathology

Regarding acute psychological states, it was shown that the CAR is associated with day-to-day variations in psychosocial, emotional and physical states (Adam et al 2006; Kuehner et al 2007; Polk et al 2005; Steptoe et al 2007). In respect to more chronic states, the CAR is positively associated with perceived chronic stress (Pruessner et al 1999; Schlotz et al 2004; Schulz et al 1998; Wüst et al 2000a) and perceived loneliness (Steptoe et al 2004a). In caregivers of dementia patients, de Vugt et al. (2005) observed significantly higher levels of cortisol at the time of morning awakening compared to controls, but a smaller awakening increase. Existing evidence on cortisol awakening levels in burnout is very inconsistent (for a recent review see Kudielka et al 2006a). Pruessner et al. (1999) observed a reduced CAR in burned out teachers and Mommersteeg et al. (2006c) reported on lowered awakening cortisol levels in burnout patients before psychotherapy, others observed higher morning cortisol levels in subjects high on burnout (Grossi et al 2005) or no group differences (Mommersteeg et al 2006a; Mommersteeg et al 2006b). Recent studies have indicated a significantly altered awakening rise in various health conditions including psychopathology. In more detail, the CAR is positively associated with present and former depressive symptomatology (Bhagwagar et al 2003; Bhagwagar et al 2005; Pruessner et al 2003). Huber et al. (2006) described higher cortisol morning levels in depressed versus non-depressed psychotherapy inpatients but a reduced awakening net increase. In contrast, Stetler et al. (2005) found a blunted cortisol response to awakening in a non-clinical sample of women from the community with mild to moderate depressive symptomatology. Lower cortisol levels immediately after awakening were found to be related to poor sleep quality in patients with primary insomnia (Backhaus et al 2004). Moreover, a reduced CAR was associated with persisting pain after surgery (Geiss et al 1997), chronic fatigue syndrome (Roberts et al 2004), posttraumatic stress disorder (de Kloet et al 2007; Rohleder et al 2004; Wessa et al 2006) and early loss experience (Meinlschmidt and Heim 2005). In patients with various forms of hippocampal damage (Buchanan et al 2004) as well as in a small sample of patients with global amnesia (Wolf et al 2005) the awakening rise was reported to be completely absent.

To conclude, our results in the sleep laboratory study (Wilhelm et al 2007) identify the CAR as a genuine response to awakening. Moreover, this data also suggest that home assessment of the CAR in saliva provides the same results as the assessment under highly controlled laboratory conditions (in this study, subjects additionally collected CAR samples at two days at home). The ease to obtain this measure, which is especially an advantage for research questions that can only be pursued in ambulant settings, in combination with the multifaceted findings reported in the literature encompassing various groups of interest including different age groups, populations exposed to chronic stress or other special circumstances, and psychiatric patients argues for its further use as a promising diagnostic marker in psychoneuroendocrinology, psychosomatic medicine, and psychiatry.

Acknowledgements

This work was supported by grants of the German Research Foundation for BMK (Grant KU140/4-1 and 4-2) and SW (Grant WU 324/3-1, 3-2 and 3-3). Moreover, BMK and SW are members of the International Research Training Group IRTG funded by the DFG (GRH 1389/1).

References

Adam EK, Hawkley LC, Kudielka BM, Cacioppo JT (2006): Day-to-day dynamics of experience-cortisol associations in a population-based sample of older adults. *Proc. Natl. Acad. Sci. U S A.* 103:17058-17063.

Backhaus J, Junghanns K, Hohagen F (2004): Sleep disturbances are correlated with decreased morning awakening salivary cortisol. *Psychoneuroendocrinology.* 29:1184-1191.

Badrick E, Kirschbaum C, Kumari M (2007): The Relationship between Smoking Status and Cortisol Secretion. *J. Clin. Endocrinol. Metab.* 92:819-824.

Bartels M, Van den Berg M, Sluyter F, Boomsma DI, de Geus EJ (2003): Heritability of cortisol levels: review and simultaneous analysis of twin studies. *Psychoneuroendocrinology.* 28:121-137.

Belanoff JK, Gross K, Yager A, Schatzberg AF (2001a): Corticosteroids and cognition. *J. Psychiatr. Res.* 35:127-145.

Belanoff JK, Kalehzan M, Sund B, Fleming Ficek SK, Schatzberg AF (2001b): Cortisol activity and cognitive changes in psychotic major depression. *Am. J. Psychiatry.* 158:1612-1616.

Bellingrath S, Weigl T, Kudielka BM (in press): Cortisol dysregulation in school teachers in relation to burnout, vital exhaustion, and effort-reward-imbalance. *Biol. Psychol. ahead of press.*

Bennett GG, Merritt MM, Wolin KY (2004): Ethnicity, education, and the cortisol response to awakening: a preliminary investigation. *Ethn. Health.* 9:337-347.

Bhagwagar Z, Hafizi S, Cowen PJ (2003): Increase in concentration of waking salivary cortisol in recovered patients with depression. *Am. J. Psychiatry.* 160:1890-1891.

Bhagwagar Z, Hafizi S, Cowen PJ (2005): Increased salivary cortisol after waking in depression. *Psychopharmacology.* (Berl) 182:54-57.

Born J, Hansen K, Marshall L, Molle M, Fehm HL (1999): Timing the end of nocturnal sleep. *Nature.* 397:29-30.

Broderick JE, Arnold D, Kudielka BM, Kirschbaum C (2004): Salivary cortisol sampling compliance: comparison of patients and healthy volunteers. *Psychoneuroendocrinology.* 29:636-650.

Buchanan TW, Kern S, Allen JS, Tranel D, Kirschbaum C (2004): Circadian regulation of cortisol after hippocampal damage in humans. *Biol. Psychiatry.* 56:651-656.

Chrousos GP (1998): Stressors, stress, and neuroendocrine integration of the adaptive response. The 1997 Hans Selye Memorial Lecture. *Ann. N Y Acad. Sci.* 851:311-335.

Chrousos GP, Gold PW (1992): The concepts of stress and stress system disorders. Overview of physical and behavioral homeostasis. *JAMA.* 267:1244-1252.

Clow A, Thorn L, Evans P, Hucklebridge F (2004): The awakening cortisol response: methodological issues and significance. *Stress.* 7:29-37.

Czeisler CA, Klerman EB (1999): Circadian and sleep-dependent regulation of hormone release in humans. *Recent Prog. Horm. Res.* 54:97-130; discussion 130-132.

Davidson JR, Moldofsky H, Lue FA (1991): Growth hormone and cortisol secretion in relation to sleep and wakefulness. *J. Psychiatry Neurosci.* 16:96-102.

de Kloet CS, Vermetten E, Heijnen CJ, Geuze E, Lentjes EG, Westenberg HG (2007): Enhanced cortisol suppression in response to dexamethasone administration in traumatized veterans with and without posttraumatic stress disorder. *Psychoneuroendocrinology.* 32:215-226.

de Vugt ME, Nicolson NA, Aalten P, Lousberg R, Jolle J, Verhey FR (2005): Behavioral problems in dementia patients and salivary cortisol patterns in caregivers. J *Neuropsychiatry Clin. Neurosci.* 17:201-207.

de Weerth C, Buitelaar JK (2005): Cortisol awakening response in pregnant women. *Psychoneuroendocrinology*. 30:902-907.

Dockray S, Bhattacharyya MR, Molloy GJ, Steptoe A (in press): The cortisol awakening response in relation to objective and subjective measures of waking in the morning. *Psychoneuroendocrinology ahead of press.*

Edwards S, Clow A, Evans P, Hucklebridge F (2001a): Exploration of the awakening cortisol response in relation to diurnal cortisol secretory activity. *Life Sci.* 68:2093-2103.

Edwards S, Evans P, Hucklebridge F, Clow A (2001b): Association between time of awakening and diurnal cortisol secretory activity. *Psychoneuroendocrinology*. 26:613-622.

Eller NH, Netterstrom B, Hansen AM (2006): Psychosocial factors at home and at work and levels of salivary cortisol. *Biol. Psychol.* 73:280-287.

Federenko I, Wüst S, Hellhammer DH, Dechoux R, Kumsta R, Kirschbaum C (2004): Free cortisol awakening responses are influenced by awakening time. *Psychoneuroendocrinology*. 29:174-184.

Federenko IS, Wolf JM, Wüst S, Schlotz W, Hellhammer J, Kudielka BM, et al (2006): Parity does not alter baseline or stimulated activity of the hypothalamus-pituitary-adrenal axis in women. *Dev. Psychobiol.* 48:703-711.

Garde AH, Hansen AM (2005): Long-term stability of salivary cortisol. *Scand. J. Clin. Lab. Invest.* 65:433-436.

Geiss A, Varadi E, Steinbach K, Bauer HW, Anton F (1997): Psychoneuroimmunological correlates of persisting sciatic pain in patients who underwent discectomy. *Neurosci. Lett.* 237:65-68.

Grossi G, Perski A, Ekstedt M, Johansson T, Lindstrom M, Holm K (2005): The morning salivary cortisol response in burnout. *J. Psychosom. Res.* 59:103-111.

Gunnar MR, Vazquez DM (2001): Low cortisol and a flattening of expected daytime rhythm: potential indices of risk in human development. *Dev. Psychopathol.* 13:515-538.

Hansen AM, Hogh A, Persson R, Karlson B, Garde AH, Orbaek P (2006): Bullying at work, health outcomes, and physiological stress response. *J. Psychosom. Res.* 60:63-72.

Harris A, Ursin H, Murison R, Eriksen HR (2007): Coffee, stress and cortisol in nursing staff. *Psychoneuroendocrinology*. 32:322-330.

Harville EW, Savitz DA, Dole N, Herring AH, Thorp JM, Light KC (2006): Patterns of salivary cortisol secretion in pregnancy and implications for assessment protocols. *Biol. Psychol.* 74:85-91.

Heim C, Ehlert U, Hellhammer DH (2000): The potential role of hypocortisolism in the pathophysiology of stress-related bodily disorders. *Psychoneuroendocrinology*. 25:1-35.

Hellhammer J, Fries E, Schweisthal OW, Schlotz W, Stone AA, Hagemann D (2007): Several daily measurements are necessary to reliably assess the cortisol rise after awakening: state- and trait components. *Psychoneuroendocrinology*. 32:80-86.

Het S, Ramlow G, Wolf OT (2005): A meta-analytic review of the effects of acute cortisol administration on human memory. *Psychoneuroendocrinology*. 30:771-784.

Holsboer F (1989): Psychiatric implications of altered limbic-hypothalamic-pituitary-adrenocortical activity. *Eur. Arch. Psychiatry Neurol. Sci.* 238:302-322.

Huber TJ, Issa K, Schik G, Wolf OT (2006): The cortisol awakening response is blunted in psychotherapy inpatients suffering from depression. *Psychoneuroendocrinology*. 31:900-904.

Hucklebridge F, Mellins J, Evans P, Clow A (2002): The awakening cortisol response: no evidence for an influence of body posture. *Life Sci*. 71:639-646.

Kirschbaum C, Hellhammer DH (1989): Salivary cortisol in psychobiological research: an overview. *Neuropsychobiology*. 22:150-169.

Kirschbaum C, Hellhammer DH (1994): Salivary cortisol in psychoneuroendocrine research: recent developments and applications. *Psychoneuroendocrinology*. 19:313-333.

Kirschbaum C, Hellhammer DH (2000): Salivary cortisol. In: Fink G editor. *Encyclopedia of stress,* Vol 3. San Diego: San Diego: Academic Press, pp 379-383.

Kudielka BM, Bellingrath S, Hellhammer DH (2006a): Cortisol in burnout and vital exhaustion: an overview. *G. Ital. Med. Lav. Ergon. [Applied Psychology to Work and Rehabilitation Medicine]* 28:34-42.

Kudielka BM, Bellingrath S, Hellhammer DH (2007a): Further support for higher salivary cortisol levels in “morning” compared to “evening” persons. *J. Psychosom. Res*. 62:595-596.

Kudielka BM, Broderick JE, Kirschbaum C (2003): Compliance with saliva sampling protocols: electronic monitoring reveals invalid cortisol daytime profiles in noncompliant subjects. *Psychosom. Med*. 65:313-319.

Kudielka BM, Buchtal J, Uhde A, Wüst S (2007b): Circadian cortisol profiles and psychological self-reports in shift workers with and without recent change in the shift rotation system. *Biol. Psychol*. 74:92-103.

Kudielka BM, Federenko IS, Hellhammer DH, Wüst S (2006b): Morningness and eveningness: The free cortisol rise after awakening in "early birds" and "night owls". *Biol. Psychol*. 72:141-146.

Kudielka BM, Hawkley LC, Adam EK, Cacioppo JT (2007c): Compliance with ambulatory saliva sampling in the chicago health, aging, and social relations study and associations with social support. *Ann. Behav. Med*. 34:209-216.

Kudielka BM, Kern S (2004): Cortisol day profiles in victims of mobbing (bullying at the work place): preliminary results of a first psychobiological field study. *J. Psychosom. Res*. 56:149-150.

Kudielka BM, Kirschbaum C (2003): Awakening cortisol responses are influenced by health status and awakening time but not by menstrual cycle phase. *Psychoneuroendocrinology*. 28:35-47.

Kudielka BM, Kirschbaum C (2007): Biological bases of the stress response. In: al'Absi M editor. Stress and addiction: biological and psychological mechanisms, section I: *Neurobiology of stress and addiction,* Vol 1. Amsterdam: Elsevier, chapter 1, pp 3-19.

Kuehner C, Holzhauer S, Huffziger S (2007): Decreased cortisol response to awakening is associated with cognitive vulnerability to depression in a nonclinical sample of young adults. *Psychoneuroendocrinology*. 32:199-209.

Kunz-Ebrecht SR, Kirschbaum C, Marmot M, Steptoe A (2004a): Differences in cortisol awakening response on work days and weekends in women and men from the Whitehall II cohort. *Psychoneuroendocrinology*. 29:516-528.

Kunz-Ebrecht SR, Kirschbaum C, Steptoe A (2004b): Work stress, socioeconomic status and neuroendocrine activation over the working day. *Soc. Sci. Med.* 58:1523-1530.

Kupper N, de Geus EJ, van den Berg M, Kirschbaum C, Boomsma DI, Willemsen G (2005): Familial influences on basal salivary cortisol in an adult population. *Psychoneuroendocrinology.* 30:857-868.

Lundberg U, Hellström B (2002): Workload and morning salivary cortisol in women. *Work Stress.* 16:356-363.

Mason JW (1968): A review of psychoendocrine research on the pituitary-adrenal cortical system. *Psychosom. Med.* 30:576-607.

McEwen BS (2003): Interacting mediators of allostasis and allostatic load: towards an understanding of resilience in aging. *Metabolism.* 52:10-16.

Meinlschmidt G, Heim C (2005): Decreased cortisol awakening response after early loss experience. *Psychoneuroendocrinology.* 30:568-576.

Mommersteeg PM, Heijnen CJ, Kavelaars A, van Doornen LJ (2006a): Immune and endocrine function in burnout syndrome. *Psychosom. Med.* 68:879-886.

Mommersteeg PM, Heijnen CJ, Verbraak MJ, van Doornen LJ (2006b): Clinical burnout is not reflected in the cortisol awakening response, the day-curve or the response to a low-dose dexamethasone suppression test. *Psychoneuroendocrinology.* 31:216-225.

Mommersteeg PM, Keijsers GP, Heijnen CJ, Verbraak MJ, van Doornen LJ (2006c): Cortisol deviations in people with burnout before and after psychotherapy: a pilot study. *Health Psychol.* 25:243-248.

Polk DE, Cohen S, Doyle WJ, Skoner DP, Kirschbaum C (2005): State and trait affect as predictors of salivary cortisol in healthy adults. *Psychoneuroendocrinology.* 30:261-272.

Portella MJ, Harmer CJ, Flint J, Cowen P, Goodwin GM (2005): Enhanced early morning salivary cortisol in neuroticism. *Am. J. Psychiatry.* 162:807-809.

Pruessner JC, Hellhammer DH, Kirschbaum C (1999): Burnout, perceived stress, and cortisol responses to awakening. *Psychosom. Med.* 61:197-204.

Pruessner JC, Wolf OT, Hellhammer DH, Buske-Kirschbaum A, von Auer K, Jobst S, et al (1997): Free cortisol levels after awakening: a reliable biological marker for the assessment of adrenocortical activity. *Life Sci.* 61:2539-2549.

Pruessner M, Hellhammer DH, Pruessner JC, Lupien SJ (2003): Self-reported depressive symptoms and stress levels in healthy young men: associations with the cortisol response to awakening. *Psychosom. Med.* 65:92-99.

Raison CL, Miller AH (2003): When not enough is too much: the role of insufficient glucocorticoid signaling in the pathophysiology of stress-related disorders. *Am. J. Psychiatry.* 160:1554-1565.

Roberts AD, Wessely S, Chalder T, Papadopoulos A, Cleare AJ (2004): Salivary cortisol response to awakening in chronic fatigue syndrome. *Br. J. Psychiatry.* 184:136-141.

Rohleder N, Joksimovic L, Wolf JM, Kirschbaum C (2004): Hypocortisolism and increased glucocorticoid sensitivity of pro-inflammatory cytokine production in Bosnian war refugees with posttraumatic stress disorder. *Biol. Psychiatry.* 55:745-751.

Rosmalen JG, Oldehinkel AJ, Ormel J, de Winter AF, Buitelaar JK, Verhulst FC (2005): Determinants of salivary cortisol levels in 10-12 year old children; a population-based study of individual differences. *Psychoneuroendocrinology.* 30:483-495.

Sapolsky RM, Romero LM, Munck AU (2000): How do glucocorticoids influence stress responses? Integrating permissive, suppressive, stimulatory, and preparative actions. *Endocr. Rev.* 21:55-89.

Scheer FA, Buijs RM (1999): Light affects morning salivary cortisol in humans. *J. Clin. Endocrinol. Metab.* 84:3395-3398.

Schlotz W, Hellhammer J, Schulz P, Stone AA (2004): Perceived work overload and chronic worrying predict weekend-weekday differences in the cortisol awakening response. *Psychosom. Med.* 66:207-214.

Schommer NC, Kudielka BM, Hellhammer DH, Kirschbaum C (1999): No evidence for a close relationship between personality traits and circadian cortisol rhythm or a single cortisol stress response. *Psychol. Rep.* 84:840-842.

Schulz P, Kirschbaum C, Pruessner JC, Hellhammer DH (1998): Increased free cortisol secretion after awakening in chronically stressed individuals due to work overload. *Stress Med.* 14:91-97.

Späth-Schwalbe E, Scholler T, Kern W, Fehm HL, Born J (1992): Nocturnal adrenocorticotropin and cortisol secretion depends on sleep duration and decreases in association with spontaneous awakening in the morning. *J. Clin. Endocrinol. Metab.* 75:1431-1435.

Steptoe A, Brydon L, Kunz-Ebrecht S (2005): Changes in financial strain over three years, ambulatory blood pressure, and cortisol responses to awakening. *Psychosom. Med.* 67:281-287.

Steptoe A, Cropley M, Griffith J, Kirschbaum C (2000): Job strain and anger expression predict early morning elevations in salivary cortisol. *Psychosom. Med.* 62:286-292.

Steptoe A, Gibson EL, Hamer M, Wardle J (2007): Neuroendocrine and cardiovascular correlates of positive affect measured by ecological momentary assessment and by questionnaire. *Psychoneuroendocrinology.* 32:56-64.

Steptoe A, Kunz-Ebrecht S, Owen N, Feldman PJ, Willemsen G, Kirschbaum C, Marmot M (2003): Socioeconomic status and stress-related biological responses over the working day. *Psychosom. Med.* 65:461-470.

Steptoe A, Owen N, Kunz-Ebrecht SR, Brydon L (2004a): Loneliness and neuroendocrine, cardiovascular, and inflammatory stress responses in middle-aged men and women. *Psychoneuroendocrinology.* 29:593-611.

Steptoe A, Siegrist J, Kirschbaum C, Marmot M (2004b): Effort-reward imbalance, overcommitment, and measures of cortisol and blood pressure over the working day. *Psychosom. Med.* 66:323-329.

Steptoe A, Ussher M (2006): Smoking, cortisol and nicotine. *Int. J. Psychophysiol.* 59:228-235.

Stetler C, Miller GE (2005): Blunted cortisol response to awakening in mild to moderate depression: regulatory influences of sleep patterns and social contacts. *J. Abnorm. Psychol.* 114:697-705.

Stratakis CA, Chrousos GP (1995): Neuroendocrinology and pathophysiology of the stress system. Annals of the New York academy of sciences. In: Chrousos GP, McCarty R, Pacâk K, Cizza G, Sternberg E, Gold PW, Kvetnanský R editors. *Stress: Basic mechanisms and clinical implications*, Vol 771. New York: The New York Academy of Sciences, pp 1-18.

Thorn L, Hucklebridge F, Esgate A, Evans P, Clow A (2004): The effect of dawn simulation on the cortisol response to awakening in healthy participants. *Psychoneuro-endocrinology*. 29:925-930.

Tsigos C, Chrousos GP (1994): Physiology of the hypothalamic-pituitary-adrenal axis in health and dysregulation in psychiatric and autoimmune disorders. *Endocrinol. Metab. Clin. North Am.* 23:451-466.

Tsigos C, Chrousos GP (2002): Hypothalamic-pituitary-adrenal axis, neuroendocrine factors and stress. *J. Psychosom. Res.* 53:865-871.

Van Cauter E (1990): Diurnal and ultradian rhythms in human endocrine function: a minireview. *Horm. Res.* 34:45-53.

Vgontzas AN, Zoumakis E, Bixler EO, Lin HM, Follett H, Kales A, Chrousos GP (2004): Adverse effects of modest sleep restriction on sleepiness, performance, and inflammatory cytokines. *J. Clin. Endocrinol. Metab.* 89:2119-2126.

Wessa M, Rohleder N, Kirschbaum C, Flor H (2006): Altered cortisol awakening response in posttraumatic stress disorder. *Psychoneuroendocrinology*. 31:209-215.

Wilhelm I, Born J, Kudielka BM, Schlotz W, Wüst S (2007): Is the cortisol awakening rise a response to awakening? *Psychoneuroendocrinology*. 32:358-366.

Williams E, Magid K, Steptoe A (2005): The impact of time of waking and concurrent subjective stress on the cortisol response to awakening. *Psychoneuroendocrinology*. 30:139-148.

Wolf OT, Fujiwara E, Luwinski G, Kirschbaum C, Markowitsch HJ (2005): No morning cortisol response in patients with severe global amnesia. *Psychoneuroendocrinology*. 30:101-105.

Wright CE, Steptoe A (2005): Subjective socioeconomic position, gender and cortisol responses to waking in an elderly population. *Psychoneuroendocrinology*. 30:582-590.

Wüst S, Federenko I, Hellhammer DH, Kirschbaum C (2000a): Genetic factors, perceived chronic stress, and the free cortisol response to awakening. *Psychoneuroendocrinology*. 25:707-720.

Wüst S, Wolf J, Hellhammer DH, Federenko I, Schommer N, Kirschbaum C (2000b): The cortisol awakening response - normal values and confounds. *Noise Health*. 2:79-88.

Index

A

B

C

E

F

G

H

I

J

K

L

M

N

T

U

V

W

X

Y

Z